# Biomarkers for Agrochemicals and Toxic Substances

ACS SYMPOSIUM SERIES **643**

# Biomarkers for Agrochemicals and Toxic Substances

## Applications and Risk Assessment

**Jerry N. Blancato,** EDITOR
*U.S. Environmental Protection Agency*

**Robert N. Brown,** EDITOR
*U.S. Environmental Protection Agency*

**Curtis C. Dary,** EDITOR
*U.S. Environmental Protection Agency*

**Mahmoud Abbas Saleh,** EDITOR
*Texas Southern University*

Developed from a symposium sponsored
by the Division of Agrochemicals

American Chemical Society, Washington, DC

**Library of Congress Cataloging-in-Publication Data**

Biomarkers for agrochemicals and toxic substance: applications and risk assessment / Jerry N. Blancato... [et al.], editor.

    p.   cm.—(ACS symposium series, ISSN 0097–6156; 643)

    "Developed from a symposium sponsored by the Division of Agrochemicals at the 209th National Meeting of the American Chemical Society, Anaheim, California, April 2–7, 1995."

    Includes bibliographical references and indexes.

    ISBN 0–8412–3449–3

    1. Agricultural chemicals—Health aspects—Congresses. 2. Biochemical markers—Congresses. 3. Endocrine toxicology—Congresses. 4. Agricultural chemicals—Environmental aspects—Congresses.

    I. Blancato, Jerry N.  II. American Chemical Society. Division of Agrochemicals.  III. American Chemical Society. Meeting (209th: 1995: Anaheim, Calif.).  IV. Series.

    [DNLM: 1. Agrochemicals—congresses. 2. Hazardous Substances—congressses. 3. Biological Markers—congresses. 4. Environmental Exposure—prevention & control—United States—congresses. WA 465 B620 1996]

RA1270.A4B55   1996
615.9—dc20

DNLM/DLC
for Library of Congress                                     96–28772
                                                           CIP

This book is printed on acid-free, recycled paper.

PRINTED IN THE UNITED STATES OF AMERICA

# Foreword

THE ACS SYMPOSIUM SERIES was first published in 1974 to provide a mechanism for publishing symposia quickly in book form. The purpose of this series is to publish comprehensive books developed from symposia, which are usually "snapshots in time" of the current research being done on a topic, plus some review material on the topic. For this reason, it is necessary that the papers be published as quickly as possible.

Before a symposium-based book is put under contract, the proposed table of contents is reviewed for appropriateness to the topic and for comprehensiveness of the collection. Some papers are excluded at this point, and others are added to round out the scope of the volume. In addition, a draft of each paper is peer-reviewed prior to final acceptance or rejection. This anonymous review process is supervised by the organizer(s) of the symposium, who become the editor(s) of the book. The authors then revise their papers according to the recommendations of both the reviewers and the editors, prepare camera-ready copy, and submit the final papers to the editors, who check that all necessary revisions have been made.

As a rule, only original research papers and original review papers are included in the volumes. Verbatim reproductions of previously published papers are not accepted.

ACS BOOKS DEPARTMENT

# Contents

Preface ..................................................................................................... xi

EXPOSURE BIOMARKERS AND RISK ASSESSMENT

1. Relationship of Biomarkers of Exposure to Risk Assessment
   and Risk Management........................................................................ 2
   Curtis C. Dary, James J. Quackenboss, Charles H. Nauman,
   and Stephen C. Hern

2. Proposed Use of Human Exposure Biomarkers in Environmental
   Legislation: Toxic Substances Control Act........................................ 24
   Larry L. Needham

3. The Use of Reference Range Concentrations in Environmental
   Health Investigations........................................................................ 39
   Robert H. Hill, Jr., Susan L. Head, Sam Baker, Carol Rubin,
   Emilio Esteban, Sandra L. Bailey, Dana B. Shealy,
   and Larry L. Needham

4. Possible Biomarkers for Assessing Health Risks in Susceptible
   Individuals Exposed to Agrochemicals with Emphasis
   on Developing Countries.................................................................... 49
   Abdel Khalek H. El-Sebae, Zaher A. Mohamed,
   Fawzia Abdel Rahman, Alaa Kamel

BIOMARKERS OF HUMAN EXPOSURE

5. Blood Cholinesterase Activity: Inhibition as an Indicator
   of Adverse Effect............................................................................... 70
   S. Padilla, L. Lassiter, K. Crofton, and V. C. Moser

6. Neurotoxic Esterase Inhibition: Predictor of Potential
   for Organophosphorus-Induced Delayed Neuropathy ...................... 79
   Marion Ehrich

7. Increased Urinary Excretion of Xanthurenic Acid as a
   Biomarker of Exposure to Organophosphorus Insecticides ........... 94
   Josef Seifert

8. Serum Protein Profile: A Possible Biomarker for Exposure
   to Insecticides .................................................................................. 106
   Mahmoud Abbas Saleh, Mohamed Abou Zeid,
   Zaher A. Mohamed, and Fawzia Abdel Rahman

9. Breast Milk as a Biomarker for Monitoring Human Exposure
   to Environmental Pollutants ......................................................... 114
   Mahmoud Abbas Saleh, Abdel Moneim Afify, Awad Ragab,
   Gamal El-Baroty, Alaa Kamel, and Abdel Khalek H. El-Sebae

10. Carnitine and Its Esters as Potential Biomarkers
    of Environmental–Toxicological Exposure to Nongenotoxic
    Tumorigens ...................................................................................... 126
    John E. Garst

11. L-Carnitine as a Detoxicant of Nonmetabolizable
    Acyl Coenzyme A ............................................................................ 140
    L. L. Bieber, Z. H. Huang, and D. A. Gage

BIOMARKERS OF ENVIRONMENTAL HEALTH

12. The Use of Human 101L Cells as a Biomarker, P450 RGS,
    for Assessing the Potential Risk of Environmental Samples ......... 150
    Jack W. Anderson, Kristen Bothner, David Edelman,
    Steven Vincent, Tien P. Vu, and Robert H. Tukey

13. Field Application of Fluorescence-Based Catalytic Assays
    for Measuring Cytochrome P450 Induction in Birds ...................... 169
    J. A. Davis, D. M. Fry, and B. W. Wilson

14. Initial Results for the Inductively Coupled Plasma–Mass
    Spectrometric Determination of Trace Elements in Organs
    of Striped Bass from Lake Mead, U.S.A. ...................................... 180
    Vernon Hodge, Klaus Stetzenbach, and Kevin Johannesson

15. Aquatic Toxicity of Chemical Agent Simulants as Determined
    by Quantitative Structure–Activity Relationships and Acute
    Bioassays ......................................................................................... 191
    N. A. Chester, G. R. Famini, M. V. Haley, C. W. Kurnas,
    P. A. Sterling, and L. Y. Wilson

16. **Use of a Multiple Pathway and Multiroute Physiologically Based Pharmacokinetic Model for Predicting Organophosphorus Pesticide Toxicity** ...... 206
    James B. Knaak, Mohammed A. Al-Bayati, Otto G. Raabe, and Jerry N. Blancato

17. **Use of Spot Urine Sample Results in Physiologically Based Pharmacokinetic Modeling of Absorbed Malathion Doses in Humans** ...... 229
    Michael H. Dong, John H. Ross, Thomas Thongsinthusak, and Robert I. Krieger

18. **Symbolic Solutions of Small Linear Physiologically Based Pharmacokinetic Models Useful in Risk Assessment and Experimental Design** ...... 242
    Robert N. Brown

19. **Comparison of Symbolic, Numerical Area Under the Concentration Curves of Small Linear Physiologically Based Pharmacokinetic Models** ...... 256
    Blaine L. Hagstrom, Robert N. Brown, and Jerry N. Blancato

INDEXES

**Author Index** ...... 273

**Affiliation Index** ...... 273

**Subject Index** ...... 274

# Preface

THE RISK ASSESSMENT PROCESS has received increased attention over the past few years. The National Research Council, the U.S. Congress, and numerous international bodies have all discussed its importance. There are considerable uncertainties with the probabilistic estimates of risk. Risk assessments consist of many components involving several scientific disciplines. Exposure assessment is often the weak link in risk assessments. While a treatise on exposure assessment is outside the scope of this volume, measurements of concentrations in environmental media and in biological fluids and tissues are important cornerstones. These measurements are often expensive and time-consuming. We need accurate, quick, and cost-effective measurement techniques. In addition, we need interpretive methods to assess exposure and risk to xenobiotics.

Public concerns about agrochemicals and toxic substances are great. The media frequently report on these areas, alerting people to newly perceived dangers, both real and imagined. The public finds pesticides in food products and in drinking water especially disconcerting. For example, we have grown more aware of the role of agrochemicals as possible endocrine disruptors.

Biomarkers and biomonitoring, the subjects of this volume, offer exciting opportunities to understand more completely environmental, chemical, and physiologic processes by providing useful measures of chemicals and their breakdown products. Many chapters in this volume show and discuss such new methods and their field applications. However, to take full advantage of any new measurement or monitoring strategies, we should understand all the mechanistic processes. Full understanding is not always possible; progress is made in stages.

This volume was developed from the work presented at the 209th National Meeting of the American Chemical Society, titled "Biomarkers for Agrochemicals and Toxic Substances," sponsored by the Division of Agrochemicals, in Anaheim, California, April 2–7, 1995. It focuses not only on measurement techniques and applications but also on new and innovative thinking regarding the interpretation of measurement results. Several chapters deal with the application of results to the exposure and risk assessment process. Other chapters discuss new physiologic end points as markers of exposure and effect. This integrated picture is more fully discussed in the first chapter.

We wish to acknowledge the efforts of all contributing authors in providing informative and insightful chapters. We also acknowledge the assistance and patience of our acquisitions editor, Michelle Althuis.

JERRY N. BLANCATO
ROBERT N. BROWN
CURTIS C. DARY
Characterization Research Division
National Exposure Research Laboratory
U.S. Environmental Protection Agency
944 East Harmon Avenue
Las Vegas, NV  89193–3478

MAHMOUD ABBAS SALEH
Environmental Chemistry and Toxicology Laboratory
Department of Chemistry
3100 Cleburne Avenue
Texas Southern University
Houston, TX  77004

May 30, 1996

# EXPOSURE BIOMARKERS AND RISK ASSESSMENT

# Chapter 1

# Relationship of Biomarkers of Exposure to Risk Assessment and Risk Management

Curtis C. Dary, James J. Quackenboss,
Charles H. Nauman, and Stephen C. Hern

Characterization Research Division, National Exposure Research
Laboratory, U.S. Environmental Protection Agency, 944 East Harmon,
Las Vegas, NV 89193-3478

An "environmental health paradigm" has been proposed for setting priorities in risk assessment for pesticides. This paradigm serves as a useful template for depicting the chain of events that relate sources of exposure with adverse environmental and human health outcomes. The paradigm is based on multiroute and multipathway exposure from contact with contaminants in multimedia as modulated by the magnitude, frequency, and duration of exposure. As part of this risk assessment paradigm, biomarkers integrate routes, media, and pathways of exposure so that scientifically defensible risk characterizations can be realized. Biomarkers of exposure are indicators of current or historical contact with an environmental agent. Moreover, biomarkers serve to evaluate the completeness of exposure assessment information by associating environmental or source information, exposure measurements, and epidemiological and human activity data with internal dose. Biomarkers may take the form of the unchanged agent itself or its metabolites, an adduct or biochemical indicator depending on the source and route of exposure, and the metabolism, storage, and excretion and physicochemical properties of the agent. This paper will review the use of biomarkers of exposure and explore the relationship biomarkers assume within the risk assessment or "environmental health" paradigm.

Risk is a function of exposure and hazard and risk assessment is the process used to evaluate the probability of exposure and the magnitude and character of the hazard posed to human health and ecosystems by environmental stressors such as pollution and alterations in the quality of habitats (1,2). The risk assessment process consists of *hazard identification, dose-response assessment, exposure assessment* and *risk characterization.* Hazard identification involves the experimental study of toxicity in test organisms and the examination and evaluation of reports of exposure incidents and epidemiological research. The potency of a stressor is determined experimentally through a *dose-response assessment.* An *exposure assessment* determines the nature and size of the population or ecosystem at risk together with the frequency, duration and magnitude of the exposure to the stressor. The data collected through the first three elements, *hazard identification, dose-response assessment* and *exposure assessment*, are used in risk characterization to qualitatively or quantitatively estimate or predict the likelihood of an adverse health outcome.

### Components of the Paradigm

Exposure assessment has been generally considered to be the weak link in hazard evaluation and the assessment of risk to toxic substances including pesticides (3). As depicted by Pirkle et al. (4) and recast to explore the pesticide exposure situation (Exhibit 1), exposure assessment represents the preponderance of elements along the spine of the environmental health paradigm that guides research in risk assessment and risk management decisions (5). The environmental health paradigm for the pesticide situation looks to a clear relationship between media concentrations, the rate of exposure and adverse environmental health outcomes. Monitoring exposure through the use of biomarkers provides a measure of the amount of the agent transferred into an organism independent of the source or medium.

**Sources.**     For the pesticide situation, the source of the exposure is generally controlled through regulation as specified under the Federal Insecticide Fungicide and Rodenticide Act (FIFRA. 7 U.S.C. 136 a-c, 136v.). The active ingredients and formulations of pesticide products have been rationally designed and packaged for the actuation of product at prescribed rates to control target pests in accordance with the lawful labeling instructions. The product label contains approved conditions for use, special restrictions, hazard warnings and general information about the product (6). Conformance with the labeling instructions would be expected to reduce exposure to the user and other non target organisms while delivering the product to the intended target. Exposure would be expected to occur to the user through the occupational activities required during handling while exposure to nontarget organisms might occur through the incidental contact with residues. Thus the source of exposure along the environmental health paradigm for pesticides (Exhibit 1) is set along parallel tracts for occupational and incidental exposure.

**Environmental Concentrations.**     Occupational exposure is clearly a human condition where the user is expected to come into contact with relatively high environmental concentrations of the product in different exposure media during mixing, loading, application, and harvesting (7,8). As depicted in Exhibit 1, the sources of occupational exposure would be expected to be the formulation concentrate, the end use product, the application rate and the commodity or target. The formulation concentrate might refer to emulsifiable concentrates, wettable powders and certain dry flowables and microencapulated formulations that are mixed with water to produce the end use product (9). The end use product may also be "ready to use" as an aerosol to be actuated at specified application rates. The

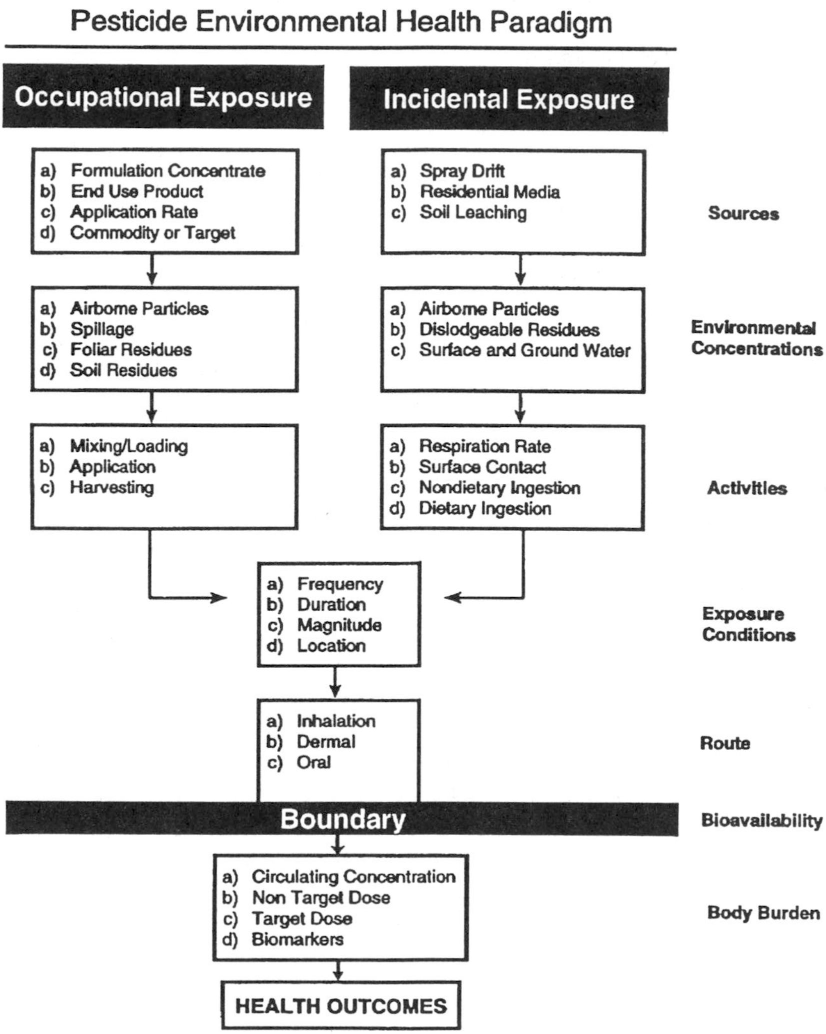

## Pesticide Environmental Health Paradigm

| Occupational Exposure | Incidental Exposure | |
|---|---|---|
| a) Formulation Concentrate<br>b) End Use Product<br>c) Application Rate<br>d) Commodity or Target | a) Spray Drift<br>b) Residential Media<br>c) Soil Leaching | **Sources** |
| a) Airborne Particles<br>b) Spillage<br>c) Foliar Residues<br>d) Soil Residues | a) Airborne Particles<br>b) Dislodgeable Residues<br>c) Surface and Ground Water | **Environmental Concentrations** |
| a) Mixing/Loading<br>b) Application<br>c) Harvesting | a) Respiration Rate<br>b) Surface Contact<br>c) Nondietary Ingestion<br>d) Dietary Ingestion | **Activities** |
| | a) Frequency<br>b) Duration<br>c) Magnitude<br>d) Location | **Exposure Conditions** |
| | a) Inhalation<br>b) Dermal<br>c) Oral | **Route** |
| **Boundary** | | **Bioavailability** |
| | a) Circulating Concentration<br>b) Non Target Dose<br>c) Target Dose<br>d) Biomarkers | **Body Burden** |
| | **HEALTH OUTCOMES** | |

Exhibit 1. Environmental Health Paradigm for the pesticide exposure as taken from Pirkle et al. (4).

application rate is generally expressed in mass of active ingredient delivered per area treated. The area treated may be a commodity or other target such as residential structures and media.

**Activity Patterns.**    The source is likely to give rise to certain environmental concentrations that may pose a hazard during the performance of activities (7,10,11). For example, wettable powder formulations may pose a hazard from the dispersion of airborne particles during mixing and loading while emulsifiable concentrates may pose a hazard from spills and splashes (9). Harvesters may contact foliar and soil residues during harvesting a certain agricultural commodity (7). The hazard would be dependent on the application rate, the re-entry interval between application and harvesting and the foliar characteristics of the commodity. For example, a greater hazard might be posed by an early unscheduled re-entry into a field where high rates of an insecticide were required to meet an emergency application on a commodity that requires a great deal of hand manipulation.

**Exposure Conditions.**    The severity of the exposure is determined by the magnitude, frequency and duration of contact as it would be for incidental exposure. The sources of incidental exposure would be expected to contribute to much lower environmental concentrations than those from occupational exposure (8). For example, spray drift may pose a potential hazard to organisms adjacent to agricultural lands. Spray drift is the movement of airborne particles  or vapors from the intended site of application to an off-target site (9). Droplet size predicts spray drift potential with mobility diminishing rapidly with droplets or particles ranging above 150 to 200$\mu$m in approximate diameter. Vapor drift might occur with highly volatile pesticides applied under environmental conditions that promote rapid volatilization and the movement of the pesticide as a gas to off-target sites.

**Routes of Exposure.**    The magnitude, frequency and duration of contact with a contaminant would be dependent on the activity patterns of the organism (12). Activities in closed environments might contribute to respiratory exposure to vapors and particles (2,13,14). Fall-out of particles on residential media at intended sites and off-target sites might contribute to contact with and dermal transfer of dislodgeable residues (15,16,17). Residues transferred to the hands may be percutaneously absorbed or ingested (17,18,19). Accumulation of residues on soil contribute to soil leaching into groundwater or run off to surface waters resulting in a persistent reservoir of contaminants for ingestion and percutaneous absorption (2,20).

**Bioavailability.**    The sources, environmental media, activities, exposure conditions and routes of exposure form the environmental pathways  which complete the exposure segment of the environmental health paradigm (Exhibit 1). While the risk of exposure can be explored and evaluated through study of the first five elements of the environmental health paradigm (21,22,23), exposure assessment is a subset of a larger problem as seen by the addition of internal dose and adverse environmental health outcomes in the total risk assessment (Exhibit 1). The weak link in the paradigm is the cusp between the elements of exposure and the perceived or realized adverse environmental health outcomes. Very few studies have been designed and performed that include the elements of exposure with the study of dose (4). In Exhibit 1, the cusp between exposure and dose is filled by the term bioavailability which may be seen as the means by which the environmental concentrations reach the biological boundary as determined by pathways and routes of exposure (3).

The route of exposure, oral, dermal and inhalation, controls the rate and magnitude of the expression of the bioavailability which describes the completeness of the uptake of a substance from the environmental media (2). This control is modulated by the conditions, locations and the exposure medium, e.g., air, water, and soil which defines the exposure pathway. The potential dose may be seen as the amount of xenobiotic that exists along an exposure pathway and can become associated with an exposure route (5). In dermal exposure of humans, the potential dose is the amount of xenobiotic in a vehicle matrix such as soil, dust, water or an organic solvent that forms a film over the surface of the skin. The potential dose may be acquired outdoors while harvesting vegetables laden with surface residues of a pesticide as part of the pathway. Ingestion of the residue laden vegetables would contribute to the potential oral dose. Inhalation of vapors and particles would contribute to the potential respiratory dose.

**Body Burden.**        Biomarkers enter the environmental health paradigm at the point where the organism acquires an internal dose (Exhibit 1). The amount of the xenobiotic that enters the organism through the dermal, oral, and inhalation routes defines the internal dose. Pirkle et al.(4) further define the portion of the internal dose that reaches a tissue of interest as the delivered dose or body burden. The tissue of interest may be the tissue site of action and that portion of the delivered dose has been referred to as the biologically effective or target dose (4). Body burden is seen here (Exhibit 1) as the circulating concentration of xenobiotic equivalents in blood and other biological fluids, e.g. cerebrospinal fluid (CSF), and the non target dose and target dose in tissues. Biomarkers represent identifiable and measurable biochemical, physiological and histopathological-pathological indicators and expressions of the body burden.

The NRC (2) defines a biological marker (biomarker) as a cellular or molecular indicator of toxic exposure, adverse health effects, or susceptibility (2). Biomarkers have been classified as biomarkers of exposure, biomarkers of effect, and biomarkers of susceptibility (2). More specifically, biomarkers are seen as biochemical, physiological, and histological indicators of change in an organism at the organismic, sub-organismic, cellular and subcellular level (24). Very simply, biomarkers are tools for detecting exposure and assessing untoward biological effects. In this context, biomarkers should have diagnostic and prognostic value. Therefore, biomarkers should be relevant to the human clinical situation as well as to veterinary medicine, horticulture, and ecological health. In this regard, humans share the same level of concern as all other organisms (24).

**Use of Biomonitoring Data.**

The Toxic Substances Control Act (TSCA - 15 U.S.C. 2601) and the Federal Insecticide Fungicide and Rodenticide Act (FIFRA - 7 U.S.C. 136, *et seq.*) were enacted to control, manage and thereby reduce exposure and injury of humans and other non-target flora and fauna to environmental chemicals. Evidence acquired from studies performed according to the TSCA test rules and Pesticide Assessment Guidelines under FIFRA would be expected to furnish the necessary toxicological and exposure data to reduce uncertainty in risk assessment. The Pesticide Assessment Guidelines (25) describe the requirements for acceptable study protocols, test conditions, and data collection and reporting. As set out in Exhibit 2, the requirements have been harmonized with the OTS TSCA Test Rules and OECD Guidelines and offered under the revised U.S. EPA Office of Pollution, Pesticides and Toxic Substances (OPPTS) nomenclature (26). For example, the requirements under the OTS TSCA Test Rules (Part 798-Health Effects Testing Guidelines) and Subpart F (Hazard Evaluation: Human and Domesticated Animal) have been harmonized with the OECD Guidelines and given the new OPPTS designation, Series 870 (26).

The OPPTS ecological effects (Series 850) and health effects (Series 870) guidelines are expected to minimize variation among and between testing procedures that must be performed to meet the data requirements under TSCA and FIFRA. In an analogous way, the other existing Pesticide Assessment Guideline Subdivisions (Exhibit 2) will be reissued as harmonized OPPTS Guidelines. The Re-entry Exposure (Subdivision K) and Application Exposure Monitoring Guidelines (Subdivision U) will be merged into the single harmonized OPPTS Series 875-Occupational and Residential Exposure Test Guidelines. Although incomplete at this juncture, the Series 875 guidelines will likely consist of individual guidelines for studies of the dissipation of foliar dislodgeable residues, soil residues and indoor surface residues, and human exposure via the dermal and inhalation routes. Included with these exposure guidelines will be the "Assessment of Dose Through Biological Monitoring Guidelines 875.2600.

**Biomarkers of Human Exposure.** As presented in this volume, Needham and in a subsequent paper, Hill et al., explore the benefits of using biomarkers and the value of biomonitoring data to assess human exposure to chemicals and metals important to TSCA and FIFRA according to methods established and performed by the National Center for Environmental Health. In this lead paper, Needham compares the virtues of biomonitoring to assess exposure and characterize risk with the traditional epidemiological approach where potential exposure to the individual or population is assessed through the determination of an exposure index (27). The exposure index is used to explain the total concentration of a substance in the environmental media that humans contact integrated over time. As explained by Needham, this "environmental approach" for assessing risk which forms the first five elements of the environmental health paradigm (Exhibit 1) is troubled by considerable uncertainty.

Under the best experimental design conditions, environmental samples, e.g., air, water, soil and food, are collected at an exposure site in conjunction with information about the frequency, duration and magnitude of contact with the contaminated environmental media along with medical histories. Difficulties in interpreting the causative relationships between exposure as determined from the environmental approach of collecting environmental samples, clinical information and activity patterns information through survey questionnaires have been recognized where multiple exposure pathways (sources, media, and routes of exposure) coincide with mixtures of causative agents, individual variability in susceptibility and behavior and the latency period of exposure (5). These factors are reviewed by Khalek and coworkers in this volume, as they explore the social, cultural, psychological, anatomical, and physiological variation among human populations that impact on human exposure and the use of biomarkers. Khalek et al., differentiate and classify geographically distinct human populations in developing countries by certain intrinsic genetic markers, e.g. blood type groupings, in combination with extrinsic markers, e.g., infectious disease distributions, age, development, and nutrition.

As observed by Needham and in a subsequent paper by Hill and colleagues in this volume, the environmental sampling approach to obtain an exposure index is best suited to identify and at times locate sources and media of concern with the intent to limit and prevent further exposure. As observed by Hill et al., the measurement of internal dose, what we choose to refer to as body burden, assesses actual exposure. We submit, that the measurement of body burden legitimizes and verifies the inferential information gained from the environmental approach.

As asserted by Needham in this volume, the best measure of exposure for assessing dose-response relationships is the biologically effective dose or target dose. For the reconstruction of dose related and health related events along the environmental health paradigm, the target dose is seldomly quantafiable although the adverse effects are generally observable (5). For example, anticholinesterase agents such as the organophosphorus (OP) and carbamate insecticides, have a clearly defined site and mode of action (28), yet, as discussed by Padilla et al., in this volume, isolation of the target tissue dose is impractical. Padilla et al., present the limitations and utility of blood cholinesterase (ChE) as a biomarker of effect as well as a biomarker of exposure. Padilla et al., further recognize the pivotal relationship that must be struck between blood cholinesterase inhibition and the onset of adverse effects and behavioral changes that follow interaction with the target site, the acetylcholinesterase enzyme (AchE: EC 3.1.1.7) that hydrolyzes the neurotransmitter, acetylcholine, that is responsible for the transmission of electrical impulses between nerve synapses and neuromuscular junctions (28).

| FIFRA 40CFR158 Subpart | Guideline Reference Numbers | OTS TSCA Test Rules | OECD Guidelines | OPPTS Number |
|---|---|---|---|---|
| D Product Chemistry | 61-64 | None | None | Series 810 & 830 |
| E Wildlife and Aquatic Organisms | 71-72, 141 | Part 795.xxxx | 201-207, 305E | Series 850.1000 |
| F Hazard Evaluation: Human and Domestic Animals | 81-85 | Part 798.xxxx | 401-416, 173-181, 117, 118, 419, 451-477 | Series 870 |
| J Hazard Evaluation: Nontarget Plants | 120-124 | Part 797.xxxx | None | Series 850.4000 |
| K Reentry Exposure | 130-134 | None | None | Series 875.xxxx |
| M Biochemical and Microbial Pesticides | 151-153 | None | None | Series 880 & 885 |
| N Chemistry: Environmental Fate | 161-167 | Part 796.xxxx | None | Series 835 |
| O Residue Chemistry | 171 | None | None | Series 860.xxxx |
| R Spray Drift | 201 | None | None | Series 835.1580 |
| U Application Exposure Monitoring | 231-236 | None | None | Series 875.xxxx |

Exhibit 2. Proposed Pesticide Assessment Guidelines as harmonized with the OTS TSCA Test Rules and OECD Guidelines and developed as the OPPTS Series Guidelines as described by Auletta (26).

This understanding is shared by Ehrich as she describes the biochemical and physiological requirements for the inhibition of another important carboxyl esterase known as Neuropathy Target Esterase (NTE) and its utility in predicting the expression of organophosphate - induced delayed neuropathy (OPIDN). This neurotoxic disorder has been described most accurately as a central peripheral distal axonopathy (29) where the long nerve fibers (axons) degenerate from a point distally along their length from a point of "chemical transection" (30). This "chemical transection" is thought to resemble the physical severing of the nerve fiber resulting in the pathologic loss of the axon with the survival of the cell body. Ehrich refers to three clinically and morphologically distinct forms of OPIDN, Type I associated with phosphoric acid, phosphonic acid, and phosphoramidic acid derivatives, Type II OPIDN associated with triphosphate compounds and a third "intermediate syndrome" associated with respiratory failure and injury to the diaphragm.

Unlike the well developed understanding of the syndrome of toxicity (28) and mode of action (31) of cholinergic poisoning following OP exposure, the relationship between NTE inhibition and the expression of OPIDN remains problematic (32). Ehrich presents a very lucid and complete discussion of the relationship and differences between OP induced acute neurotoxicity resulting from AchE inhibition and the delayed and often progressive neurotoxicity resulting from NTE inhibition. She further examines the value of the use of NTE as a field marker of OP exposure compared with the use of blood ChE. In addition to these comparisons, Ehrich explains the realized value of NTE as an indicator of OPDIN in premarket testing.

As the papers by Needham, Hill et.al, Padilla et al., and Ehrich reveal, the internal dose is more easily determined from an examination of the delivered dose in tissues apart from the target dose. The exposure can be detected and the agent identified in easily accessible tissues such as blood, and excreta. The mildly invasive collection and examination of blood and the noninvasive examination of urine and to a lesser extent human milk form the core biological matrices that have been used to develop biochemical and metabolic biomarker assays (3). One novel approach to the investigation of possible OP insecticide exposure as presented by Seifert in this volume, is the increased urinary excretion of xanthurenic acid. This method would be expected to be complimentary with the measurement of blood ChE inhibition as described by Padilla et al., in this volume and possibly provide an important link between cholinesterase activity and urinary metabolite excretion.

As an alternative method for detecting contaminants in blood, Saleh and his colleagues explore the binding of pesticides and two industrial chemicals, trichlorophenol (TCP), and polychlorinated biphenyls (PCB's 1221), to serum proteins and the resultant shifts in the associated serum protein profile. Saleh et al., demonstrate the utility of Fast Protein Liquid Chromatography (FPLC) as a tool for rapid screening of human plasma for protein adducts. In another study, Saleh and coworkers examine the value of human mothers milk as an important biological matrix in exposure surveillance of three distinct populations at different levels of risk of exposure to lead and certain chlorinated insecticides. These studies serve to illustrate the linkage between human time-location activity patterns and the bioaccumulation of xenobiotics in accessible human fluids of human populations in developing countries as presented by Khalek et al.

These biomarkers may be categorized as delivered dose biomarkers whereas blood cholinesterase and NTE are clear examples of biomarkers of effect (5). Sexton et al. (5) further categorize biomarkers as effective dose markers, e.g. DNA, protein and hemoglobin adducts, and adverse effects (subclinical disease) biomarkers which may be seen as cellular or tissue changes. We prefer to lump these subclassifications into a single category as "endogenous response biomarkers". With respect to this classification, Garst and Bieber et al., present the potential use of L-carnitine and carnitine esters as biomarkers of metabolism and cellular and subcellular injury. These "endogenous response biomarkers" are important products of mitochondrial β-oxidation of long-chain fatty acids. Interference with fatty acid β-oxidation would have dire effects on energy production as expressed through a secondary carnitine deficiency. In his paper, Garst theorizes the role that acylcarnitine ratios play in the expression of hepatomegaly and the production of non-genotoxic hepatic tumors in response to structurally diverse peroxisomal proliferating agents. Beiber et al., in a subsequent paper, explain how the urinary acylcarnitine profile can be used for diagnostic purposes.

**Biomarkers of Environmental Health.**   Biomarkers find importance as biomarkers of environmental health just as they do in human health risk assessment (33). Biomarkers assume a broader importance beyond the human health situation as biomarkers or indicators of exposure, susceptibility, and toxic effect that impact on numerous species in diverse communities of which humans are a part and where humans can act as benign participants or stressors (24). Deleterious changes in ecosystem health that impact risk prediction are often indicated by sharp reductions in the population densities of certain species through injury and lethality that contribute to declines in species diversity and biomass (34). Thus, the morbidity and mortality of populations of certain species, such as nontarget beneficial insect species, assume the unfortunate role as biomarkers where intervention or r emmediation are acutely indicated. Under certain circumstances, populations and species will likely remain "biomarkers of stress" through injury and lethality until external and endogenous response biomarkers are found.

We apply the term "external biomarker" to evidence of exposure or stress of a species as determined by the detection and measurement of physical, chemical and biological changes in biological remains such as excreta, exuviae, eggs, feathers, leaves and seeds that can be obtained without stress to the organism (35). Such biological remains might lend themselves to traditional residue analysis of persistent chemicals, e.g. polychlorinated biphenyls (PCB's), as logical compliments to residue analysis of co-located environmental media for monitoring the status and trends of contaminated sites and for the surveillance of sensitive and endangered species. An external biomarker might also be traced back to the living organisms to be detected and measured as an "endogenous response biomarker" with immediate importance for intervention.

As detailed in the reviews of McCarthy and Shugart (36), Peakall (34) and Huggett et al. (24), the list of currently recognized biomarkers ranges over several levels of subcellular, cellular and tissue organization (35). The majority of these biomarkers and others recently recognized by Butterworth et al. (37) may be classified as endogenous response biomarkers. Direct analysis of blood and urine of wild fauna for xenobiotic equivalents and metabolites is not practiced to the same extent as it is in human clinical surveillance or biological monitoring. A very limited number of species are suitable for biological monitoring of status and trends. Large mammals and certain avian and reptile species might be suitable for routine nondestructive blood and tissue monitoring while small and delicate species may require whole-body analysis (34,37). The destructive use of any species carries with it profound ethical considerations (35) and therefore a primary aim of biomarker development in ecotoxicology is the development of noninvasive and nondestructive biomarkers (34,35).

As observed by Peakall (34), the wealth of biomarker research in ecotoxicology has grown larger than that which a single treatise can convey. The growth in the volume of scientific papers devoted to ecotoxicology has very recently escalated as compared to the more established record of proliferation of papers in environmental health and other related disciplines that impact on the development of the environmental health paradigm (Exhibit 3). The reader is invited to refer to the extensive reviews of McCarty and Shugart (36), Jeffrey and Madden (38), Huggett et al. (24), Peakall (34), Peakall and Shugart (39) and Butterworth et al. (37). In this volume, Anderson and coworkers and Davis and colleagues review the use of two novel techniques, the reporter gene system (RGS) assay and fluorescence-based catalytic (EROD) assay respectively, to measure one of the most well researched endogenous response biomarkers, the induction of P450 enzymes (24,34). Genetic engineering is used to derive the RGS assay upon which the measurement of inducers of P450 in environmental media and animal tissues is related to relative toxicity and potential carcinogenicity. Thus, the RGS assay may be seen as a genetically designed external biomarker of effect that can be used to study exposure to environmental concentrations of P450 inducers in the initial stages of the environmental health paradigm where as Sexton et al. (5) point out, the opportunities for intervention and prevention of unacceptable risks would be most timely and effective.

The RGS and EROD assays serve another important field application as surveillance and monitoring tools for the study of status and trends of exposure of marine animals (see Anderson et al.) and birds (see Davis et al.). Davis and coworkers describe the value of the EROD assay for the longitudinal study of exposure of double-crested cormorants to TCDD-like compounds. With this example, Davis et al., explain the importance and perhaps essential relationship that can and possibly should always be established between controlled laboratory studies of biomarker responses and responses expected in the field. The successful application of these biomarker assays can be attributed to the extensive research in toxicology and hazard evaluation of the analytes, chlorinated insecticides, dioxins, PCB's, and polycyclic aromatic hydrocarbons (PAH's) to document and support mechanisms of action, metabolism, disposition and potential harm (35).

The establishment of endogenous biomarker assays for field application must necessarily lag behind the discovery of unusual events that signal possible ecological imbalances such as the disproportionate bioaccumulation of rare earth elements in the organs of fresh water fish from Lake Mead as seen by Hodge and colleagues. Such ecological field studies have traditionally been followed by toxicity testing as part of the risk assessment process. Another approach that has gained acceptance is the prediction of toxicity of chemical agent simulants (CAS) by quantitative structure activity relationships and the use of acute bioassays (see Chester et al.). In this paper, Chester et al., describe the predictive power of QSAR for the generation of environmental toxicity data for chemical structures with missing values. The use of statistical models such as the QSAR model from TOPKAT° (Health Designs, Inc.) described by Chester and coworkers have become important and valuable tools in risk assessment particularly as budgets for toxicity testing and hazard evaluation continue to decline (5).

**Mathematical Modeling in Risk Assessment.** Mathematical modeling holds the promise of assembling data bases and algorithms associated with each element of the environmental health paradigm. Advances in computer programing have lead to the creation of robust information based systems that facilitate the interaction between models and data in a personal computer (PC) environment (40,41). One such PC driven data and modeling platform, the Total Human Exposure Relational database and Advanced Simulation Environment (THERdbASE°), manages and processes information relating to sources, environmental concentrations, activity patterns, and dose assessment parameters and data (40). Physiologically based pharmacokinetic (PBPK) models represent the most well developed system of algorithms for processing dose assessment parameters and data. PBPK models are systems of mass balance equations (42) used to unravel the acquired dose into compartments of mass equivalents. PBPK models are built on assumptions that depend on the use of plausible estimates that relate to physiological processes and on direct measurements of mass and expression of mass action as taken from biomarker data bases.

One of the most powerful and complete PBPK models is the one presented in this volume by Knaak and coworkers for the examination of multiple routes of exposure and metabolic pathways in three animal species to isofenphos. The model has been expanded from previous work (43) to include metabolism studies via intravenous (i.v.) infusion and single i.v. bolus in the rat, i.v. bolus and dermal routes in the dog and intraperitoneal (i.p.) route (single bolus and infusion) in the guinea pig. One of the important features of this model is the linkage between tissue concentrations of $[^{14}C]$-ring stabilized and $[^{3}H]$-ethyl equivalents of isofenphos and the expression of enzyme markers of exposure, e.g., inhibition of Ache, and metabolism, e.g. induction of P450. The reader is advised to study this model in

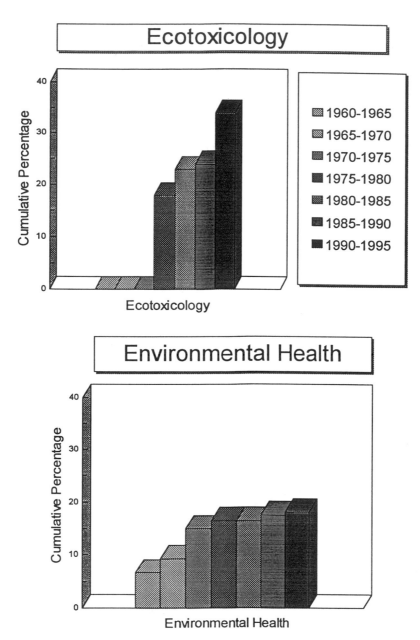

Exhibit 3.  Growth in the percentage of scientific papers dedicated to Pesticide Ecotoxicology and Environmental Health as contrasted against scientific papers reporting findings on Pesticides and Epidemiology, Chemistry, Human Exposure, and Toxicology.

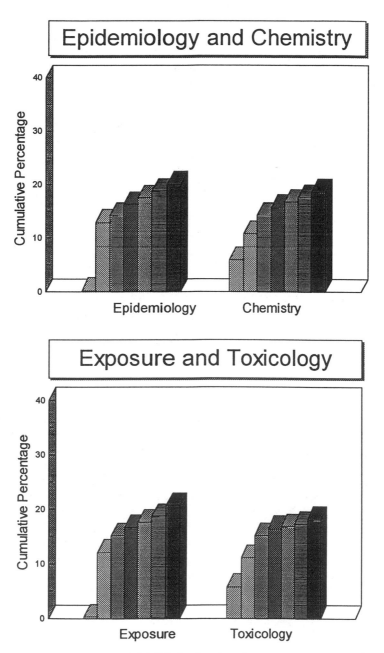

Exhibit 3. *Continued*

relation to the use of these marker enzymes in exposure assessment as presented by Padilla et al., Ehrich, and Anderson et al., and  Davis et al., in this volume.

Another important feature of the Knaak model is the complete accounting of mass balance in blood and excreta. The relationships struck between tissue concentrations, metabolism, receptor interactions and clearance in animal models may form the pharmacokinetic basis for the reconstruction of dose in humans where exposure is most often determined through the measurement of parent compounds in blood and metabolites in urine (3). The power of PBPK modeling for interpretation of internal dose in humans from the measurement of metabolites in urine is convincingly presented by Dong and coworkers. In their paper, Dong et al., validate the reconstruction of dose from spot urine samples from individuals incidentally exposed to malathion  against controlled laboratory studies.

The success of model validation is dependent on the reliability (accuracy, precision and representativeness) of the data and the symbolic and analytical methods incorporated into the model. These symbolic and analytical tools are detailed by Brown  for the expansion of a four compartment model to an eight compartment model with solutions given in a subsequent paper by Hagstrom, Brown and Blancato. These linear PBPK models provide comparative route-to-route formulas for the extrapolation of dose  between animal systems and humans to identify dominant tissue and organ pathways. Properly validated PBPK models may be seen as the mathematical and statistical means to test default exposure assumptions, measure uncertainty of specific physiological and biochemical parameters, and predict tissue uptake and response biomarker expression without animal experimentation.

**Application of the Environmental Health Paradigm.**

Putting the environmental health paradigm into action requires a system of postulates associated with each of the elements that can be supported by data or inferences (44). The aim of an environmental risk assessment is to obtain likelihood projections for present and future intermediate and final environmental health outcomes (45,46). An intermediate health outcome is the appearance of biological changes in an organism caused by stress or illness that cannot be felt by the organism but can be measured as a biomarker (46). A health outcome may be the physiological stress resulting from continual or repeated insults to cells, organs and organ systems and the development of a disease state. Eddy et al. (46) describe a human health outcome as an outcome of disease or injury that impacts on the length and quality of life such as pain, anxiety and disfigurement. In humans, a health outcome may be the psychological stress or dread of  uncontrollable exposure or loss of income with the advance of a particularly onerous disease state. An ecological health outcome may be identified as the alteration or destruction of habitat or the reduction or loss of reproductive potential (24,34,35).

The successful prediction of present and future health outcomes depends on the reliability of the data, inferences and postulates that enter the linked elements of the environmental health paradigm (5). Rarely if ever will a single study be perfectly designed and conducted to obtain results with the level of precision that could be used as the best estimates of parameters for each of the linked elements of an environmental health assessment (46). The evidence is generally acquired piecemeal from studies designed at various levels of control and consistency (45). Studies performed to satisfy regulatory requirements would be expected to conform with regulatory and Good Laboratory Practices (GLP) guidelines (47). Studies performed in the interest of exploratory research and development may not follow such regulatory and GLP guidelines although the data obtained may be in retrospect used in exposure and risk assessment.

Standards of good laboratory practice (GLP) as specified under Good Laboratory Practice Standards Final Rule (48) are expected to control the quality of animal experiments under guidelines set by TSCA and FIFRA. The aim of guideline harmonization is to arrive at a single, consistent and suitably detailed guidance document that can assist researchers, pesticide registrants and applicants. Guideline harmonization is also expected to reduce the data collection burden by providing a common set of procedures. In the case of pesticide registration,  toxicity data is required under 40 CFR 158.340 while dissipation and post-application exposure data for re-entry protection to pesticides is conditionally required under 40 CFR 158.390. The decision to require  post-application exposure data is dependent on toxicity criteria detailed in 40 CFR 158.340. Thus exposure data for certain substances may not be obtained. This situation is not unreasonable given the expense of conducting occupational exposure and environmental monitoring studies. However, evidence of untoward effects from a previously tested substance may surface years later as evidenced by the interest in non-hodgkins lymphoma (49,50) and estrogen mimics or endocrine disruption (51,52).

The challenge that faces research planners and practitioners of the environmental health paradigm is the assessment of uncertainty and sensitivity of parameters and data bases associated with the elements of the paradigm. Such an assessment might identify extreme sources of variation in certain elements that might render more stable or conservative distributions meaningless (5,44). This can be accomplished through the use of complete modeling systems like THERdbASE* where prior distributions of exposure and dose related data can be tested against incoming data to estimate in a Bayesian fashion, joint posterior distributions for parameters of interest (44,46,53). A great deal of data has been amassed in various environmental databases that could be used to establish prior distributions and test the status and trends in environmental and human health and predict future environmental health outcomes (54-58). Moreover, with the reality of dwindling budgetary resources for basic and applied research, it becomes essential to assess existing databases to establish prior distributions from which to test incoming data from proposed national probability health surveys in an iterative fashion.

National probability health surveys such as the National Health and Nutrition Examination Survey (NHANES) (59) and the National Human Exposure Assessment Survey (NHEXAS) (60-66) are expected to provide a large amount of information on the prevalence of health conditions and the distributions of physiological, biochemical, and exposure related parameters for the assessment of current status and historical trends and predict future human health outcomes. Underlying these extensive surveys are numerous databases (52) from which information can be drawn to test prior assumptions. In this way, the cost of obtaining additional data may be evaluated against the statistical significance of the outcomes.

## Literature Cited

1.  Risk Assessment in the Federal Government: Managing the Process. National Academy Press: Washington, D.C., pp. 191. 1983. NRC (National Research Council).

2.  Monitoring Human Tissues for Toxic Substances. National Academy Press: Washington, D.C. 1991. Committee on National Monitoring of Human Tissues.

3.  Nauman, C.H.; Santolucito, J.A.; Dary, C.C. In: Biomonitoring for Pesticide Exposure. Saleh, M.A.; Blancato, J.N.; Nauman, C.H. ACS Symposium Series 542; Washington, D.C., 1994.

4.  Pirkle, J.L.; Needham, L.L.; Sexton, K. Journal of Exposure Analysis and Environmental Epidemiology, 1995, 5, 405-424.

5.  Sexton, K.; Callahan, M.A.; Bryan, E.F. Environmental Health Perspectives, 1995, 103.

6.  Baker, S.R.; Wilkinson, C.F. In: The Effects of Pesticides on Human Health: Introduction and Overview. Wilkinson, C.F. 1990, 18, pp. 5-33.

7.  Knaak, J.; Iwata, I.; Maddy, K.T. In: The Risk Assessment of Environmental and Human Health Hazards: A Textbook of Case Studies; Paustenbach, D.J., Ed., John Wiley & Sons, New York, 1989.

8.  Nigg, H.N.; Beier, R.C.; Carter, O.; Chaisson, C.; Franklin, C.; Lavy, T.; Lewis, R.G.; Lombardo, P.; McCarthy, J.F.; Maddy, K.T.; Moses, M.; Norris, D.; Peck, C.; Skinner, K.; Tardiff, R.G. In: The Effects of Pesticides on Human Health: Exposure to Pesticides. Baker, S.R.; Wilkinson, C.F., 1990.

9.  Meister, R.T.; Fitzgerald, G.T.; Sine, C.; Miller, W.J. (II); Meister Publishing Company. In: 1993 Farm Chemicals Handbook, 1993, 79.

10. Reinert, J.C.; Severn, D.J. Dermal Exposure to Pesticides: The Environmental Protection Agency's Viewpoint. ACS Symposium Series 273, Washington, D.C. American Chemical Society, pp. 357-68. 1985.

11. U.S. EPA Nonoccupational Pesticide Exposure Study (NOPES). Final Report. Office of Research and Development. Washington, D.C. 1990.

12. Nelson, W.C.; Ott, W.R.; Robinson, J.P. The National Human Activity Pattern Survey (NHAPS): Use of Nationwide Activity Data for Human Exposure Assessment. 94-WA75A.01, Proceedings of the A&WMA 87th Annual Meeting, Cincinnati, OH, June, 1994.

13. Lewis, R.G. Human Exposure to Pesticides Used in and Around the Household. Baker, S.R.; Wilkinson, C.F. In: The Effects of Pesticides on Human Health. Advances in Modern Environmental Toxicology. Princeton Scientific, 18, 1990.

14. Ott, W.R. Human Exposure Assessment: The Birth of a New Science. Journal of Exposure Analysis and Environmental Epidemiology, 5, 1995.

15. Berteau, P.E.; Knaak, J.B.; Mengle, D.C.; Schreider, J.B. Insecticide Absorption from Indoor Surfaces. In: Biological Monitoring for Pesticide Exposure. ACS Symposium Series 382. Wang, R.G.; Franklin, C.A.; Honeycutt, R.C.; Reinert, J.C. pp. 315-326. American Chemical Society, Washington, D.C.

16. Ross, J.; Fong, H.R.; Thongsinthusak, T.; Margetich, S.; Kreiger, R. HS. 1603, Interim Report, California Department of Food and Agriculture. 1990.

17. Finley, B.L.; Scott, P.K.; Mayhall, D.A. Development of a Standard Soil-to-Skin Adherence Probability Density Function for Use in Monte Carlo Analyses of Dermal Exposure. Risk Analysis, 14, 1994.

18. Sheppard, S.C. Parameter Values to Model the Soil Ingestion Pathway. Environmental Monitoring and Assessment, pp. 27-44, 34, 1995.

19. Simcox, N.J.; Fenske, R.A.; Wolz, S.A.; Lee, I.C.; Kalman, D.A. Pesticides in Household Dust and Soil: Exposure Pathways for Children of Agricultural Families. Environmental Health Perspectives, 103, 1995.

20. Turkall, R.M.; Skowronski, G.A.; Abdel-Rahman, M.S. Differences in Kinetics of Pure and Soil-Adsorbed Toluene in Orally Exposed Male Rats. Arch. Environ. Contam. Tolicol., 20, pp. 155-160, 1991.

21. Kopfler, F.C. and Craun, G.F.. Assessment of Human Exposure to Chemicals Through Biological Monitoring. In: Environmental Epidemiology. Bernard, A.M. and Lauwerys, R.R. pp. 17-27, 1986.

22. Paustenbach, D.J. The Practice of Health Risk Assessment in the United States (1975-1995): How the U.S. and Other Countries Can Benefit From That Experience. Human and Ecological Risk Assessment, 1, pp. 29-79, 1995.

23. Walker, C.H. Biomarkers and Environmental Risk Assessment. pp. 7-12, 7, 1996.

24. Huggett, R.J.; Kimerle, R.A.; Mehrle, P.M. (Jr.); Bergman, H.L. Biomarkers: Biochemical, Physiological, and Histological Markers of Anthropogenic Stress. SETAC Special Publication Series, 1989.

25. U.S. EPA (1984a) U.S. Environmental Protection Agency. Pesticide assessment guidelines. Subdivision K. Exposure: Reentry Protection. U.S. Environmental Protection Agency, Washington, DC.

26. Auletta, A.; OECD Test Guidelines Program. U.S. Environmental Protection Agency, 1996.

27. Needham, L.L.; W.M. Draper (ed.), Environmental Epidemiology: Effects of Environmental Chemicals on Human Health, Advances in Chemistry Series, 241, American Chemical Society, Washington, D.C., 1995, pp. 121-135.

28. Ecobichon, D.J. Toxic Effects of Pesticides. In: Casarett and Doull's Toxicology: The Basic Science of Poisons. Amdur, Mary O.; Doull, J. and Klaassen, C.D. Pergamon Press, 1992.

29. Spencer, P.S. and Schaumburg, H.H. Centralperipheral Distal Axonopathy: The Pathology of Dying-back Polyneuropathies. In Progress in Neuropathology. Zimmerman, H. (ed.):, 3, pp. 253-295, 1976.

30. Anthony, D.C. and Graham, D.G. Toxic Responses of the Nervous System. In: Casarett and Doull's Toxicology: The Basic Science of Poisons. Amdur, M.O., Doull, J. and Klaassen, C.D. Pergamon Press, pp. 407-429, 1992.

31. Eto, M. Organophosphorus Pesticides: Organic and Biological Chemistry. CRC Press, Cleveland, Ohio, 1974.

32. Johnson, M.K. Chem.-Biol. Interactions, pp. 339-346, 87 1993.

33. Melancon, M.J.; Alscher, R.; Benson, W.; Kruzynski, G.; Lee, R.F.; Sikka, H.C. and Spies, R.B. Metabolic Products as Biomarkers. In: Biomarkers: Biochemical, Physiological, and Histological Markers of Anthropogenic Stress. Huggett, R.J.; Kimerle, R.A.; Mehrle, P.M. (Jr.) and Bergman, H.L. Lewis Publishers, 1989.

34. Peakall, D.B. Animal Biomarkers as Pollution Indicators. Chapman and Hall, London, 1992.

35. Shugart, L.R. In: Use of Biomarkers in Assessing Health and Environmental Impacts of Chemical Pollutants. State of the Art - Ecological Biomarkers. Plenum Press, New York, 1993.

36. McCarthy, J.F. and Shugart, L.R. Biomarkers of Environmental Contamination. Lewis Publishers, Inc., Boca Raton, FL. 1990.

37. Butterworth, F.M.; Corkum, L.D. and Guzmán-Rincón, J. Biomonitors and Biomarkers as Indicators of Environmental Change, pp. 341. Plenum Press, New York and London.

38. Jeffrey, D.W. and Madden, B. Bioindicators and Environmental Management, pp. 445. Academic Press.

39. Peakall, D.B. and Shugart L.R.  Biomarkers:  Research and Application in the Assessment of Environmental Health.  NATO ASI Series H:  Cell Biology, 68, Springer-Verlag, Heidelberg, 1993.

40. Pandian, M.D.; Dary, C.C. and Behar, J.V.  The International Toxicologist:  An Official Publication of the International Congress of Toxicology, 50-PD-8, 7, 1995.

41. Pandian, M.D.; Dary, C.C.; Behar, J.V. and Hern, S.C.  Integrating Pesticide Handlers Exposure Database (PHED) with a Dermal-Driven PBPK Model.  Picogram and Abstracts, American Chemical Society, 1996.

42. Blancato, J.N.  Pharmacokinetics, Chemical Interactions, and Toxicological Risk Assessment in Perspective.  102, Supplement 9, November 1994.

43. Knaak, J.B.; Al-Bayati, M.A.; Raabe, O.G. and Blancato, J.N.  In:  Biomarkers of Human Exposure to Pesticides.  Saleh, M.A.; Blancato, J.N. and Nauman, C.H.  ACS Symposium Series 542.  American Chemical Society, Washington, D.C.  1994.

44. Brand, K.P. and Small, M.J.  Updating Uncertainty in an Integrated Risk Assessment: Conceptual Framework and Methods.  Risk Analysis, 15, 1995.

45. Eddy, D.M.  The Confidence Profile Method:  A Bayesian Method for Assessing Health Technologies.  Oper. Res. 37(2), pp. 210-228, 1989.

46. Eddy, D.M.; Hasselblad, V. and Shachter, R.  An Introduction to a Bayesian Method for Meta-Analysis:  The Confidence Profile Method.  Med. Dec. Making 10(1), pp. 15-23, 1990.

47. Hawkins, N.C.; Jayjock, M.A. and Lynch, J.  A Rationale and Framework for Establishing the Quality of Human Exposure Assessments.  Am. Ind. Hyg. Assoc. J. 53(1), pp. 34-41, 1992.

48. U.S. EPA (1992b) Good Laboratory Practice Standards Final Rule.  54 FR 34052.  1989.

49. Hoar, S.K.; Blair, A.; Holmes, F.F.; Boysen, C.D.; Robel, R.J.; Hoover, R. and Fraumeni, J.F. (Jr.)  Agricultural Herbicide Use and Risk of Lymphoma and Soft-Tissue Sarcoma, JAMA, 256, 1986.

50. Wigle, D.T.; Semenciw, R.M.; Wilkins, K.; Riedel, D., Ritter, L.; Morrison, H.I. and Mao, Y.  Mortality Study of Canadian Male Farm Operators:  Non-Hodgkin's Lymphoma Mortality and Agricultural Practices in Saskatchewan.  Journal of the National Cancer Institute, pp. 575-582, 82, 1990.

51. Davis, D.L.; Bradlow, H.L.; Wolff, M.; Woodruff, T.; Hoel, D.G. and Anton-Culver, H.  Medical Hypothesis:  Xenoestrogens As Preventable Causes of Breast Cancer. Environmental Health Perspectives, 101, 1993.

52. Colborn, T.; vom Saal, F.S. and Soto, A.M.  Developmental Effects of Endocrine-Disrupting Chemicals in Wildlife and Humans.  Environmental Health Perspectives, 101, 1993.

53. Hasselblad, V.; Eddy, D.M. and Kotchmar, D.J.  Synthesis of Environmental Evidence: Nitrogen Dioxide Epidemiology Studies.  Journal of the Air & Waste Management Association, 42, 1992.

54. Sexton, K.; Selevan, S.G.; Wagener, D.K. and Lybarger, J.A.  Estimating Human Exposures to Environmental Pollutants: Availability and Utility of Existing Databases.  Archives of Environmental Health, 47, December 1992.

55. Graham, J.; Walker, K.D.; Berry, M.; Bryan, E.F.; Callahan, M.A.; Fan, A.; Finley, B.; Lynch, J.; McKone, T.; Ozkaynak, H. and Sexton, K.  Role of Exposure Databases in Risk Assessment.  Archives of Environmental Health, 47, December 1992.

56. Burke, T.; Anderson, H.; Beach, N.; Colome, S.; Drew, R.T.; Firestone, M.; Hauchman, F.S.; Miller, T.O.; Wagener, D.K.; Zeise, L. and Tran, N.  Role of Exposure Databases in Risk Management.  Archives of Environmental Health, 47, December 1992.

57. Goldman, L.R.; Gomez, M.; Greenfield, S.; Hall, L.; Hulka, B.S.; Kaye, W.E.; Lybarger, J.A.; McKenzie, D.H.; Murphy, R.S.; Wellington, D.G. and Woodruff, T.  Use of Exposure Databases for Status and Trends Analysis.  Archives of Environmental Health, 47, December 1992.

58. Matanoski, G.; Selevan, S.G.; Akland, G.; Bornschein, R.L.; Dockery, D.; Edmonds, L.; Greife, A.; Mehlman, M.; Shaw, G.M. and Elliott, E.  Role of Exposure Databases in Epidemiology.  Archives of Environmental Health, 47, December 1992.

59. Ezzati-Rice, T.M. and Murphy, R.S.  Issues Associated with the Design of a National Probability Sample for Human Exposure Assessment.  Environmental Health Perspectives, 103, Supplement 3, 1995.

60. Sexton, K.; Kleffman, D.E. and Callahan, M.A.  An Introduction to the National Human Exposure Assessment Survey (NHEXAS) and Related Phase 1 Field Studies.  Journal of Exposure Analysis and Environmental Epidemiology, 5, Number 3.

61. Sexton, K.; Callahan, M.A.; Bryan, E.F.; Saint, C.G. and Wood, W.P.  Informed Decisions About Protecting and Promoting Public Health: Rationale for a National Human Exposure Assessment Survey.  Journal of Exposure Analysis and Environmental Epidemiology, 5, Number 3.

62. Callahan, M.A.; Clickner, R.P.; Whitmore, R.W.; Kalton, G. and Sexton, K.  Overview of Important Design Issues for a National Human Exposure Assessment Survey.  Journal of Exposure Analysis and Environmental Epidemiology, 5, Number 3.

63. Burke, T.A. and Sexton, K. <u>Integrating Science and Policy in a National Human Exposure Assessment Survey</u>. Journal of Exposure Analysis and Environmental Epidemiology, <u>5</u>, Number 3.

64. Lebowitz, M.D.; O'Rourke, M.K.; Gordon, S.; Moschandreas, D.J.; Buckley, T. and Nishioka, M. <u>Population-Based Exposure Measurements in Arizona: A Phase 1 Field Study in Support of the National Human Exposure Assessment Survey</u>. Journal of Exposure Analysis and Environmental Epidemiology, <u>5</u>, Number 3.

65. Pellizzari, E.; Lioy, P.; Quackenboss, J.; Whitmore, R.; Clayton, A.; Freeman, N.; Waldman, J.; Thomas, K.; Rodes, C. and Wilcosky, T. <u>Population-Based Exposure Measurements in EPA Region 5: A Phase 1 Field Study in Support of the National Human Exposure Assessment Survey</u>. Journal of Exposure Analysis and Environmental Epidemiology, <u>5</u>, Number 3.

66. Buck, R.J.; Hammerstrom, K.A. and Ryan, P.B. <u>Estimating Long-Term Exposures from Short-Term Measurements</u>. Journal of Exposure Analysis and Environmental Epidemiology, <u>5</u>, Number 3.

Chapter 2

# Proposed Use of Human Exposure Biomarkers in Environmental Legislation: Toxic Substances Control Act

Larry L. Needham

National Center for Environmental Health, Centers for Disease Control and Prevention, Public Health Service, U.S. Department of Health and Human Services, 4770 Buford Highway, Atlanta, GA 30341–3724

In the United States, many laws have been enacted to eliminate or to at least reduce human exposure to environmental chemicals. One such law is the Toxic Substances Control Act (TSCA) which includes approximately 72,000 chemicals. Recently, the U.S. Congress expressed interest in examining available technologies for screening all or a selected portion of these chemicals for health effects in humans and to a lesser degree, their effects on the ecologic system. In terms of risk assessment, such screening procedures would yield data for use in hazard identification. Obviously, in order to screen this immense number of chemicals, the chemicals must be prioritized. One such focus could be on the 14,000 nonpolymeric TSCA Inventory chemicals produced in amounts greater than 10,000 pounds per year. Nonetheless, screening all of these 14,000 chemicals for various health endpoints still requires further prioritization. No doubt, quantitative structure activity relationships will be used to set priorities. However, we submit that priority setting could also be based at least in part on another aspect of risk assessment--human exposure assessment--for without human exposure, there are no adverse health effects, and no need would exist for further risk characterization. Human exposure has been assessed by a variety of means. We prefer to assess human exposure by measuring biomarkers of exposure in human specimens, when such measurements are possible.

In this chapter, I give examples of how using such biomarkers provided qualitative and quantitative exposure information that proved useful in conducting epidemiologic studies. I also discuss how reference-range levels of exposure biomarkers in humans (that are determined through biomonitoring programs) have been extremely beneficial in conducting exposure assessments and how

expansion of such programs would directly benefit the TSCA.
Programs, such as the National Health and Nutrition Examination
Survey (NHANES) and the National Human Exposure Assessment
Survey (NHEXAS), could be available to collect and bank the
needed specimens. Analytical methods would then be used in these
programs to determine whether, and to what extent, humans were
being exposed to particular TSCA-related substances. If they were,
more extensive "effect screening" methodologies would be used to
characterize risk associated with these substances; if no, or if little
exposure was detected, these substances may be given a low
priority for "effect screening", and further risk characterization.

Humans are exposed daily to a variety of chemicals that are present in the
environment as pollutants or that are in commercial products. For this chapter, I
shall assume that these chemicals are included in the Toxic Substances Control
Act (TSCA) list of some 72,000 chemicals. Humans are exposed when they come
into contact for an interval of time with such chemicals in an environmental
medium--soil, water, air, food, or in another medium, such as a commercial
product, or in an occupational setting (1-4). Epidemiologists, risk assessors, and
others often classify the degree of exposure or potential exposure by using the
concentration of a given chemical in the medium that humans contact, integrated
over the time of contact; this is then the basis of an exposure index (5). When
humans have contact with these environmental media, the chemical may enter the
body via inhalation, ingestion, or skin absorption. Once in the body, the chemical
may be distributed to tissues, and adverse health effects may result.

The amount of chemical absorbed in body tissue is called the internal dose.
Common measures used to determine internal dose are the blood and urine levels
of chemicals or their metabolites (3). A portion of this internal dose may reach
and interact with a target site over a given period so as to alter physiologic
function; this portion is called the biologically effective dose (3). Other exposure
and dose terms are further defined by Sexton et al. (4).

Various methods have been used to assess human exposure to xenobiotics--
one such approach is the use of biomarkers. I present here case studies that
demonstrate the benefits of using biomarkers of exposure compared with using
other estimates of exposure. I then describe biomonitoring programs and
analytical methods that may be beneficial for setting priorities for TSCA. In
addition, I evaluate various methods that have been used to assess human
exposure.

**Exposure Index (without biomarkers).** Traditionally, exposure has been
assessed by estimating a person's or population's potential for exposure. If the
concentration of a given chemical in various media is known, then the total
concentration of that chemical in the environmental media that humans contact,
integrated over the time of contact, usually forms the basis of an exposure index.
The concentration of the chemical in the environmental media is sometimes based
on analytical measurements of environmental samples -- water, air, soil, food --
collected at the exposure site near or as to close to the time of exposure as

possible.  Depending on the pathway of exposure, all of these environmental media, and perhaps multiple samples of each, may have to be analyzed (sometimes at a high cost) and yet may not be representative of the concentration of the chemical in the media at the time of human exposure.  For example, is the average level of a pollutant in fish caught in a river representative of all such fish in that river?  Perhaps the "best" environmental sample for an airborne chemical would be a personal air sample collected at the time of exposure by an organic vapor badge; in one experiment, this technique correlated more highly with blood levels of selected volatile organic compounds (VOCs) than did VOC levels in breathing-zone air that was collected by charcoal tubes (6).  Clearly, such techniques are only available in designed experiments.  Information about the estimated time of contact, including frequency and duration, with the environmental medium containing the chemical is generally collected by questionnaire.  This combination of questionnaire/history information and environmental measurements are then weighted into an exposure model, which is used as an estimate of exposure for each person.   We call this the "environmental approach" for assessing exposure.

This approach may be useful in human-exposure assessment as a preliminary screen to help ascertain the potential for human exposure.  Various models have been used for both qualitative and quantitative predictions.  However, they are based on many assumptions that may contain several potential problems, such as the inability to adjust for individual factors that relate to how much chemical enters the body and how much is absorbed (e.g., individual metabolism differences, individual nutritional status during exposure, individual differences in surface area or body mass, and personal habits such as hand-to-mouth activities).  In addition, the frequency and duration of contact with the environment that contains the chemical are difficult to estimate because of uncertainty of recall or bias in administering and answering the questionnaire.  This bias may arise whenever noncomparable information is obtained from the different study groups, a factor that may be the result of the interviewer eliciting or interpreting the information differently from the interviewee's intent (interviewer bias) or of the participants either intentionally or unintentionally reporting the events in a noncomparable manner (recall bias).  For example, participants may have problems recalling the frequency of playing on contaminated soil or consuming a certain food.  Thus, we believe that such exposure indices are useful but are not the best means to assess human exposure to environmental chemicals.

**Types of Biomarkers of Exposure**

By definition, the best measure of exposure for assessing dose-response relationships is the biologically effective dose.  Ideally, environmental health scientists would like to have sensitive and specific measurements of the biologically effective dose.  However, identifying the target site(s) of the chemical is a major impediment to using measures of the biologically effective dose to quantify exposure.   Even when the target site is known, an invasive procedure may be required to sample that site (e.g., biopsies of liver or brain tissue).  Some organic toxicants or their metabolites covalently bond to DNA, thus forming a DNA adduct; most notably, carcinogens and mutagens form such adducts.  The measurement of such adducts is an example of a biologically effective dose, but the levels of these adducts may reflect only recent exposure because of DNA repair.  On the other hand, measurements of adducts with hemoglobin and other proteins, such as albumin, have been considered measurements of the internal dose, and as exemplified by 4-aminobiphenyl, the hemoglobin adduct has been shown to be significantly associated with DNA adduct concentration in the epithelial cells of the  human bladder (7).  Some of these adducts are specific markers for a toxicant (e.g., benzo(a)pyrene in lymphocytes), whereas others are much less specific (e.g., DNA adducts with alkyl groups).  The measurement of adducts in humans is still in the developmental stage, and for most chemicals, much more information is needed before they can be used as a quantitative measurement of exposure (8).  Nonetheless, it can be used as a marker of exposure.  Other disadvantages to be considered in these measurements are that sample throughput may be too slow for moderate-size epidemiologic studies, and many adducts may arise from a single chemical.

Other examples of internal dose are the direct measurements of chemicals or their metabolites in blood or urine.  Such measurements have significantly improved human exposure assessment and thus have improved assessing the risk to humans of many important chemicals.  For example, it is fair to say that without blood lead measurements, most of the central nervous system effects of low-level lead exposure could not have been detected.

To interpret blood or urine chemical levels accurately, analysts must know the pharmacokinetics of the chemical and also must have a knowledge of the background levels found in the general population.  For example, some chemicals, such as VOCs, are rapidly eliminated, whereas others, such as the chlorinated hydrocarbon pesticides, may have a half-life in humans of greater than 5 years.  Thus, such information is critical for interpreting whether the measured concentration of a chemical reflects recent exposures, long-term (chronic) exposures, or both.  Of course, to the extent possible, it is still of great importance for the epidemiologist to collect, nonbiased information from study participants regarding their potential exposure.

Additional biomarkers that have been monitored in humans include biomarkers of susceptibility and effect.  Biomarkers of response, such as cytogenetic markers, stress proteins, and enzyme induction, are sometimes classified as exposure biomarkers and sometimes as effect biomarkers.  We will not consider them in this presentation because of space limitations but also because these biomarkers are very nonspecific (i.e., abnormalities of these biomarkers would not specify to what TSCA chemical, if any, a person may have been overtly exposed).

## Examples of Use of Biomarkers in Exposure Assessment

I will now demonstrate how biomarkers of exposure have been used in exposure assessment in epidemiologic studies and the advantages of this approach over the "environmental approach" for assessing human exposure.  In so doing, I do not wish to imply that the environmental approach is meaningless, but that the biologic approach is preferred as a marker of human exposure.  Certainly, in risk management when the objective is to reduce the potential exposure, the "environmental approach" is useful for identifying the source of pollution.

## Dioxin:  Operation Ranch Hand Study

From 1962 through 1970 during the Vietnam Conflict, the main mission of
the U.S. Air Force's Operation Ranch Hand was to spray defoliants, such as
Agent Orange, over densely vegetated areas of South Vietnam.  Agent Orange
consisted of an equal mixture of 2,4-D and 2,4,5-T in diesel oil; the 2,4,5-T was
contaminated with 2,3,7,8-TCDD (dioxin) in the parts-per-million range.  Dioxin
is lipid soluble and thus tends to be stored in the lipid-rich depots of the human
body.  Dioxin has a long half-life--more than 7 years in humans (9,10).  In 1982,
the Air Force began a prospective cohort study, specifically looking at health,
reproductive, and mortality outcomes that might be associated with exposure to
Agent Orange and other herbicides containing dioxin.  These health studies are
examining the veterans of Operation Ranch Hand every 5 years through the year
2002.  One of the first tasks was to develop an exposure index in order to classify
each veteran's exposure; this index would then be used as the measure of
exposure and a way of correlating exposure with any health effects.

This exposure scenario was similar to that of exposure in an occupational
setting in that the primary exposure was thought to be direct exposure to the
herbicide itself, rather than indirect exposure through an environmental pathway.
The exposure index consisted of the average concentration of dioxin in Agent
Orange during one's tour of duty multiplied by the number of gallons of Agent
Orange sprayed during one's tour divided by the number of men in one's specialty
during that period.  The total number of eligible men in the study was limited to
the 1200 to 1300 survivors of the Operation.  The U.S. Air Force and various
review boards believed that this exposure index not only could serve as a reliable
basis for assessing exposure to dioxin but that any noted adverse health effects
could be related to this index.

In 1987, the U.S. Air Force contracted with our laboratory to analyze 150
serum samples from Operation Ranch Hand veterans in order to compare the Air
Force's exposure index with the measured internal dose of dioxin in the veterans.
There was essentially no correlation between the exposure index and the serum
dioxin level (11).  Because of this finding, the Air Force further contracted with
CDC to analyze the serum of all surviving members of Operation Ranch Hand,
and this serum-dioxin level became the exposure index used to correlate with any
adverse health effects (12).  Had the Air Force used its original exposure index
for the Operation Ranch Hand study, a great deal of misclassification would have
resulted, and any health effect conclusions of the study would have been invalid.

Thus, the use of the serum dioxin measurement, the biomarker, was preferred
over the exposure index that was derived without the biomarker.

## Dioxin:   U.S. Army Ground Troops in Vietnam

The chemical of concern for U.S. Army Ground troops in Vietnam was again
the dioxin in Agent Orange.  The potential environmental pathways were skin
contact with and inhalation of the spray containing the herbicide, skin contact with
sprayed vegetation and soil, and ingestion of water and food that had been

sprayed.  The amount of dioxin in Agent Orange from 1966-1969 was known. The duration of contact was gathered from questionnaires given to the veterans and from U.S. military records containing the locations of military units, the locations where herbicide was sprayed, and the dates when the herbicide was sprayed.

Six exposure indices were generated from this information; four of the indices were based on a soldier's potential for exposure from direct spray or on his being located in an area that had been sprayed within the previous 6 days; the other two exposure indices used self-reported data and included an index that was based on the veteran's perception of how much herbicide he has been exposed to.  To test the validity of these exposure indices, CDC measured serum dioxin levels in 646 enlisted ground-troop veterans who had served in III Corps a heavily sprayed area, for an average of 300 days during 1966 to 1969.  For comparison, serum dioxin levels in 97 non-Vietnam U.S. Army veterans who served during the same time were also measured (13).

The results showed no meaningful association between dioxin levels and any of the exposure indices.  The mean, median, and frequency distributions for both the Vietnam and non-Vietnam veterans were remarkably similar, indicating that there was little, if any,  increased exposure to dioxin in this population.  The study had a 95% statistical power to detect a difference of only 0.6 parts-per-trillion (ppt) in the medians, but this difference was not found.  This finding exemplifies the value of measurements of internal dose in exposure assessment. It also points out the need to develop specific and sensitive methods, for if the detection limit for dioxin had been 20 ppt (lipid adjusted), then most all the results would have been nondetectable.  Furthermore, because elevated exposures could not be documented, plans for a prospective cohort health study were dropped.

### Dioxin:  Occupational Setting

CDC's National Institute of Occupational Safety and Health (NIOSH) conducted a retrospective study to evaluate health outcomes, including mortality from cancer, among more than 5000 workers who may have been occupationally exposed to dioxin as a result, for example, of the production of 2,4,5-trichlorophenol (14).  Many of these workers were deceased.  Because many were deceased and because of the large number of potentially exposed men, NIOSH epidemiologists had to develop an exposure index for use in correlating the health outcomes (the effect).  Serum dioxin measurements were performed on 253 workers; the results were compared with various exposure indices.  From this analysis, epidemiologists determined that the best exposure index was years of work in a job with potential exposure.  Since this exposure index had been validated to, and calibrated with, serum dioxin levels, it could be used as the exposure index in this study, and exposure and effects could be compared directly with those found in other studies.  This process again demonstrates the need for measuring the internal dose in exposure-assessment or health-effect studies.

## Lead

Toxicity associated with high levels of lead in humans has long been recognized. However, biochemical and epidemiologic studies have noted hematologic and neurologic damage among children with relatively low levels of lead in their blood and teeth. The second National Health and Nutrition Examination Survey (NHANES II), conducted by the National Center for Health Statistics (NCHS), provided blood lead measurements, which were the basis for estimating the degree of exposure of the general U.S. population to lead (15). As a result of federal regulations requiring the removal of lead from gasoline, the amount of lead in gasoline decreased about 55% from early 1976 to early 1980. The population-based NHANES II showed that the predicted mean blood level in the U.S. population had decreased 37% during that same period, from 14.6 µg/dL to 9.2 µg/dL. Environmental modeling did not accurately predict the magnitude of the impact of decreasing the amount of lead in gasoline because the contribution of lead from gasoline to humans via the soil was not well characterized. These NHANES II data were major factors in the Environmental Protection Agency's (EPA's) decision to implement a more rapid removal of lead from gasoline. This implementation and the banning of the lead-soldering of cans produced in the U.S. have been major factors in reducing blood lead levels in the United States. NHANES II (1988-91) found that the mean blood level in the U.S. had declined even further, to 2.8 µg/dL (16).

Thus, exposure assessment by measuring blood lead levels has been a public health success story. It helped identify lead in gasoline as a major preventable source and showed that removing lead from gasoline was an effective prevention strategy. However, the latest data indicate that 8.9%, or approximately 1.7 million children, aged 1-5 years, still have blood lead levels equal to or greater than 10 µg/dL, which is the level of concern under the 1991 CDC guidelines. The population at risk for excessive lead exposure comprises primarily black, inner-city children and has been targeted for more extensive lead poisoning prevention efforts (17). This example again shows the value of biomarkers of exposure for quantifying the amount of human exposure and documenting the effectiveness of regulatory actions to reduce exposure.

## Volatile Organic Compounds (VOCs)

Many volatile organic compounds (VOCs) are ubiquitous in the environment and many have been shown to exist in higher concentrations in indoor rather than outdoor air (18). Reported health effects from exposure to VOCs have included eye irritation, sick-building syndrome, neurologic effects, and cancer. CDC developed an isotope-dilution purge-and-trap gas chromatography/mass spectrometry method to quantify 32 VOCs (see Table) in 10 mL of blood with detection limits in the ppt range (19). This method is a full-scan method at 3000 resolving power, so that in addition to acquiring quantitative data on these 32 VOCs, many additional VOCs can be qualitatively identified and in many cases, quantified (20).

CDC, with financial support from the Agency for Toxic Substances and Disease Registry (ATSDR), selected a 1,000-person subset of the NHANES III

population to determine reference ranges for these 32 VOCs. The 1,000 people
were chosen from both sexes, all regions of the contiguous U.S., urban and rural
residents, and were adults from 20 through 59 years of age (21). The data
showed that 11 of these VOCs were measured above the limit of detection in
more than 75% of the people, with the nonchlorinated aromatics being the most
prevalent group. These VOCs included styrene, toluene, ethylbenzene, o-xylene,
m,p-xylene, and benzene, which is a known human carcinogen. The primary
sources of these compounds are tobacco smoke and exhaust from internal
combustion engines. The nonendogenous compound found at the highest
concentration and highest frequency was 1,4-dichlorobenzene (22). The blood-
exposure data for this moth repellant and room deodorizer correlated highly with
urinary levels of its primary metabolite, 2,5-dichlorophenol (23). This high
correlation indicated that either blood 1,4-dichlorobenzene or urinary 2,5-
dichlorophenol levels could be used as a biomarker of exposure to 1,4-
dichlorobenzene.

Five of the VOCs were found in 10%-75% of the selected population,
whereas the remainder of the VOCs were in less than 10% (22). Thus, this latter
group would be of lower priority for inclusion in human effect studies, but if
these VOCs were found in exposure studies, their presence would likely indicate
an exposure to a point source. These analytical methods and reference range
studies have been applied to a variety of case studies and population studies,
including exposure-assessment studies of toxic waste sites, oil-well fires (24), sick-
building syndrome (22), multiple chemical sensitivity, and oxygenated fuels
involving methyl tertiary-butyl ether (MTBE) (25). In each of these examples, the
blood concentrations of VOCs were compared with the reference-range population
data. However, pharmacokinetic data are needed to properly interpret blood levels
of VOCs. Scientists from CDC and EPA have collaborated in determining the
elimination kinetics of many VOCs in humans subjected to low level mixtures of
VOCs in well-controlled chamber studies (26). The blood half-lives were less
than one-half hour, but the elimination kinetics were multiexponential, suggesting
multiple storage sites within the body. The blood-uptake portion of the 4-hour
exposure curve exhibited a rapid uptake that reached a plateau after about 50
minutes; the uptake rate was not concentration dependent, but the blood
concentration was directly dependent on the air concentration. When exposure
ceased after 4 hours, the decay was rapid, but the decay rate also reached a
plateau after about 1 hour; however, the VOC levels remained elevated even 24
hours after exposure as compared with the pre-exposure blood levels. Thus, like
those compounds with long biologic half-lives, such as dioxin, VOCs also can be
the focus of exposure assessment studies, if the blood samples are collected within
a specified time frame after exposure. This time frame should be based on the
half-life of the VOC and should be selected so that the sampling period would
correspond to the plateau region of the decay curve; e.g., from 5 - 100 times the
half-life of the VOC.

## Collecting and Banking of Human Specimens

We have presented examples of the benefits of biomarkers of exposure; now
we focus on the mechanisms of collecting and banking human specimens for such

biomonitoring. The first U.S. program of biological monitoring tissue specimens for environmental pollutants and also for human tissue specimen banking was the National Human Monitoring Program (NHMP), which began in 1967 and was conducted by the U.S. Public Health Service. When the U.S. Environmental Protection Agency (EPA) was created in 1970, the NHMP was transferred to it. One of the major activities of the NHMP was the National Human Adipose Tissue Survey (NHATS), which was designed to be a continuously operating survey that would collect, store, and analyze autopsy and surgical specimens of human adipose tissue from the major U.S. metropolitan areas. However, during the 1980s, budget cuts restricted the NHMP to a reduced and modified NHATS, which continued until 1990. In 1991, the National Research Council published its findings that programs that provide more useful data based on probability samples for the entire U.S. population should be designed and properly funded (27).

One program that is based on a national probability sample is the NHANES, which is conducted by the CDC's NCHS. Data from NHANES I, II, and III have provided important information on the prevalence of various health conditions and distributions of physical and biochemical characteristics of the U.S. population. As previously mentioned, data on blood lead levels in NHANES II and III provided longitudinal trend data on human levels and the effect of legislation on that trend. The data also pinpointed a subpopulation still at risk for excessive lead exposure. Serum levels of cotinine, the major metabolite of nicotine, are being measured in NHANES III in order to ascertain exposure levels as a result of both active and passive smoking (28). As mentioned previously, CDC measured blood VOCs and selected urinary pesticide residues in a subset of the NHANES population in order to assess human exposure to these compounds. In addition, blood, urine, and DNA have been banked from NHANES III.

Phase I of the National Human Exposure Assessment Survey (NHEXAS), which is conducted under cooperative agreements with the EPA, began in 1995. These Phase I studies are population based surveys for exposure assessment to selected environmental pollutants in the state of Arizona and in EPA's Region V (28).

Many issues must be considered in designing and implementing national probability sampling surveys for human exposure assessment (29). However, certainly NHANES, and now NHEXAS, have addressed these issues. Therefore, the mechanism is in place to collect and bank specimens needed to assess biomarkers of exposure in human specimens for many of the chemicals included in TSCA.

## Prioritizing Chemicals

We have presented examples of the benefits of biomarkers of exposure and the ability of programs, like NHANES, to collect and bank the needed biologic specimens for assessing human exposure to many of the chemicals included in TSCA. This does not argue that the entire number of probability based samples have to be analyzed but that mechanisms are in place to collect such samples. Assuming the needed biologic samples are available, the list of TSCA chemicals must be prioritized for the effective application of biomarkers for human exposure assessment. The following factors would be included in such prioritization:

. Potential for human exposure.
  - Degree of exposure.
    . Pounds produced per year.
    . Physical/chemical characteristics of chemical.
    . How the chemical is made, used, and its fate.
  - Number of people, especially susceptible people, potentially exposed.
. Hazard identification and information about the severity of effect.
. Dose response information in both animals and humans.
. Possibility of measuring biomarkers.

Such prioritization of this chemical list would therefore involve developing a model that would include the following factors: the potential for human exposure (degree and number), severity of adverse effects in a dose-response manner, and the possibility of the biomarkers existing and ability of the laboratory to develop the needed analytical methods to evaluate exposure. For those chemicals that lack the needed information, quantitative structure activity relationship data, if available, would also be used. Exposure databases (30) would be used in this process.

**Analytical Methods**

As mentioned previously, one of the criteria for prioritizing the list of chemicals for the development of biomarkers is the possibility of measuring biomarkers of exposure (i.e., does a biomarker exist, and can the laboratory develop the needed analytical methods to measure the biomarker?). Unless the biomarker exists, there is no need for the analytical method. Assuming the biomarker exists, the analytical methods should have the following characteristics:

. Multianalyte (several biomarkers sequentially or simultaneously).
. Compatible with sample matrix.
. Demonstrated acceptable sensitivity.
. Demonstrated acceptable specificity.
. Demonstrated acceptable precision.
. Demonstrated acceptable accuracy.
. Cost effective.
. High sample throughput.

These characteristics, except for cost effectiveness, can be defined in objective terms. Certainly, the methods used in our examples meet the needed objective criteria. For measuring organic biomarkers of exposure, the analytical methods that are atop the method hierarchy include high-resolution mass spectrometry and tandem mass spectrometry using the isotope dilution technique for quantification. Whether a particular analyses is cost effective is more subjective. For example, the cost for measuring 32 VOCs in 10 mL of blood is about $500 per sample or less than $20 per analyte. Commercial prices for measuring the 17 polychlorinated dibenzo-p-dioxins and furans plus 4 co-planar polychlorinated biphenyls that are in human serum are about $1000 per sample or less than $50 per analyte. One can decide if this is too costly for the intended purpose.

Historically, mass spectrometric methods have suffered in the area of rapidity or high throughput, but this is not always the case. For example, for the measurement of cotinine in NHANES III, serum extracts are analyzed at the rate of 1 every 2 minutes by using high-performance liquid-chromatography\ atmospheric-pressure ionization tandem mass spectrometry. This technique also requires less sample preparation than traditional methods although sample preparation is the rate-limiting step because of the speed of mass spectrometric analysis.

Other methods, such as immunoassays that may appear to be more amenable to screening methodologies, have been developed for many chemicals, primarily pesticides, in the environmental area (31,32). To expand this list to many of the TSCA chemicals in biological specimens would require much work in developing both the antiserum and the methods. Many of the current immunoassays have high levels of false positives (because of cross-reactivity or matrix effects) and false negatives (because of matrix effects unless sufficient sample preparation procedures are followed). Therefore, to meet the objective requirements of the desired analytical methods, one frequently must employ methods of higher specificity for many of the samples. One new technique that employs many of the advantages of immunoassays with the specificity and multianalyte capability of the mass spectrometer is a mass spectrometric immunoassay (33). Such combinations of techniques will be used increasingly for biomonitoring. In addition, for screening individual compound or mixtures, bioassays will be used. Such techniques may have the advantage of using small volumes of serum and little, if any sample preparation (34).

I believe that the bottom line is that following some prioritization of the chemicals, if the biomarker of exposure exists in a readily accessible biologic specimen, such as blood or urine, this biomarker can be measured effectively to assess human exposure and thus be used to help prioritize TSCA chemicals for health effect screening. Biomarkers of effect can also help prioritize chemicals for exposure studies. I also believe that this biomarker approach for exposure assessment should be more widely used in other environmental legislation such as the Comprehensive Environmental Responses, Compensation, and Liability Act; Safe Drinking Water Act; Clean Water Act; and Federal Insecticides, Fungicide, Rodenticide Act.

## Summary

I have attempted to show that a biomonitoring program would be beneficial in assessing human exposure to many of the chemicals on the TSCA list. Such a program might be also a way to 1) establish reference ranges in the general population; 2) identify subpopulations potentially at risk; 3) establish trends in exposure and, hence, judge the effectiveness of pollution prevention practices and regulations; 4) provide dose assessment over total exposure; 5) and provide a data base for comparison with other data sets such as ecologic data sets. The needed sample-collection programs and analytical procedures are now available for conducting such a program. These procedures incorporate the benefits of having the required sensitivity, specificity, and multi-chemical measurements and are cost effective. National population-based programs such as NHANES or NHEXAS could be used to collect the specimens. Each of these would offer certain advantages. The TSCA list would have to be prioritized by using an algorithm consisting of the potential for human exposure, severity of adverse human effects, and the possibility of measuring the required biomarker. Once this model is formed, it could be validated by the biomonitoring program.

I have also included a list of the chemicals (see Table), for which CDC has national human internal-dose data, the biologic specimen needed and the amount, and the lower detection limits; these data are from various sources and are of varying quality for predicting national mean and ranges of human levels. Nonetheless, they do show whether exposure is common for particular chemicals. In addition, many of these chemicals, such as the pesticides, are not on the TSCA list.

I believe that there is a hierarchy of means to assess human exposure. This hierarchy includes self reports, professionally-developed exposure questionnaires, measurements of external dose, and modelling of all or portions of these data. All of this information may be useful, but I believe that the "gold standard" for assessing human body burden (internal dose) is the measurement of a biomarker of exposure in human specimens. Thus, if exposure data and classification from any of the other techniques are to be used, they should be both validated and calibrated to human biomonitoring data. However, programs such as NHANES or NHEXAS and many of the analytical methods are available to gather exposure information on many TSCA chemicals as well as on other environmental pollutants. This exposure information would then be used to determine which chemicals should be examined for health effects.

## Acknowledgment

This paper was presented in part at the Office of Technology Assessment's (OTA) Workshop on Testing and Screening Technologies for Review of Chemicals in Commerce by Dr. Larry L. Needham in Washington, D.C., on April 24, 1995; other members of the Exposure Assessment Panel were Dr. James Bond of the Chemical Industry Institute of Toxicology and Dr. Steve Tannenbaum of the Massachusetts Institute of Technology.

**Biological Monitoring Measurements Currently Performed at the National Center for Environmental Health, of the Centers for Disease Control and Prevention.**

**Metals** (typical urine or blood sample - 3 mL; typical limit of detection - low parts per-billion

| | |
|---|---|
| Lead | Beryllium |
| Mercury | Chromium |
| Cadmium | Nickel |
| Arsenic | Thallium |
| Vanadium | |

**Polychlorinated dibenzo-dioxins, polychlorinated dibenzo-furans, coplanar polychlorinated biphenyls (PCBs).** All analytes measured in serum from one 25 mL blood sample if exposure is near background levels; smaller samples are adequate for higher exposures. Typical limit of detection - low parts-per-trillion on a lipid-weight basis, low parts-per-quadrillion on a whole-weight basis.

2,3,7,8-Tetrachlorodibenzo-p-dioxin
1,2,3,7,8-Pentachlorodibenzo-p-dioxin
1,2,3,4,7,8-Hexachlorodibenzo-p-dioxin
1,2,3,6,7,8-Hexachlorodibenzo-p-dioxin
1,2,3,7,8,9-Hexachlorodibenzo-p-dioxin
1,2,3,4,6,7,8-Heptachlorodibenzo-p-dioxin
1,2,3,4,6,7,9-Heptachlorodibenzo-p-dioxin
1,2,3,4,6,7,8,9-Octachlorodibenzo-p-dioxin
2,3,7,8-Tetrachlorodibenzofuran
1,2,3,7,8-Pentachlorodibenzofuran
2,3,4,7,8-Pentachlorodibenzofuran

1,2,3,4,7,8-Hexachlorodibenzofuran
1,2,3,6,7,8-Hexachlorodibenzofuran
1,2,3,7,8,9-Hexachlorodibenzofuran
2,3,4,6,7,8-Hexachlorodibenzofuran
1,2,3,4,6,7,8-Heptachlorodibenzofuran
1,2,3,4,7,8,9-Heptachlorodibenzofuran
1,2,3,4,6,7,8,9-Octachlorodibenzofuran
3,3',4,4'-Tetrachlorobiphenyl
3,4,4',5-Tetrachlorobiphenyl
3,3',4,4',5-Pentachlorobiphenyl
3,3',4,4',5,5'-Hexachlorobiphenyl

**Volatile organic compounds (VOCs)** All analytes measured in one 10 mL blood sample; typical limit of detection - low parts-per-trillion.

1,1,1-Trichloroethane
1,1,2,2-Tetrachloroethane
1,1,2-Trichloroethane
1,1-Dichloroethane
1,1-Dichloroethene
1,2-Dichlorobenzene
1,2-Dichloroethane
1,2,-Dichloropropane
1,3-Dichlorobenzene
1,4-Dichlorobenzene
2-Butanone
Acetone
Benzene
Bromodichloromethane
Bromoform
Carbon Tetrachloride

Chlorobenzene
Chloroform
cis-1,2-Dichloroethene
Dibromochloromethane
Dibromomethane
Ethylbenzene
Hexachloroethane
m-/p-Xylene
Methylene chloride
o-Xylene
Styrene
Tetrachloroethene
Toluene
trans-1,2-dichloroethene
Trichloroethene

**Chlorinated pesticides and non-coplanar polychlorinated biphenyls** (all analytes measured in serum from one 5 mL blood sample; typical limits of detection - low parts-per-billion).

Aldrin
Chlordane, alpha
Chlordane, gamma
beta-Hexachlorocyclohexane
gamma-hexachlorocyclohexane
Biphenyls, Polychlorinated (total)
Biphenyls, Polychlorinated (individual congeners)
DDD
Trans-Nonachlor

DDE
DDT
Dieldrin
Endrin
Heptachlor
Heptachlor epoxide
Hexachlorobenzene
Mirex
Oxychlorodane

**Non-persistent pesticides** (all analytes measured in one 10 mL urine sample typical limits of detection - low parts-per-billion).

| Urine metabolites | Parent Pesticides |
|---|---|
| 2-Isopropoxyphenol | Propoxur |
| 2,5-Dichlorophenol | 1,4-Dichlorobenzene |
| 2,4-Dichlorophenol | 1,3-Dichlorobenzene, Dichlofention, Prothiofos, Phosdiphen |
| Carbofuranphenol | Carbonfuran, Benfuracarb, Carbosulfan, Furanthiocarb |
| 2,4,6-Trichlorophenol | 1,3,5-Trichlorobenzene, Hexachlorobenzene, Lindane |
| 3,5,6-Trichloro-2-pyridinol | Chloropyrifos, Chlorpyrifos-methyl |
| 4-Nitrophenol | Parathion, Methyl parathion, Nitrobenzene, EPN |
| 2,4,5-Trichlorophenol | 1,2,4-Trichlorobenzene, Fenchlorphos, Trichloronate, Lindane |
| 1-Naphthol | Naphthalene, Carbaryl |
| 2-Naphthol | Naphthalene |
| 2,4-Dichlorophenoxyacetic acid | 2,4-D |
| Pentachlorophenol | Pentachlorophenol |

## Literature Cited

1. National Research Council (NRC) Subcommittee, Committee on Advances in Assessing Human Exposure to Airborne Pollutants. *National Academy Press,* Washington, D.C., **1991**, pp. 15-37.
2. Needham, L.L.; Pirkle, J.L.; Burse, V.W.; Patterson, D.G. Jr.; Holler, J.S. *J Exp. Anal. Environ. Epidem.* Supplement **1992,** 1:209-221.
3. Lioy, P.J. *Environ. Sci. Technol.* **1990,** 24:938-945.
4. Sexton, K.; Callahan, M.A.; Bryan, E.F. *Environ. Health Persp.* **1995,** 103(3):13-29.
5. Needham, L.L.; W.M. Draper (ed.), Environmental Epidemiology: Effects of Environmental Chemicals on Human Health, Advances in Chemistry Series Vol. 241, American Chemical Society, Washington, D.C., **1995,** 121-135.
6. Mannino, D.M.; Schreiber, J.; Aldous, K.; Ashley, D.; Moolenaar, R.; Almaguer, D. *Int. Arch. Occup Environ. Health* **1995,** 67:59-64.
7. Skipper, P.L.; Tannenbaum, S.R. *Environ. Health Persp.* **1994,** 102:17-21.
8. Skipper, P.L.; Tannenbaum, S.K. *Carcinogenesis* **1990,** 11:507-518 (1990).
9. Pirkle, J.L.; Wolfe, W.H.; Patterson, D.G. Jr.; Needham, L.L.; Michalek, J.E. Miner, J.C.; Peterson, M.R.; Phillips, D.L. *J. Toxicol. Environ. Health* **1989,** 27:165-171.
10. Needham, L.L.; Gerthoux, P.; Patterson, D.G. Jr.; Brambilla, P.; Pirkle, J.L.; Tramacere, P.L.; Turner, W.E.; Beretta, C.; Sampson, E.J.; Mocarelli, P. *Organohalogen Compounds* **1994,** 21:81-86.
11. Michalek, J.E. *Appl.Inc. Hyg.* **1989,** 12:68-72.
12. Wolfe, W.E.; Michalek, J.E.; Miner, J.C.; Rahe, A.; Silva, J.; Thomas, W.F.; Grubbs, W.D.; Lustik, M.B.; Karison, T.G.; Roegner, R.H.; Williams, D.E. *JAMA* **1990,** 264:1824-1831.
13. Centers for Disease Control (CDC). *JAMA* **1988,** 260:1249-1254.

14. Fingerhut, M.A.; Halperin, W.E.; Marlow, D.A.; Piacitelli, L.A.; Honchar, P.A.; Sweeney, M.H.; Greife, A.L.; Dill, P.A.; Steenland, K.; Suruda, A.J. *N. Engl. J. Med.* **1991**, 324:212-218.
15. Annest, J.L.; Pirkle, J.L.; Makuc, D.; Neese, J.W.; Bayse, D.D.; M.G. Kovar, *N. Engl. J. Med.* **1983**, 308:373-1377.
16. Pirkle, J.L.; Brody, D.J.; Gunter, E.W.; Kramer, R.A.; Paschal, D.C.; Flegal, K.M.; Matte, T.D. *JAMA* **1994**, 272:284-291.
17. Brody, D.J.; Pirkle, J.L.; Kramer, R.A.; Glegal, K.M.; Matte, T.D.; Gunter, E.W.; Paschal, D.C. *JAMA* **1994**, 272:277-283.
18. Wallace, L.A.; Pellizzari, E.D.; Hartwell, T.D.; Sparacino, C.; Whitmore, R.; Sheldon, H.; Zelon, L.; Perritt, R. *Enivron. Res.* **1987**, 43:2990-307.
19. Ashley, D.L.; Bonin, M.A.; Cardinali, F.L.; McCraw, J.M.; Holler, J.S.; Needham, L.L.; Patterson, D.G. Jr. *Anal. Chem.* **1992**, 64:1021-1029.
20. Bonin, M.A.; Ashley, D.L.; Cardinali, F.L.; McCraw, J.M.; Wooten, J.V. *J. Anal. Toxicol.* **1995**, 19:187-191.
21. Needham, L.L.; Hill, R.H. Jr.; Ashley, D.L.; Pirkle, J.L.; Sampson, E.J. *Environ Health Persp.* **1995**, 103(3):89-94.
22. Ashley, D.L.; Bonin, M.A.; Cardinali, F.L.; McCraw, J.M.; Wooten, J.V. *Clin. Chem.* **1994**, 40:1401-1404.
23. Hill, R.H. Jr.; Ashley, D.L.; Head, S.L.; Needham, L.L.; Pirkle, J.L. *Arch. Environ. Health* **1995**, 50:277-280.
24. Etzel, R.A.; Ashley, D.L. *Int. Arch. Environ. Health* **1994**, 47:1-5.
25. Moolenarr, R.L.; Heflin, B.J.; Ashley, D.L.; Middaugh, J.P.; Etzel, R.A. *Arch. Environ. Health* **1994**, 49:402-409.
26. Ashley, D.L.; Prah, J.D. CDC, personal communication.
27. National Research Council. Monitoring Human Tissues for Toxic Substances. Washington: National Academy Press, **1991**
28. Centers for Disease Control and Prevention. "Preliminary Data: Exposure of Persons Aged Greater Than or Equal for 4 Years - United States, 1988-1991. *MMWR* **1993**, 42:37-39.
29. Ezzoti-Rice, T.M.; Murphy, R.S. *Environ. Health Persp.* **1995**, 103(3):55-59.
30. Graham, J.; Walker, K.D.; Berry, M.; Bryan, E.F.; Callahan, M.A.; Fan, A.; Finley, B.; Lynch, J.; McKone, T.; Ozkaynak, H.; Sexton, K. *Arch. Environ. Health* **1992**, 47:408-420.
31. Van Emon, J.M.; Lopex-Avilla, V. *Anal. Chem.* **1992**, 64:79-88.
32. Meulenbert, E.P.; Mulder, E.H.; Stoks, P.G. *Environ. Sci. & Tech.* **1995**, 29:553-561.
33. Nelson, R.W.; Krone, J.R.; Bieber, A.L.; Williams, P. *Anal. Chem.* **1995**, 67:1153-1158.
34. Balaguer, P.; Denison, M.; Zacharewski, T. *Organohalogen Compounds* **1995**, 23:215-220.

# Chapter 3

# The Use of Reference Range Concentrations in Environmental Health Investigations

Robert H. Hill, Jr., Susan L. Head, Sam Baker, Carol Rubin,
Emilio Esteban, Sandra L. Bailey, Dana B. Shealy,
and Larry L. Needham

National Center for Environmental Health, Centers for Disease Control
and Prevention, Public Health Service, U.S. Department of Health and
Human Services, 4770 Buford Highway, Atlanta, GA 30341–3724

Reference range concentrations are obtained by measuring
xenobiotics, their residues, or their metabolites in human specimens
from the general population. These reference range concentrations
provide a foundation for assessments of exposure to xenobiotics in
specific exposure situations and also provide information about the
extent and magnitude of xenobiotic exposure in the general
population. Reference range concentrations for $p$-nitrophenol served as
a basis for comparison for residents exposed to methyl parathion
following inappropriate residential exposure. Reference range
concentrations for 2,5-dichlorophenol and 3,5,6-trichloro-2-pyridinol
also provide information about the magnitude and extent of exposure
to environmental contaminants, $p$-dichlorobenzene, and chlorpyrifos,
respectively.

Biological monitoring, sometimes called biomonitoring, evaluates exposure to
xenobiotics by measuring the concentration of that xenobiotic in a biological sample,
usually urine, serum, blood, or tissue. The xenobiotic measured may be the toxicant
itself, a metabolite, a product of the toxicant and a bio-molecule such as DNA,
hemoglobin, or other proteins, or a change in a naturally occurring biochemical
within the body, such as cholinesterase depression. These are known as biomarkers
of exposure. The concentration of the xenobiotic or its metabolite(s) is known as the
internal dose and reflects toxicant exposure from all routes and sources, providing a
summary or integrated index of exposure. Air monitoring, wipe sampling, and food
and water analyses are typically used to estimate exposure from a single source, and
these environmental monitoring methods are useful in risk management to locate and
prevent exposure from a source or primary route of exposure. The objective of
measuring internal dose is to assess actual human exposure and to correlate this
internal dose with biomedical changes or adverse health effects.

We have extensive experience in biological monitoring, particularly in support of various epidemiologic investigations involving known or possible exposures to toxic compounds. One important finding is that often a measurement of internal dose provides a much better assessment of exposure than all exposure indices (*1,2*). Exposure indices are estimates of exposure based upon information collected from questionnaires, medical history, and environmental measures. However, the paucity of biomarker data makes interpreting biological monitoring difficult. Detection of a xenobiotic in urine, blood, or serum indicates exposure; nevertheless, the magnitude and extent of exposure is at the heart of exposure assessment and risk analysis. In an effort to improve exposure assessments, we conducted a program to establish reference range concentrations for 32 volatile organic compounds in blood and 12 pesticide metabolites or residues in urine (*3*).

Reference range concentrations are those concentrations of a toxicant, its residue, or its metabolite that are found in the general population. We assume that the general population is not occupationally exposed to the parent compound and is not otherwise overtly exposed. Of course, most people in the general population are exposed to many xenobiotics – for instance, to pesticide residues in foods; generally however, these residue concentrations are usually low. Reference range concentrations can be used as baseline or normal concentrations for the "unexposed," general population. These values are analogous to normal values of clinical laboratory measurements, but they are not normal biochemical constituents of the body. Reference range concentrations provide a foundation for assessments of exposure to xenobiotics by serving as a basis of comparison for possibly exposed populations. Reference range concentrations also provide information about the extent and magnitude of exposure of the general population to xenobiotics (e.g. pesticide residues), and they can suggest areas for future research. The uses for reference range concentrations for pesticide residues or metabolites will be illustrated below.

## Methods and Materials

Reference range concentrations were determined in samples collected from approximately 1000 people – a subset of the National Health and Nutrition Examination Survey III [NHANES III], a national survey of the general population of the United States (*4*). Twelve pesticide residues were measured in urine samples by using an isotope dilution technique, enzyme hydrolysis and extraction, derivatization and concentration, and finally capillary gas chromatography combined with tandem mass spectrometry (*5*). This method was also used to measure *p*-nitrophenol in urine samples collected from Lorain County residents.

## Non-Occupational Exposure to Methyl Parathion

Methyl parathion is a highly toxic pesticide and its use, which is principally to spray cotton, is restricted by the Environmental Protection Agency (EPA) (*6*). Furthermore, methyl parathion's highly toxic properties require that field re-entry standards be observed with this pesticide (that is, workers may not re-enter fields that have been sprayed with methyl parathion for at least 48 hours). In late 1994, state officials from Ohio discovered that this pesticide had been used to exterminate pests from a home– a use not permitted by federal or state regulations and laws.  The ensuing investigation showed that an unlicensed pesticide operator had illegally used this pesticide in hundreds of homes, thus exposing several hundred people, including children, expectant mothers, and the elderly.  Local, state, and federal health officials working together quickly went to selected homes to assess whether the people in those homes had suffered exposure or adverse health effects from methyl parathion. This population not only had exposure that could cause acute illness but also had experienced chronic, long term exposure to methyl parathion.

Exposure was determined by measuring methyl parathion in air and wipe samples from these homes and by measuring *p*-nitrophenol (PNP), a metabolite of methyl parathion, in urine.  Figure 1 shows an ion chromatogram (top) of a urine extract from one of the residents selected for evaluation; the ion chromatogram (bottom) of a urine extract from the reference range population represents a PNP concentration of  16 $\mu$g/L– the 99th percentile of the PNP reference range concentrations.  The PNP urinary concentration (4000 $\mu$g/L) from  the resident was 250 times greater than the 99th percentile concentration.  To accurately measure such a high concentration, the sample had to be diluted by a factor of 100.  All residents of this household had PNP concentrations greater than or equal to 4000 $\mu$g/L; a child had a concentration of 4800 $\mu$g/L – almost 1000 times greater than the 95th percentile concentration of the reference range.

Table I shows a comparison of PNP concentrations from 131 Lorain County residents to PNP reference range concentrations.  For every category of comparison, these residents had concentrations 75 to 250 times greater than the reference concentrations.  Comparisons with reference range data can be made using frequencies of detection, mean, and various percentile concentrations.   Usually we consider concentrations less than or equal to the 95th percentile concentration of the

**Table I.  Comparison between *p*-Nitrophenol Urinary Concentrations Found in Lorain County Residents Exposed to Methyl Parathion and in the Reference Range Subjects**.

| *Measures of Comparison* | *Lorain County*<br>*(N = 131)* | *Reference Range*<br>*(N = 974)* |
|---|---|---|
| Frequency of detection | 86% | 41% |
| Mean concentration, $\mu$g/L | 240 | 1.6 |
| Median concentration, $\mu$g/L | 28 | <1.0 |
| 90th Percentile concentration, $\mu$g/L | 280 | 3.3 |
| 95th Percentile concentration, $\mu$g/L | 910 | 5.2 |
| 99th Percentile concentration, $\mu$g/L | 4,000 | 16 |
| Maximum observed concentration, $\mu$g/L | 4,800 | 63 |

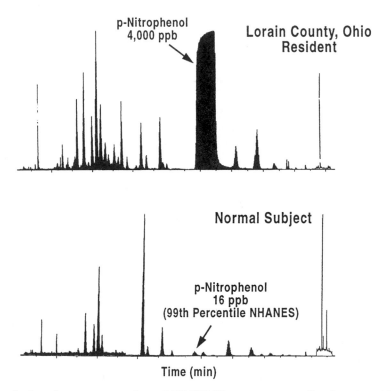

Figure 1. Ion chromatograms from GC/MS/MS measurement of *p*-nitrophenol in the urine of a Lorain County, Ohio, resident (top) and in the urine of a normal subject from the NHANES III study (bottom).

reference range as normal or expected.  Concentrations greater than the 95th percentile are considered elevated– indicating probable overt exposure to a toxicant that results in that residue or metabolite in urine.  In Lorain County, 78% of the methyl parathion-exposed residents had PNP concentrations greater than or equal to the 95th percentile concentration of the reference range.  Clearly the majority of residents living in homes sprayed with methyl parathion had greater than normal exposure to this pesticide.  Cholinesterase measurements of the Lorain County residents did not show cholinesterase depression, even though urinary PNP concentrations showed that methyl parathion exposure was often quite significant.  This perhaps illustrates the relative insensitivity of cholinesterase measurements in evaluating exposures to cholinesterase-inhibiting pesticides.  That is, when cholinesterase is depressed, the patient has been acutely poisoned; however, the absence of cholinesterase depression does not rule out exposure to pesticide concentrations that could produce adverse health effects, particularly in the very young, the very old, or expectant mothers.

Based upon the information concerning elevated urinary PNP levels  and the results of air and wipe samples, the homes in this area were declared a Superfund cleanup site by EPA, with the result being a full effort to identify and decontaminate all homes in which methyl parathion presented a hazard.  Data from urinary PNP concentrations of Lorain County residents were combined with data from air and wipe sample concentrations of methyl parathion in their homes to produce a matrix decision tree for evaluation of exposure (Milton Clark, EPA, Chicago, Il. Personal communication, 1995).  These evaluations were used to determine which priority action was required for residents:

*Priority 1*)     Exposed, must be evacuated immediately, must remain out of the home until it is decontaminated;
*Priority 2*)     Exposed but can remain in the home until decontamination  begins;
*Priority 3*)     Exposed at a very low level and  can remain in the home without further assistance;
*Priority 4*)     Not exposed.

The data in Table I illustrate how reference range concentrations are used to determine if elevated exposure to a toxic agent has occurred and to provide a basis for decision-making by public health officials.  In the past, biological monitoring was used principally to assess occupational exposures.  Often, occupational exposures result in very high concentrations of the toxicant or its metabolite in human specimens.  Judging the importance of environmental, nonoccupational exposures is difficult without a basis for comparison.  Reference range concentrations provide this basis and help bridge the intermediate zone between non-exposed and occupationally exposed populations.  Table II shows PNP concentrations resulting from exposure to parathion (ethyl or methyl) among workers and victims of acute poisonings (*7-10*).  The studies reported in Table II utilized a colorimetric method with a detection limit of approximately 100 parts per billion (*11*).  The detection limit of the method used in the Lorain County investigation was 1 $\mu$g/L or 1 part per billion (*5*).

The PNP reference range concentrations (shown in Table I) fall far below the concentrations found among those exposed in the situations listed in Table II.

**Table II. Urinary concentrations of *p*-Nitrophenol from Parathion (ethyl and methyl) Exposure Reported Among Pesticide Workers and Among Persons with Acute Poisoning**

| Source of Exposure | p-Nitrophenol Concentrations |
|---|---|
| Spraying (7) | Mean = 1,600 $\mu$g/L; Highest = 8,000 $\mu$g/L [n=43] |
| Spraying (8) | Highest = 2,400 $\mu$g/L [n=16] |
| Careful spraying (9) | Means = 800 $\mu$g/L- 900 $\mu$g/L [n=9] |
| Careless spraying (9) | Mean = 4,300 $\mu$g/L [n=7] |
| Mixing (7) | Mean = 1,200 $\mu$g/L; Highest = 11,300 $\mu$g/L [n=33] |
| Aerial spraying (10) | Range = 400 $\mu$g/L-12,300 $\mu$g/L [n=23] |
| Non-fatal poisonings (9) | Mean = 10,800 $\mu$g/L; Range = 700 $\mu$g/L-22,000 $\mu$g/L [n=9] |
| Non-fatal poisonings (7) | Means = 1,000 $\mu$g/L-8,400 $\mu$g/L; Ranges = 570 $\mu$g/L-32,200 $\mu$g/L [n=13] |
| Fatal poisonings (9) | Mean = 40,300 $\mu$g/L; Range = 2,400 $\mu$g/L-122,000 $\mu$g/L [n=14] |

However, many Lorain County residents had PNP concentrations within the uppermost part of the reference range and some concentrations were within the range of concentrations experienced by pesticide workers and victims of poisonings, as illustrated in Figure 2. The urinary PNP concentrations shown in Figure 2 from selected occupational exposures to ethyl and methyl parathion were reported by Roan et al. (10); those resulting from poisonings are from selected parathion (ethyl) intoxications (7,9). Although ethyl parathion is approximately twice as toxic as methyl parathion, the PNP concentrations in Lorain County were high enough to be of concern. Davies et al. (9) observed that urinary PNP concentrations from occupational exposure to parathion of 3,000 $\mu$g/L to 4,000 $\mu$g/L were indicative of excessive exposure and indicated a careless operator. Although adult exposure to toxic agents has been generally accepted– as long as adverse effects are not detected– exposure of children to the same toxic agents is not acceptable. Indeed the report of the National Research Council (12) emphasized that children are not small adults and their developing physiological systems make them more susceptible than adults to toxicants. Our knowledge of subtle health effects on children by these toxicants is minimal. The Lorain County exposed population should be followed to identify adverse health effects that may arise from this chronic exposure.

The methyl parathion exposure in Lorain County, Ohio, is one example that illustrates the value of the reference range concentrations. We have been involved in several other studies involving possible exposure to pesticides that used the reference range concentrations for pesticide residues or metabolites as a basis for comparison.

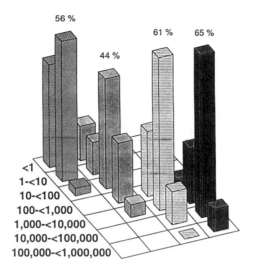

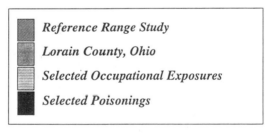

Figure 2. Frequency distribution plots of urinary *p*-nitrophenol (PNP) concentrations in the reference range population [n=974], Lorain County, Ohio residents [n=131], selected pesticide workers [n=23] (10), and selected persons who suffered pesticide poisonings [n=23] (9). The percent of the largest group in each frequency plot is shown above that bar (e.g. 56% of the reference range population had PNP concentrations from 1 to <10 $\mu$g/L).

These include studies of a group of people living near an abandoned pesticide manufacturing facility in Arkansas; a group of people living near Brownsville, Texas, where a cluster of birth defects occurred; farmers in Kansas exposed to a herbicide, 2,4-D; consumers of fish in the Great Lakes area; and Ohio infants with pulmonary hemosiderosis - a serious medical condition.

## Environmental Exposure to *p*-Dichlorobenzene and Chlorpyrifos

The study to determine reference range concentrations also revealed interesting information involving the extent and magnitude of exposure to several pesticides. One of those pesticides was *p*-dichlorobenzene (DCB). DCB is used in the United States (and other parts of the world) as a moth repellent, a toilet deodorizer, and a precursor for a polymer. DCB is metabolized to 2,5-dichlorophenol (DCP) and excreted in urine. The results of the reference range study showed that 98% of the 1000 adults had detectable concentrations of DCP (*13*). Concentrations ranged as high as 8,700 $\mu$g/L, and the 95th percentile concentration for DCP was 790 $\mu$g/L. This analyte was at least one to two orders of magnitude higher than the other analytes in the study. The information derived from this reference range study showed that everyone in the United States is exposed to DCB, with concentrations for those in the upper range nearing occupational types of exposure.

Data from the reference range study provided the information for urinary DCP concentrations and DCB blood concentrations. As a result we found an excellent correlation [Pearson correlation coefficient = 0.82, p <0.0001] between urinary DCP concentrations and DCB blood concentrations (*13*). We previously reported a similar correlation (r = 0.92) between pentachlorophenol concentrations in urine and in blood (*14*). These observations are particularly interesting in view of many reports that there is often a lack of correlation between urinary toxicant concentrations and estimated exposure.

Another analyte in the reference range study was 3,5,6-trichloro-2-pyridinol (TCPY), the metabolite of the pesticide chlorpyrifos. TCPY was detected in 82% of the reference range population with a maximum concentration of 77 $\mu$g/L and 95th percentile concentration of 13 $\mu$g/L. Although the concentrations were considerably lower than those for DCP, the frequency or prevalence of TCPY in urine specimens indicates that many people in the United States are exposed to chlorpyrifos. Chlorpyrifos is an insecticide that is commonly used in homes. It has become even more popular as a termiticide, replacing the banned, persistent chlorinated termiticides like chlordane. This same analyte was measured in samples from NHANES II (1976-1980) with a finding that 5.8% of the U.S. population had detectable TCPY concentrations (*15*) at a detection limit of 5 $\mu$g/L, which was higher than the reference range study detection limit of 1 $\mu$g/L. The difference does not account for the large change in the frequency of detection, since more than 31% of the reference range population had detectable TCPY concentrations of 5 $\mu$g/L or greater. Although the reference range study was not a population- based study, the data suggest that exposure to chlorpyrifos has increased since the period of 1976-1980 (NHANES II).

## Conclusion

The results of the reference range study for PNP, DCP, DCB, and TCPY illustrate how biological monitoring measurements provide valuable information regarding human exposure to environmental toxicants. The primary purpose of the reference range concentrations is to provide a basis for comparison for exposure assessments. In addition, this information suggests several future studies. Studies to determine the half-life of DCP and TCPY could provide information that may explain part of the widespread detection of these analytes. For example, if DCB is stored in fat tissues, as one report indicates (*16*), perhaps intermittent exposure with DCB retention and subsequent slow release of DCP account for part of our observations. Half-life information about TCPY could also be used in a similar fashion. Studies to indicate where and how most exposures to DCB or chlorpyrifos occur would help us identify ways to reduce exposure.

The establishment of reference range concentrations for biological monitoring of common toxicants is an important step in developing our understanding of exposure assessment and environment health. Concentrations of blood lead found in samples from NHANES II and NHANES III have demonstrated a substantial decline in blood lead levels of the entire U.S. population as a result of removal of lead from gasoline and from soldered cans (*17*). Ashley et al.(*18*) has recently reported reference range concentrations for selected volatile organics in blood samples from the NHANES III population, and these reference range concentrations have played a key role in identifying elevated exposure to some of these volatile organics (*19*). Selected chlorophenol concentrations in urine from the general population in Germany have been reported (*20*); in that study, the investigators also used reference range values to show exposure to organic substances among workers at municipal incinerators (*21*). Alessio (*22*) discussed the value and need for reference range values. Most recently Danish researchers reported reference range values for 1-hydroxypyrene and α-naphthol (*23*).

Reference range concentrations for biological monitoring serve as a sentinel for evaluating the community's exposure to xenobiotics; and for judging the effects of decisions and efforts at prevention. These points are illustrated by the banning of chlordane– a persistent pesticide– and its replacement by chlorpyrifos– a non-persistent pesticide– and by the decrease in blood lead concentrations after the removal of lead from gasoline and food cans. When used with exposure assessment data, the reference range values directly assist in risk analysis and management and also suggest areas for future research and investigation. Reference range concentrations provide important information that is a basis for decision making in the public health community and, in turn, should provide for a safer and more healthful living environment for everyone.

**Acknowledgments.** This work was supported in part by funds from the Comprehensive Environmental Response, Compensation, and Liability Act trust fund through an interagency agreement with the Agency for Toxic Substances and Disease Registry, U.S. Public Health Service.

**Literature Cited.**
1. Needham, L.L.; Pirkle, J.L.; Burse, V.W.; Patterson, D.G.; Holler, J.S. *J. Expo. Anal. Environ. Epidemiol.*, **1992**, Suppl. 1, pp. 209-221.
2. Needham, L.L. In *Environmental Epidemiology: Effects of Environmental Chemicals on Human Health*; Draper, W.M. Ed.; Advances in Chemistry Series 241; American Chemical Society: Washington, DC, 1994, Chapter 10, pp. 121-135.
3. Needham, L.L.; Hill, R.H.; Ashley, D.L.; Pirkle, J.L.; Sampson, E.J. *Environ. Health Perspect.*, **1995**, *103*, pp. 89-94.
4. National Center for Health Statistics. *Plan and Operation of NHANES III (1988-1994)*; Vital and Health Statistics Series 1, Number 32; National Center for Health Statistics, Hyattsville, MD, 1994.
5. Hill, R.H.; Shealy, D.B.; Head, S.L.; Williams, C.C.; Bailey, S.L.; Gregg, M.; Baker, S.E.; Needham, L.L. *J. Anal. Toxicol.* **1995**, 19, pp. 323-329.
6. Agency for Toxic Substances and Disease Registry. *Toxicological Profile for Methyl Parathion,* TP-91/21; Agency for Toxic Substances and Disease Registry, Atlanta, GA, 1992.
7. Arteberry, J.D.; Durham, W.F.; Elliot, J.W.; Wolfe, H.R. *Arch. Environ. Health* **1961**, 3, pp. 476-485.
8. Durham, W.F.; Wolfe, H.R.; Elliot, J.W. *Arch. Environ. Health* **1972**, 24, pp. 381-387.
9. Davies, J.E.; Davis, J.H.; Frazier, D.E.; Mann, J.B.; Welke, J.O. In *Organic Pesticides in the Environment: A Symposium*; Rosen, A.A.; Kraybill, A.F., Eds.; Advances in Chemistry Series 60; American Chemical Society, Washington, DC, 1966, Chapter 5, pp. 67-78.
10. Roan, C.C.; Morgan, D.P.; Cook, N.; Paschal, E.H. *Bull. Environ. Contam. Toxicol.* **1969**, 4, pp.362-369.
11. Elliot, J.W.; Walker, K.C.; Penick, A.E.; Durham, W.F. *J. Agr. Food Chem.***1960**, 8, pp. 111-113.
12. National Research Council. *Pesticides in the Diets of Infants and Children;* National Academy Press, Washington, DC, 1993.
13. Hill, R.H.; Ashley, D.L.; Head, S.L.; Needham, L.L.; Pirkle, J.L. *Arch. Environ. Health* **1995**, 50, pp.277-280.
14. Cline, R.E.; Hill, R.H.; Phillips, D.L.; Needham, L.L. *Arch. Environ. Contam. Toxicol.* **1989**, 18, pp. 475-481.
15. Kutz, F.W.; Cook, B.T.; Carter-Pokras, O.D.; Brody, D.; Murphy, R.S. *J. Toxicol. Environ. Health*, **1992**, 37, pp. 277-291.
16. Morita, M.; Ohi, G. *Environ. Pollut.* **1975**, 8, pp. 269-274.
17. Pirkle, J.L.; Brody, D.J.; Gunter, E.W.; Kramer, R.A.; Paschal, D.C.; Flegal, K.M.; Matte, T.D. *JAMA* **1994**, 272, pp. 284-291.
18. Ashley, D.L.; Bonin, M.A.; Cardinali, F.L.; McCraw, J.M.; Wooten, J.V. *Clin. Chem.* **1994**, 40, pp. 1401-1404.
19. Etzel, R.A.; Ashley, D.L. *Int. Arch. Occup. Environ. Health* **1994**, 66, pp. 125-129.
20. Angerer, J.; Heinzow, B.; Schaller, K.H.; Welte, D.; Lehnert, G. *Fresenius J. Anal. Chem.* **1992**, 342, pp. 433-438.
21. Angerer, J.; Heinzow B.; Reimann, D.O.; Knorz, W.; Lehnert, G. *Int. Arch. Occup. Environ. Health* **1992**, 64, pp. 265-273.
22. Alessio, L. *Int. Arch. Occup. Environ. Health* **1993**, 65, pp. S23-S27.
23. Hansen, A.M.; Christensen, J.M.; Sherson, D. *Sci. Total Environ.* **1995**, 163, pp. 211-219.

# Chapter 4

# Possible Biomarkers for Assessing Health Risks in Susceptible Individuals Exposed to Agrochemicals with Emphasis on Developing Countries

Abdel Khalek H. El-Sebae[1], Zaher A. Mohamed[2], Fawzia Abdel Rahman, and Alaa Kamel

Environmental Chemistry and Toxicology Laboratory, Department of Chemistry, Texas Southern University, 3100 Cleburne Avenue, Houston, TX 77004

Interindividual variation in response to agrochemicals and their potential toxic effects is mediated by intrinsic and extrinsic factors. The intrinsic variations stem from inherited predispositions exemplified in different blood groupings, immunoantigens, metabolic activation and detoxification, and DNA repair capacity, in addition to age, development and nutrition. The extrinsic factors include the geographical climatic variations, background of infectious diseases, health sanitation, socioenvironmental stresses and standard of living. Populations in developing countries have their own genetical trait characteristics where some genetic diseases are more dominant e.g. sickle cell anemia and other metabolic deficiencies. Parasitic endemic diseases such as schistosomiasis and hepatitis viruses are more abundant in the poor sanitation conditions of the third world countries. These unfavorable genetic and ecological factors when interacting synergistically together will render the people in the developing countries at higher risk because of high susceptibility to exposure and health hazards of agrochemicals. Highly susceptible population groups in general and those in the third world in particular, require specific biomarkers to measure and assess the inherited susceptibility, exposure dose and health effects of the agrochemicals. In this review, attempts are made to address such specific biomarkers needed to detect and to evaluate the levels of susceptibility, exposure and adverse health effects of agrochemicals with special reference to pesticides in developing countries.

[1]Current Address: Department of Pesticides Chemistry and Toxicology, Faculty of Agriculture, Alexandria University, Chatby, Alexandria, Egypt
[2]Current Address: Department of Biochemistry, Faculty of Agriculture, Cairo University, Fayoum, Egypt

Poisoning by agrochemicals particularly pesticides occur during their manufacturing, transportation, testing and application. Environmental hazards take place through air, water and food pollution. Exposure to sublethal doses of pesticides for long periods cause chronic and long term toxicities such as neurological disorders, sensitization, respiratory intoxication, hepatic and kidney lesions, genotoxicity including fetotoxicity, teratogenesis, mutagenesis and carcinogenesis. The mode of toxic action of most of the pesticides is not fully elucidated and there are only few known antidotes to combat acute toxicity. The chronic effects of the pesticides are irreversible in most cases and seldom can be medically cured (1). The World Health Organization (WHO) estimated that two million cases of human poisonings occur annually, of which 40,000 are fatalities occurring mostly in the third world (1,2). The high incidents of pesticide poisonings in the developing countries is a reflection of the developing countries conditions where they suffer from abuse of pesticides, inefficient regulations and enforcement, lack of pollution awareness, low standard of living and poor sanitation.

The present review is devoted to address the need for specific biomarkers to detect the genetically high susceptible individuals to toxic chemicals exposure. Such inherited susceptibility is expected to be more intensified when those individuals are simultaneously exposed to residues of other toxicants such as air pollutants, toxins, heavy metals and other organics which can interact synergistically to intensify the toxic burden, particularly in developing countries where the people are more vulnerable. Recently, pesticides proved to cause immunodepression, a fact which is of great concern in our fight against infectious viral diseases including AIDS (3). The more vulnerable people are more susceptible to pesticides exposure due to their genetic characteristics and/or environmental exposure background. There are indications that people with different blood groupings will vary in their susceptibility to diseases and hazards of toxic agents (4). It should be anticipated also that people in the tropical and subtropical developing countries will be at higher risk because their liver and kidney functions are already impaired due to the endemic diseases caused by schistosomiasis parasite and hepatitis virus, etc (5,6). Therefore this review is actually devoted to address the possible molecular biomarkers which can be helpful in detecting the more susceptible population and quantify the risk of exposure and internal toxic effects at the biochemical targets with specific reference to the population in developing countries.

## Intrinsic Factors For Variation in Human Susceptibility to Toxic Chemicals

**Genetic variation in human population.** Our everyday experience reveals that there is an extraordinary amount of variation among us as human beings. We differ in height, weight, hair and skin color, texture, facial features, customs, and expressions. Our generalizations such as "women are more sensitive than men" obscure the fact that there is immense variation within sexes in all physiological, psychological, and anatomical traits. Moreover, there is a great deal of overlap between these traits. We are all aware that people from different geographic regions differ not only socially, but also biologically. For example, the red blood cells in all human beings contain hemoglobin. However, all people do not have exactly the same form of this molecule. Most people possess the common form of hemoglobin A, but in Western and Central Africa, in parts of Southern India, and in the Arabian peninsula, about one-quarter of the population carries another molecular form, hemoglobin S. People with only hemoglobin S suffer severe anemia because this molecule forms large crystalline structures inside the red blood cells, causing them to take on a characteristic sickle shape and then to break down. This condition is known as sickle-cell anemia. In population with hemoglobin S, only about 1-3% of the people suffer from this anemia while about 25-30% of such population have both hemoglobin S and hemoglobin A and those will suffer only very mild anemia and are significantly more resistant to malaria. The human population is clearly differentiated along the geographical lines. Currently, 17 polymorphic human genes have been studied in enough nations and tribes to make possible a calculation of comparative diversity. These polymorphic forms can be inherited as homozygote or heterozygote allelic genes in different human traits *(7)*. Karl Steiner (1900-1902) discovered the blood type polymorphism expressed as A, B, O, blood groups system as shown in Table 1. Humans are remarkably polymorphic for this ABO blood type system. In most European populations 45% of the people have type O blood group, 35% have type A, 15% have type B, and 5% have type AB. There is some variation from one major region of the world to another, but type O is always the most common followed by type A.

**Table 1. ABO Blood Types, Their Antigens, Antibodies, and Compatible Transfusions**

| Blood group type | Cell antigen type | Specific antibody | Compatibletransfusions |
|---|---|---|---|
| A | A | Anti- B | Transfuse to A & AB Receive from A &O |
| B | B | Anti-A | Transfuse to B & AB Receive from B &O |
| AB | AB | None | Transfuse to AB Receive from all types |
| O | None | Anti-A & Anti-B | Transfuse to all types Receive from O |

Adapted from Lewontin, R. *Human Diversity*, **1982** *(7)*.

The blood groups are genetically determined antigens of the erythrocytes. At least 15 different blood group systems, each determined by a separate locus on the red blood cells, have been identified. Furthermore, at these separate loci multiple alleles are now known. Table 2 shows the frequencies of alleles for seven polymorphic enzyme loci chosen in red blood cells in a number of European and African populations.

**Table 2.   Variation in Allelic Frequencies in Red Blood Cells of European and Black Africans**

| Red Cell Locus | Europeans | | | Black Africans | | |
|---|---|---|---|---|---|---|
| | Allele 1 | Allele 2 | Allele 3 | Allele 1 | Allele 2 | Allele3 |
| Acid phosphatase | 0.36 | 0.60 | 0.04 | 0.17 | 0.83 | - |
| Phosphoglucomutase-1 | 0.77 | 0.23 | - | 0.79 | 0.21 | - |
| Phosphoglucomutase-3 | 0.74 | 0.26 | - | 0.37 | 0.63 | - |
| Adenylate kinase | 0.95 | 0.05 | - | 1.00 | - | - |
| Peptidase A | 0.76 | - | 0.24 | 0.90 | 0.10 | - |
| Peptidase D | 0.99 | 0.01 | - | 0.95 | 0.03 | 0.02 |
| Adenosine deaminase | 0.94 | 0.06 | - | 0.97 | 0.03 | - |
| Average heterozygosity per individual | $0.068 \pm 0.028$ | | | $0.052 \pm 0.023$ | | |

Adapted from Lewontin, R. *Human Diversity*, **1982** *(7)*.

The reported data revealed that the genetical variation between the two populations was also reflected in the variation of the enzyme activities regulating metabolic, hormonal and oxidative phosphorylation processes including detoxification and activation of toxic chemicals *(7)*.  A correlation was suggested between blood group type and human susceptibility to infectious diseases.  A survey of cholera epidemic, indicated high attack rates in Peru, and determined the association between blood group O and severe cholera. The data revealed strong association between blood group O and the severe cholera that infected individuals. Since prevalence of blood group O in Latin - America may be the world's highest: treatment requirements were increased to combat the disease *(8)*. It was also reported that variations in corneal epithelium of subjects with different blood groups can be detected with the use of a monoclonal antibody and the lectins  DBA, GSL-1B-4, and UEA-1. Previous reports of lectin binding to ocular surface epithelium should be reevaluated since they did not take into account blood-group specific binding *(9)*. Distribution of blood groups ABO, Rh, P-1, and  MN  were  studied in 85 patients with multi modular euthyroid colloid goiter. An association has been revealed between the presence of this disease and MN blood group, as well as the absence  of  P-1 antigen. These results permit us considering MN blood group and P-1 antigen absence as factors of risk for multi modular euthyroid colloid goiter disease *(10)*. Accordingly, human population is heterogenous and consists of subpopulations with a considerable range of susceptibilities to chemical-induced toxic effects.   Parameters such as metabolic variability, DNA repair capacity and genetic predisposition can influence the severity of a given exposure to a known toxic or genotoxic agent *(11-13)*.

Several reports emphasized that the blood group type is also the main limiting factor  for  deciding  the  immune  response  induced  by  the  blood  antigens.  The immunoglobulin E   binding protein, epsilon BP is a betagalactoside protein of approximately 30-KDa and a number of the human soluble lactins which act by mediation of IgE binding to cellular surface proteins. Recently, the specificity of binding of the radio iodinated epsilon-BP to a series of lipid-linked oligosaccharide sequences of the lacto/neolacto family was investigated. The results showed that the minimum lipid-linked oligosaccharides that can support epsilon-BP binding are the pentasaccharides of the lacto/neolacto series and that the lectin binds more strongly to oligosaccharides of this family that bear blood group A, B, or B-like determinants than to those bearing blood group O. Similarly, EBP was preferentially bound to erythrocytes of blood groups A, B, and O *(14)*. These results imply that biological activity of the immunoglobulin E binding protein epsilon BP, in its degree of cell association and thresholds for activation of the cellular immune system and the allergic response, depends on the blood group type prevailing in the human subject.

**Immuno-Antigens and Human Genetic Diversity.**   Human lymphocyte antigens (HLA) were discovered in the white blood cells and were shown to have specific antigenic properties which are similar to those of the human red blood cells in being also polymorphic. Distribution of HLA antigens, haptoglobin phenotypes (HP), ABO blood groups and rhesus factor were investigated in 60 patients suffering from tuberculosis. Tuberculosis reactivation and postoperative recurrences occurred more often in carriers of HLA antigen DR2 *(15)*. HLA and ABO genetic markers were determined in 406 workers occupationally exposed to industrial biotechnology. The study proved the importance of a genotype in determination of individual sensitivity in the working places *(16)*. It was also reported that pesticide field workers vary in their acute response to pesticide exposure according to the blood group type *(17)*.

**Glucose -6- Phosphate Dehydrogenase ( G-6-PD) Deficiency.**   In the case of glucose-6-phosphate dehydrogenase (G-6-PD) deficiency, a number of observations suggested that such individuals will be more susceptible to oxidizing agents. This G-6-PD deficiency is due to many X-linked genetic variants abnormality which occurs most frequently in two forms. The mild more common form is the African variant (A-) which affect approximately 12% of the Afro-American males. The more severe form is the Mediterranean variety (B-). On exposure to hazardous oxidizing agents the G-6-PD deficient red cell accumulates peroxides which damage the cell membrane and shorten its life span. Exposure to trinitrotoluene (TNT) has been associated with hemolytic episodes in workers. The relationship between TNT exposure and severe hemolysis in G-6-PD deficient individuals strongly suggests a cause and effect relationship *(18)*.

**Acetylated Phenotype.** The activity of liver N-acetyltransferase differs widely among humans. A single gene determines the rate of acetylation and rapid acetylation is an autosomal dominant trait. Population studies revealed that about 60% of Caucasians and Negroes and about 10-20% of Orientals are slow acetylators. A number of drugs are metabolized by the acetylation of amino or hydrazino groups. These include isoniazid, hydralazine, procarbinamide, and a number of sulfa drugs. With long term administration of isoniazid, slow acetylators develop higher blood levels of the drug due to the slow detoxification. Besides, the rate of acetylation is of critical importance in industrial chemical exposure. The detoxification of the arylamine bladder carcinogens e.g. 2-naphthylamine and benzidine requires rapid acetylation. Slow acetylation is expected to lead to bladder cancer due to continuous exposure to arylamines in such cases *(19)*.

**Arylhydrocabon hydroxylase (AHH).** Cytochrome P-450 dependent monoxiginase enzymes are a large family with various substrate specificities. Naturally, it is useful in oxidative detoxification of xenobiotics leading to more polar metabolites which can be excreted. However, in some cases, these metabolites can be cytotoxic even more than the parent compound. The activity of this enzyme AHH, which is abundant in mammalian tissues, is stimulated by induction after exposure to polycyclic aromatic hydrocarbons, insecticides, steroids, and drugs. AHH catalyzes the metabolism of polycyclic aromatic hydrocarbons to some products which were shown to be carcinogenic. Several studies indicated that differences in AHH levels in humans are controlled genetically. The high inducibility AHH phenotype is thought to represent about one tenth of the U.S. population *(20)*.

**Serum $\alpha_1$–antitrypsin.** $\alpha_1$-antitrypsin is produced in the liver, and is responsible for producing 90% of the $\alpha$-globulin proteins in human serum. Its main protective function is to prevent the proteolytic enzymes from attacking the alveolar structures in the lungs. In case of $\alpha_1$-antitrypsin deficiency, the destruction of alveolar tissue produces emphysema which takes place in humans during their forties and fifties. The level of $\alpha_1$-antitrypsin is under genetic control. Individuals with homozygous deficiencies are at high risk of having emphysema. They are highly susceptible to respiratory poisons and pesticidal fumigants *(21)*.

**Age, Development, and Nutrition**. Children and elders are more susceptible to the hazardous chemicals than the middle-aged people. The newly born children lack some detoxifying enzymes such as glutathione-s-transferase. Chemicals may cause developmental toxicity to the embryo at levels not harmful to the mother. Malformation, growth retardation, and fetotoxicity might result from fetus exposure due to less efficient immune system. On the other hand, the aged people are at high risk for exposure to the environmental pollutants because of the deteriorating immune system and the high sensitivity especially when they suffer from chronic diseases *(22)*.

**Hormone disruptors**. It was recently postulated that environmental chemicals e.g. chlorinated insecticides and other polychlorinated cyclic compounds, can bind to steroid hormone receptor. Such interference might induce abnormal hormonal activities in females, leading to hormonal imbalance or hyperactivity which results in cancer and developmental toxicity in pregnant mothers *(23)*. Recent epidemiological studies revealed that breast milk fat and serum lipids of women with breast cancer contain significantly higher amounts of chlorinated organic compounds including pesticides. It was concluded that cancer in most cases may be initiated by the interaction between host factors such as genetic susceptibility together with simultaneous exposure to environmental toxicants such as chlorinated pesticides *(24)*. In recent years, chlorinated pesticides, dioxins and dibenzofurans have been recorded in breast milk from different countries. DDT, DDE, and other chlorinated pesticides were detected in mother's milk in developing countries at higher levels, e.g. Thailand, Egypt, and Vietnam; while polychlorinated biphenyls (PCB's), dioxins, and dibenzofurans were detected at higher levels in mothers' milk in industrialized countries, e.g., Germany and U.S.A. more than in the third world countries *(25,26)*.

**Nutritional and Food Habits.**   Changes in nutrition may affect the toxicity of environmental chemicals by altering the defensive immuno response, and the other processes of detoxication, biotransformation, and excretion. Impairment of the host defenses plays a role in the unique susceptibility of the young as well as the elders to foreign chemicals. Antioxidant action of vitamins E, A, and C was found to protect the lungs against oxidant's damage. Some food components were found to afford protection against chemically induced neoplasia, e.g., phenols, indoles, isothiocyanates, and flavones *(27)*.

**Extrinsic Factors for Variation in Human Susceptibility to Toxic Chemicals**

**Variation in Background of Infectious Diseases Distribution.**   It is reasonable to assume that humans infected with parasitic diseases and suffering from impaired biological systems will be highly vulnerable to poisoning with toxic chemicals such as pesticides. Developing countries located mostly in tropical and subtropical regions have the climatic conditions suitable for the reproduction and spreading of the organisms responsible for causing and transmitting many of the infectious human diseases. Table 3 presents examples of the more abundant human transmitted diseases in developing countries *(28)*. The low standard of living and sanitation intensifies the continuous spreading of such diseases and explains its epidemic and endemic status in the third world. Individuals with impaired liver will have a reduced capacity to detoxify certain pollutants or to excrete them in the bile, and those suffering from renal failure may not be able to excrete toxic metabolites. Those suffering from cardiovascular diseases would be at greater risk when exposed to carbon monoxide. It was found that asthmatic patients are more susceptible to exposure to sulfur dioxide *(29)*. It was also reported that those suffering from liver damage due to schistosomiasis were proved to be unable to detoxify DDT when they were occupationally exposed as field spraymen *(30)*. This demonstrates the impact of infectious diseases in reducing the resistance of humans to toxic chemicals.

**Ecogenetics and Enhanced Susceptibility to Toxic Chemicals.**   Few studies have investigated the relationship between inherited traits and enhanced susceptibility to pollutant exposure. There is already known evidence that individuals with immotile cilia syndrome, and $\alpha$-antitrypsin deficiencies are unusually vulnerable to cigarette smoke and other respiratory poisons.

**Hazards of Exposure to Chemical Mixtures.** Chemical agents commonly cited as inducing symptoms in susceptible individuals include tobacco smoke, diesel and vehicle exhaust, paints, organic solvents, pesticides, natural gas fuel, plastics, perfumes, synthetic fibers, household cleaners, new building materials, and poorly ventilated new buildings.

**Table 3.     Some Parasitic and Enteric Infectious Human Diseases**

| Disease | Causing Organism | Way of Infection |
|---------|------------------|------------------|
| Schistosomiasis | Blood fluke/snails | Contaminated water |
| Fasiolopsiasis | Fluke/snails | Contam. water and food |
| Ascariasis | Round worm | Soil, water, and food |
| Ancylostomiasis | Hook worm | Soil, water, and food |
| Fliariais | Nematode | Mosquito bite |
| Malaria | Plasmodium | Mosquito bite |
| Trichinosis | Round worm | Raw pork |
| Mycotoxins | *Aspergillus* Sp. | Food |
| Paratyphoid and typhoid fever | *Salmonella* Sp. | Food and water |
| Diarrhea | *E. coli* | Food and water |
| Cholera | *Vibro cholerae* | Food and water |
| AIDS | HIV virus | Blood transfusion and Sexual contact |
| Hepatitis | Entero virus (diff. Strains) | Blood transfusion, water and food |

Clinical ecology is based on the total environmental load from a wide range of low level environmental exposures *(31)*.

Malathion acute toxicity was found to be potentiated by the presence of its two main impurities resulting as rearrangement and breakdown products: - isomalathion, O, S, S-trimethylphosphoro-dithioate (OSS-TMP), and O, O, S-trimethylphosphorothioate (OOS-TMP) *(32)*. These same impurities are also capable of causing an unusual delayed toxicity that has been studied. OOS-TMP has an $LD_{50}$ of 15-20 mg/kg in rat. However, death occurs no sooner than 22 days after exposure, in contrast to the acute lethal AChE inhibition which occurs in few hours. This delayed toxicity was not a cholinergic effect, but was due to hepatic hemostatic disorders reducing the production of the plasma clotting factors *(33)*. OOS-TMP was also shown to cause glutathione depletion in lung and other tissues. The immunotoxic potential of these impurities is probably the most serious concern *(34)*. Pentachlorophenol and its salts which are widely used as pesticides all over the world, are known to be contaminated with the extremely noxious Dioxins. Such mixtures are of extreme occupational health hazard *(35)*. The toxicities of parathion and paraoxon were potentiated by pretreatment with dietary DDE in Japanese Quail *(36)*. In mice, the acute toxicities of five N-methylcarbamates were potentiated by low-dose treatment with some OP insecticides *(37)*. A pretreatment of mice with EPN or DEF organophosphate enhanced the toxicity of the subsequent treatment with the pyrethroid fenvalerate and the OP malathion *(38)*. Pretreatment of rats with phenobarbital enhanced the acute toxicity of heptachlor, chlordane and DDT in neonatal rats *(39)*. Also pretreatment with carbon disulfide, a cytochrome P450 inhibitor, potentiated parathion and EPN but antagonized diazinon and dimethoate *(40)*. In an animal study, disulfiram (the drug used frequently in the management of alcoholism), increased tumor development after exposure to the fumigant ethylene dibromide (EDB) in rats *(41)*. A fact which caused a lot of concern for workers exposed to EDB who are currently using disulfiram as a medication *(42)*. Sequential administration of two different carcinogens has been demonstrated to result in synergistic effects *(43)*. Phenobarbital and polychlorinated biphenyls (PCBs), which are inducers of the liver mixed function microsomal enzymes, inhibit the carcinogenic effects of 2-acetylaminofluorene (2-AAF) or other hepatocarcinogens when administered simultaneously, whereas they are well known to enhance hepatocarcinogenecity when given after carcinogen treatment *(44)*. Similar timing impacts were reported for other combinations *(45)*. Polychlorinated biphenyls (PCBs) are themselves a mixture of analogous compounds polluting the environment. Although their production was banned since 1970, yet because of their persistence they are still in the environment bioaccumulating in the food chain. They cause health adverse effects mainly in reproduction through inhibition of spermatogenesis and exerting ovarian histopathological effects in additional to hormonal imbalance and other cytotoxic effects *(46)*. It can be concluded that pesticide mixtures with other chemicals may be synergistic, additive or antagonistic in their toxicity. Therefore, the occupationally exposed workers may be at higher risk than the general public in particular cases *(47)*.

**Epidemiological Biomarkers of Exposure, Effect and Susceptibility to Pesticides and Agrochemicals**

**Biomarkers of Exposure.** Measurement of activity of target enzymes, hormones, receptors, and neurotransmitters are commonly used as biomarkers for acute and short term exposure to agrochemicals including pesticides as well as other toxic chemicals. However, few biomarkers have been validated as tools for environmental epidemiology. Validation of any epidemiological biomarker must involve the following four stages:- association between the marker and a preceding exposure or subsequent effect; dosimetry of the exposure/marker relationship; the threshold of "no observed effect level" if any; and the positive value of the biomarker for exposure or adverse effect *(48)*. Besides, the biomarker must have the capacity to include and identify the genetically and occupationally highly susceptible individuals in the exposed group. The successful biomarker must finally be applicable under both laboratory and population conditions *(49-51)*. To determine whether the biomarker is measuring the degree of exposure, the adverse effect (disease) or the level of susceptibility in an exposed population depends on the available knowledge and the possibly quantified biological processors or phenomena under exposure. DNA and various protein adducts are promising as biomarkers for epidemiological and long term continuous exposure to low levels of pollutants including the hazardous agrochemicals. Increased rate of hepatic biosynthesis of protein, DNA, and RNA are indicative of the inductive effect which activates cytotoxicants *(52)*. The specific potential of agrochemicals to bind with DNA and RNA can be used as a criterion to predict for their cytotoxicity. DNA and protein adducts also can be used to measure exposures that cause other non-genotoxic adverse health effects. Protein adducts could be a better reliable measurement of exposure dose levels than DNA adducts because the protein adducts are not repaired and tend to persist for the life of the protein which reaches 17 weeks in the case of hemoglobin in humans. The protein adducts are therefore useful as a biomarker of exposure because they fulfil the criteria of linear dose-response relationships for single exposures *(53)*. Hemoglobin adducts were successful biomarkers for blood levels of exposure to ethylene oxide and lead *(54)*. Recently, there was evidence that rats' serum protein profile reflects electrophoretic gradient changes induced by the dermal administration of malathion to rats *(55)*. Similar data were reported showing specific alterations in the serum protein profile as a result of *in vivo* application of a sublethal dose of cyanophos. *In-vitro* incubation of rat serum with a sublethal level of a number of cyclodiene and lindane insecticides, PCB's and trichloroethane showed characteristic changes in the rat serum protein profile as shown by SDS-PAGE followed by laser integration scanning and by FPLC chromatography *(56)*. Preferential binding of rat serum transferrin with aluminum cations showed another example of using specific blood protein as a specific biomarker for detection and quantification of $Al^{3+}$ cation intake during occupational or medical treatments *(57)*. It can thus be concluded that most DNA, hemoglobin, albumin, and other blood proteins are selective sensitive biomarkers of exposure and effective dose *(58)*.

**Immune Dysfunction as a Biomarker for Exposure to Toxic Chemicals**. Immunotoxicology is the study of adverse effects on the immune system that result from exposure to toxic chemicals including industrial, agrochemicals, pharmaceuticals, and environmental pollutants. In these instances, the immune system acts as a nonspecific target of the xenobiotic, which may lead to an increased incidence or severity of infectious disease or neoplasia because of an inability to respond to an invading agent. The sensitivity of the immune system to reflect the environmental stress and particularly the xenobiotics, suggests using the immune response and its suppression as a biomarker for long term health adverse effects subsequent to exposure to hazardous agrochemicals including pesticides. However, it is a real challenge to distinguish homeostatic changes from pathogenomic ones *(48,59)*. Several investigators in the last decade have used the immune system markers to indicate biological response for low doses of toxic substances *(60)*. In one of the studies, some of the strengths and limitations in using markers are the effect on the immune activation and autoantibodies in persons who have long term inhalation exposure to formaldehyde. Four groups of exposed patients were compared with controls. The four exposed groups have shown higher antibody titer to antibodies to formaldehyde-human serum albumin (HCHO-HSA) conjugate; and increases in Tal+; IL2+, B cells, and autoantibodies were recorded *(61)*. It was found to be much better to use a collection of immune system markers because no single marker will accurately reflect the state of the immune system as a whole. Two pesticide fertilizer mixtures were examined for their ability to induce immunotoxicity in mice. These mixtures were made to resemble groundwater contaminations from agrochemicals in either California (aldicarb, atrazine, dibromochloropropane, dichloropropane, ethylene dibromide, simazine, and ammonium nitrate); or Iowa (alachlor, atrazine, cyanazine, metolachlor, meribuzin, and ammonium nitrate. No consistent suppression was shown for either mixture, however, a slight but significant suppression of the bone marrow progenitor cells occurred after 90 days exposure to the high dose of the California mixture *(62)*. Chemical-induced immunosuppression may result through multiple mechanisms including cellular depletion or functional alteration in which the cells do not respond adequately to an antigen. Hypersensitivity is the result of an appropriate immune response directed against chemical agents that bind to host tissues and are recognized as foreign antigenic agents by the immune system. Such chemical-host tissue complex may lead to respiratory tract allergies (e.g. asthma or rhinitis) or allergic disorders. The chemical agents known to cause allergic disorders include anhydrides, aldehydes, isocyanates, organophosphate insecticides and some antibiotics *(63,64)*. Besides, there is an increasing interest in chemical-mediated autoimmune disease. Autoimmunity, which has similarities to the hypersensitivity response is characterized by an immune response against self-antigens. However, several environmental chemicals e.g. vinyl chloride or trichloro ethylene and certain metals e.g. mercury, generate immune responses that may lead to the production of autoantibodies, although not necessarily to symptomatic autoimmune diseases *(65)*.

**Immune Dysfunction as a Biomarker for Chronic Exposure to Organophosphate**.
A large number of epidemiological studies of organophosphate (OP) exposed populations revealed that farmers succumb more frequently lymphomas or hematopoietic tumors than would be expected in healthy nonfarming populations *(66,67)*. Such data have been documented not only for farmers residing in the USA, but also for those in other countries *(68)*. An excess of lymphomas has been recently reported in Italian farmers *(69)*. Studies by epidemiologists from the National Cancer Institute in the USA have reported that the farm workers who acquire liver tumors in excess have been exposed to pesticides, which are postulated as the cause for the development of such tumors in exposed workers in the fields or grain mills treated with pesticides. Malathion, an OP insecticide was one of the identified chemicals by workers exposed to pesticide-treated grain. Several recent studies have shown OP-induced abnormalities in human cytotoxic T-lymphocyte, natural killer cell and cytotoxic monocyte functions that regulate human immune surveillance *(69)*. Both chromosomal aberrations and cell-mediated immunity can be readily assessed using peripheral blood samples from populations at risk. In addition to OP-induced defects in natural killer cell and cytotoxic T-lymphocyte activities; the inhibition constants of human monocyte carboxyl esterase and plasma butyl cholinesterase, were compared. The data suggest that immune cell esterases may be a more sensitive marker of organophosphate exposure than plasma butyryl-cholinesterase, which is a standard biomarker currently used to detect OP exposure *(69)*.

Recent studies also showed that chronic OP exposures causing an impairment in esterase-dependent immune surveillance systems, may facilitate oncogenic virus infection, an increase in the frequency of lymphocyte mutations and the subsequent development of lymphomas *(70)*. This is one of the good examples to demonstrate that chronic exposure to pesticides might lead to higher susceptibility to genotoxic and infectious diseases. Several *in-vitro* studies have also shown OP-induced mutations in human peripheral blood lymphocytes and human lymphoid cell lines of B-cell origin

*(71)*. The frequency of hybrid antigen-receptor genes in peripheral blood lymphocytes of agricultural workers has been determined to be higher during pesticide application times than before or after the pesticide use, and significantly higher than controls *(72)*. Thus, such functional OP-induced cytotoxic immuno-toxic changes may represent sensitive biomarkers of chronic OP exposure and effect when integrated with duration and intensity of exposure.

**Biomarkers of Adverse Health Effects.** Another type of molecular biomarker for adverse effects can be the cytogenetic investigation with peripheral blood lymphocytes, splenocytes, and pulmonary cell culture to monitor genotoxic effects of low level mixture or single toxicant. However, this research area is not yet fully developed and needs further exploration for standardization *(73)*.

In a recent review on biomarkers in toxicology, biomarkers of effect were emphasized to include several examples of different aspects. Inhibition of blood enzymes such as δ-aminoleuvolenic acid dehydratase (ALAD) by lead. Number and damage of blood cells, e.g., presence of sister chromatic exchanges due to potential damage of chromosomes in workers exposed to ethylene oxide. Lack of particular lymphocytes indicating immune suppression caused by dioxin (TCDD). Similarly, induction of cytochrome P450 isozymes, as a biomarker of the effect of many types of chemicals particularly organochlorine pesticides, polycyclic hydrocarbons and related chemicals. As a biomarker for long term exposure, it is suggested that we can make use of the urinary markers for cytochrome P450 induction by measuring the increased excretion of D-glucaric acid and the 6-β-hydroxycortisol/17-hydroxycorticosteroid ratio *(58)*. A recently suggested biomarker for toxic effects are the stress proteins. Immunocytochemistry might be one of the best methods to quantify the specific heat shock protein HSP 72 which is found normally in minute levels. Heat shock proteins proved to be induced by various chemical stresses, and any increase can be easily related to the type of environmental stress *(74)*. Although NMR has been used only recently to study biological systems, it has been proven to be a powerful technique and direct proton NMR of biological fluids such as serum bile and urine has been developed. The use of sophisticated techniques such as two dimensional correlation spectroscopy (COSY) analysis allows separation of overlapping resonances and greatly improved resolution of spectra. This new technique has the advantages of providing structure and quantitative information, it is rapid and not preselective. Therefore, any xenobiotic in body fluids can be detected and quantified *(75)*.

**Biomarkers for Susceptibility.** There is an increasing interest in the role of genetic variation in toxic responses. Therefore, variation in human susceptibility and development of relevant biomarkers are of increased importance. An example of the variation in metabolism can be described under the acetylator phenotype. The ability to acetylate and detoxify amines, hydrazines, and sulphoamides vary between individuals. Slow acetylators have mutations responsible for less functional N-acetyltransferase enzyme. There are correlations between slow acetylator phenotype, bladder cancer, and occupational exposure to aromatic amines. On the other hand, different types of cancers (colo-rectal cancer) have been linked to the fast acetylator phenotype *(58)*. With the advancement of molecular biology we will be able to differentiate between these phenotypes by gene analysis, rather than by biochemical reactions. It has been suggested also that taurine levels may be a useful biomarker of susceptibility. Urine taurine level reflects taurine status in the liver as shown in animal studies *(58)*. Taurine in the liver is protective and when its levels are reduced, the liver will be rendered more susceptible to toxicants. This finding is crucial because humans normally do not biosynthesize taurine and depend on its dietal sources. Individuals with relatively less taurine in their liver will be at higher risk. Current risk assessment models fail to consider genetic predisposition that make people more sensitive or resistant to exogenous exposures and endogenous processes. Cytochrome P450 enzymes activating carcinogens show wide genetic polymorphisms and differ widely among various ethnic and genetic groups. In addition to the inherited factors for detection of susceptible individuals, determination of the effective dose of carcinogen will help in setting the risk assessment evaluation. DNA adducts will be a cumulative measure of exposure, absorption, metabolic activation, detoxification and DNA repair process. The combination of type and abundance of adduct and genotyping assays will reflect the interaction between exogenous and endogenous factors along the long term exposure to carcinogens. Such criteria for risk assessment will be useful to differentiate between the level of cancer risks between individuals, and genetic traits by marking the more vulnerable people *(76)*. A biological marker of susceptibility, the alpha-1-antitrypsin ZZ allele, has been found to be associated with the disease emphysema. People with the ZZ homozygote are 30 times at the risk of developing emphysema than the general population. However, because emphysema has various causes, the heterozygous state-although genetically not susceptible can be at increased risk when hazardous environmental factors exist: e.g. pollution by cadmium, ozone, or cigarette smoke. Generally, the use of biomarkers of susceptibility in environmental epidemiology has the advantage of increasing both the precision and reliability of risk assessment and correlation between exposures and health diseases in populations without neglecting the highly susceptible section in the population which might be diluted by the majority of the nonsusceptible *(77)*.

Future extensive research is still needed to reach more sensitive measures and critical biomarkers that define the wide variation in susceptibility in a population, to protect those who are at high risk because they are more susceptible for one reason or another.

## Educational, Socio-economic, and Regulatory Constraints in Developing Countries.

**Socio-economic Status.** The developing countries have their own historical, genetical, and environmental characteristics. Being located in tropical and subtropical regions, they have the suitable ecological conditions for spreading and reproduction of not only agricultural pests but also the infectious medical diseases and their vectors. This explains their increasing demand for pesticides and drugs to combat agricultural pests and human diseases vectors as well. This implies higher exposure to pesticides.

**Environmental Education and Extension Services Needed.** Environmental education as defined by the International Union of Conservation of Nature and National Resources (IUCN) is based on a long term strategy to raise the level of awareness among the population so that local people will be able to take over the management of the environment by themselves. Extensive extension or services using media communication facilities including pictograms and television will have an enormous influence for the dissemination of proper instructions and knowledge to create public awareness regarding hazards of toxic chemicals and the best way for safer handling, storing, and application *(78,79)*. The Nongovernmental Organizations (NGO's) provide different in-service training programs for people who work in the field of handling and use of agrochemicals. We believe that, the success of any environmental educational program in developing countries is closely linked to a number of challenges: financial constraints, social awareness, cooperation between governmental and NGO's, public education system, interaction between media and public, and least, not last, the implementation of laws concerning environmental protection and human health.

**Regulatory and Management Constraints in Developing Countries.** Developing countries need updated legislation for chemical safety. Parallel to that, there must be an upgraded enforcement system with sufficient capabilities to test and evaluate the toxicological data submitted by the industry. The regulations should take into account the long-term cytotoxic effects in addition to acute and short-term effects. Cooperation between industrialized countries and the third world in advanced technology transfer will improve and upgrade the chemical industries in the developing countries toward cleaner production where the working place and the products quality will be up to the international standards. The hazards of unwise use of pesticides including the ones banned and the residues of the highly toxic pesticides and agrochemicals on exported food and those disposed in the environment should not be the concern of the developing countries alone, it should be an international responsibility. Hence assistance will be needed to provide the third world countries with the helpful techniques for setting the reliable biomarkers in order to measure the exposure levels and their effects. These measures will reduce the hazardous exposure to agrochemicals in the developing countries.

## Conclusions

1. People in the developing countries have their own characteristic genetic traits, historical behaviors, and traditions in addition to their warm tropical and subtropical ecological conditions. All these factors should be considered when trying to generalize risk assessment criteria, and when suggesting universal biomarkers for monitoring levels of exposure and to estimate degrees of respective adverse health effects at the biochemical targets.
2. Population in the developing countries vary in their susceptibility to hazardous agrochemicals as do all humans according to their ABO blood group type, and the genetic antigenic variations. However, the following genetic deficiencies were reported to be more abundant in the third world populations:
   * lower glucose -6- phosphate dehydrogenase,
   * more sickle cell anemia, and sickle cell trait,
   * lower arylhydrocarbon hydroxylase induction, and
   * slower acetylation genetic type.

Such genetic deficiencies make people more susceptible to the toxic chemicals including pesticides and other agrochemicals.

3. People in developing countries being liable to schistosomiasis and other parasitic infectious diseases, will suffer from impaired liver and kidneys in addition to deterioration of the defensive immune system. In such conditions they are more vulnerable to be intoxified with agrochemicals.

4. The general public in the third world is exposed to relatively higher levels of environmental pollutants. In the urban areas and crowded cities, they suffer from higher levels of air pollutants, e.g., tetraethyl lead, sulfur dioxide, nitrogen oxide, carbon monoxide, ozone, and benzopyrines. In addition, the rural areas will suffer from contaminated drinking water and heavy residues of agrochemicals. Such higher loads of mixture of pollutants render the population highly sensitive for subsequent exposure to hazardous agrochemicals.

5. Although pesticides had an undenied advantage of reducing vectors transmitted diseases, in addition to the protection and rise of the agricultural production in developing countries; yet the severe unwise application of pesticides and their abuse especially those which had been banned or restricted in the developed countries, make pesticides and agrochemicals very harmful to humans in the third world. The hazardous types and the unprotected exposure to excessive amounts of pesticides makes the people of the developing countries at higher risks.

6. The known molecular biomarkers that can be adapted to the developing countries population include:

    * Protein and DNA adducts as biomarkers to detect levels of exposure,
    * depression of immune system as a criteria for the health adverse effect induced
      by such exposures.

7. Other pathological lesions, physiological deficiencies, and behavioral habits should be regarded when monitoring exposure levels or effects and when setting the threshold limit values (TLV's) for the working places, or the no observable effect level (NOEL) for ADI's, and the amount of the safety factor should be higher to compensate for the highly susceptible individuals.

8. The third world countries lack the advanced regulatory enforcement capabilities needed to cope with the international schemes for safe handling and use of toxic chemicals including pesticides and other agrochemicals. The hazards of all these chemicals and their residues will not be kept locally but they are transportable to other parts on the globe in polluted exported food products, discharged water wastes in rivers, seas, and as vapor in the atmosphere to many remote places. Being a universal problem, it might necessitate UN agencies and international organizations to help under the known Global Monitoring System Programme.

## Literature Cited

1. El-Sebae, A.H. *Environ. Abstract Annual.* **1988**, *18*, pp. A47-A50.
2. Forget, G. *J. Toxicol. Environ. Health.* **1991**, *32*, pp. 11-31.
3. El-Sebae, A.H., *Genetic Toxicology of Environmental Chemicals; Part B Genetic Effects and Applied Mutagenesis*, Alan R., Liss. Inc., Sweden, **1986**, pp. 273-281.

4.  El-Sebae, A.H. In *Appropriate Waste Managment for Developing Countries*; Curi, K. Ed.; Plenum Pub. Comp, N.Y., **1985**, pp. 563-577.

5.  Dieter, M.P.; J.L. Ludke *Bull. Environ. Contam. Toxicol.* **1975**, *13*, pp. 257-261.

6.  Wasserman, M.; Prata, A.; Tomatis, L.; Wasserman, D.; Oliviera, V. *Rev. Soc. Braz. Med. Trop.* **1975**, *VIII*, pp. 271-273.

7.  Lewontin, R., *Human Diversity*; Scientific American Library Inc., N.Y. 1982, p.179.

8.  Swerdlow, D.L.; Mintz, E.D.; Rodriguez, M.; Tejada, E.; Ocampo, C.; Espejo, L.; Seminario, L.; Tauxe, R.V., *J.Infectious Diseases*, **1994**, *170,* pp. 468-472.

9.  Watanabe, R.; Gipson, I.K., *Arch. Of Ophthalmology*, **1994**, 112, pp. 667-673.

10. Makarov, A.D.; Bazarova, E.N.; Koslov, G.I., *Problemy Endocrinologii*, **1993**, *39*, 25-26,  in Russian, English summary, Biol.Abstracts 1994, *98*, iss.1 , ref. 4016.

11. *Principles and Methods of Toxicology*, Hayes A.W., Ed. Raven Press, N.Y., N.Y., 1994,  3rd Edition.

12. Shields, P.; Harris, C.C. *JAMA* **1991**, 266, pp. 681-687.

13. Harris, C.C., *Carcinogenesis*, **1989**, 10, pp. 1563-1566.

14. Feizi, T.; Solomon, J.C.; Yuen, C.T.; Jeng, K.C.G.; Frigeri, L.G.; Hsu, D.K.; Liu, F.T.; *Biochemistry*, **1994**, *33*, pp. 6342-6349.

15. Naumov, V.N.; Shaikhaev, A.Ya.; Paspelov, L.E.; Testov, V.V., *Problemy Tuberculeza*, **1993**, 0(4), 17-19, in Russian, English summary, Biol.Abstracts 1994,  *98*, 2:21746.

16. Kuznetsov, N.F.; Artamonova, V.G., *Meditsina Truda I Promyshlennaya Ekologiya*, **1993**, 0(9-10), 12-15, in Russian, English summary, Biol. Abstracts, 1994, *98*, *155.2*:26799.

17. El-Sebae, A.H., In *Ecotoxicology and Climate*; Bourdeau, P.; Haines, J.A.; Klein, W.; Krishna, C.R. Murti, Eds.; SCOPE, John Wiley & Sons Ltd., N.Y., 1989, pp. 359-371.

18. Luzzatto, L.; Mehta, A., In *The Metabolic Basis of Inherited Diseases*; C.R., Scriver, Beauder, A.L., Sly, W.S., Eds., McGraw-Hill, N.Y., 1989, p. 2237.

19. Cartwright, R.A.; Glashau, R.W.; Rogers, J.H., *Lancet*, **1982**, *2*, p. 824.

20. *Detection of Cancer Predisposition, Lab. Approach*; Spatz, L., Bloom, A,D., Paul, N.W. Eds.; Monograph no. 3, March of Dimes Birth Defects Foundation, White Plains, N.Y., 1990.

21. Cox, D.M., In *The Metabolic Basis of Inherited Diseases*; Scriver, C.R, Bauder, A.L., Sly, W.S., Eds.; McGraw-Hill, N.Y., 1989, p. 2409.

22. Wilson, J.E., *Environment and Birth Defects*, Academic Press, N.Y., 1973.

23. Colborn, T.; Voon Saal, F.S.; Soto, A.M., *Environ. Health Perspectives*, **1993**, *101*, pp. 378-384.

24. Davis, D.L.; Bradlow, H.L.; Wolff, M.; Woodruff, T.; Hoel, D.G.; Culver, H.A., *Environ. Health Perspectives*, **1993**, *101*, pp. 372-377.

25. Schecter, A.; Purst, P.; Kruger, C.; Meemken, H.A.; Groebel, W.; Constable, J.D., *Chemosphere,* **1989**, *18*, pp. 445-454.

26. Dogheim, S.M.; El-Saafeey, M.; Afifi, A.M.H.; Abdel Aleem, F.E., *J.Assoc. Off. Anal. Chem.* **1991**, *74*, pp. 2-4.

27. Wattenberg, L.W., *Environ. Pathol. Toxicol.* **1980**, *3*, p. 35.
28. *Envirovmental Health*; Purdom, W., Ed.; Academic Press, N.Y., 1980, pp. 12-20.
29. Tarcher, A.B.; Calabrese, E.J., In *Principles and Practice of Environmental Medicine*; Tarcher, A.B., Ed.; Plenum Publ. Co., N.Y., 1992, pp. 189-213.
30. El-Sebae, A.H., In *Proceedings of the International Workshop on Chemical Safety Communication. Special Needs for Developing Countries*; Alex. Univ., Egypt, 1994, pp. 1-10.
31. American College of Physicians, Position Paper, Clinical Ecology, *Am. Intern. Med.*, **1989**, 111, p. 168.
32. Ryan, D.L.; Fukuto, T.R., *Pesticide Biochem. Physiol.*, **1985**, *23*, pp. 413-424.
33. Umesto, N.; Mallipudi, N.M.; Tola, R.F.; March, R.B.; Fukuto, T.R., *J. Toxicol. Environ. Health,* **1981**, *7*, pp. 481-497.
34. Rodgers, K.E.; Grayson, M.H.; Imamura, T.; Devens, B.H., *Pesticide Biochem. Physiol.* **1985**, *24*, pp. 160-166.
35. Williams, P.L., *Occup. Health Safety*, **1983**, pp. 14-16.
36. Ludke, J.L., *Pesticide Biochem. Physiol.* **1977**, *7*, pp. 28-33.
37. Takahashi, H.; Kato, A.; Yamashita, E.; Naito, Y.; Tsuda, S.; Shiraso, Y., *Fundam. Appl. Toxicol.* **1987**, *8*, pp. 139-146.
38. Gaugham, L.C.; Engel, J.E.; Casida, J.E., *Pesticide Biochem. Physiol.* **1980**, *14*, pp. 81-85.
39. Harbison, R.D., *Toxicol. Appl. Pharmacol.* **1974**, *32*, pp. 443-446.
40. Yasoshima, M.; Masuda, Y., *Toxicol. Lett.* **1986**, *32*, pp. 179-184.
41. Plotnick, H. B., *J. Amer. Med. Assoc.* **1978**, *239*, p. 1609..
42. Yodaiken, R. E., *J. Amer. Med. Assoc.* **1978**, p. 2783.
43. Williams, G. M.; Katayama, S.; Ohmari, T., *Carcinogenesis*, **1981**, *2*, pp. 1111-1117.
44. Tatematsu, M.; Nakanishi, K.; Morasaki, G.; Miyata,Y.; Hirose, M.; Ho, N., *J. Natl. Cancer Inst.*, **1979**, *63*, pp.1411-1416.
45. Wade, G.G.; Mande, R.; Ryser, H. J., *Cancer Res.* **1987**, *47*, pp. 6606-6613.
46. Gray, L. E.; Ostby, J.; Sigmon, R.; Andrews,J., *Fundam. Appl. Toxicol.* **1993**, *20*, 288-294.
47. Chambers, J. E.; Dorough, G. In *Toxicology of Chemicals Mixtures, Case Studies, Mechanisms and Novel Approaches.* Yand, R.S.H., Ed.; Academic Press, N.Y., 1994, pp.135-155.
48.*Environmental Epidemiology*, Committee on Environ. Epidemiology, Board on Environ. Studies and Toxicology, Commission on Life Sciences, US National Research Council, 1991, *1*, 282.
49. Gann, P. In *Environmental Epidemiology*, Topfler, F.C.; Craun, G.F., Eds.; Lewis Publisher, Chelsia, 1986, pp. 109-122.
50. Hatch, M. C.; Stein, Z. A. In *Mechanisms of Cell Injury Implications for Human Health.* Flower, B. A., Ed. John Wiley and Sons, N.Y. 1987, pp. 303-314.
51. Grifith, J.; Doncan, R.C.; Holka, B.S. *Arch. Environ. Health,* **1989**, *44*, pp. 375-381.
52. El-Sebae, A. H.; Salem, M. H.; El-Assar; M.R.S.; Enan, E. E., *J. Environ. Health,* **1988**, *B23*, pp. 439-451.
53. Schell, F. C.; Chiang, T. C., *Protein Adducts Forming Chemicals for Exposure*

*Monitoring.Literature Summary and Recommendations.* EPA 60014-901007, Las Vegas, US EPA, 1990.

54. Osterman-Golkar, S. And Bergmark, E. *Scand. J. Work Environ. Health*, **1988**, *14*, pp. 372-377.

55. Abou Zeid, M. M.; El-Barouty; G.; Abdel Rahim, E.; Blancato, J.; Dary, C.; El-Sebae, A. H.; Saleh, M. A. *J. Environ. Sci. Health*, **1993**, *B28*, pp. 413-430.

56. Saleh, M.A.; Abou Zeid, M. M.; Kamel, A.; Blancato, J.; and El-Sebae, A. H. *ACS Biomarkers For Human Exposure.* In press.

57. El-Sebae, A. H.; Abou Zeid, M. M.; Abdel Rahim, E.; and Saleh, M. A., *J. Environ. Sci. Health,* **1994**, *B29*, pp. 303-321.

58. Timberell, J. A.; Draper, R.; Waterfield, C. J. *Toxicology and Ecotoxicology News.* **1994**, pp. 4-14.

59. *Occupational Lung Diseases*; Weill, H. ; Turner-Warwick, M. Eds.; Decker, N.Y. 1981.

60. *Environmental Chemical Exposure and Immune System Integrity*; Burger, E.J.; Tardiff, E. J.; Bellanti, J.A. Eds.; Princeton Scientific, Princeton, 1987.

61. Thasher, J. D.; Broughton, A.; Madison, R. *Arch. Environ. Health*, **1990**, *45*, pp. 217-223.

62. Thomas, P. T.; Basse, W. W.; Kirkvliet, N. I.; Luster, M. I.; Munsos, A. E.; Morray, M.; Roberts, D.; Robinson. M.; Silkworth, J.; Sjoblad, R.; and Snialowicz, R. In: *The Effects of Pesticides on Human Health*; Baker, S. R.; Wilkinson, C. F., Eds.; Princeton Sci. Publ., N.J., 1990, pp. 216-295.

63. Luster, M.I.; Wierds, D.; Rosenthal, *G.J. Med. Clin. North Am.* **1990**, 74, pp. 425-439.

64. Dean, J.H.; Cornacoff, J.B.; Rosenthal, G.J; Luster, M.I. In *Principles and Methods of Toxicology*; Hayes, A.W. Ed.; Raven Press, N.Y., 1989, pp. 741-760.

65. Rosenthal, G.J. and Luster, M.I.In *Xenobiotics induced inflamation. The role of growth factors and cytokines*; Schook, L. and Laskin, D., Eds.; Academic Press, N.Y. 1994.

66. Burmeister, L.F. *J.Natl. Cancer Inst.* **1981**, 66, pp. 461-464.

67. Pearce, N.E.; Smith, A.H.; Fisher, D.O. *Amer. J. Epidemiol.* **1985**, *121*, 225-237.

68. Alvauga, M.C-R; Blair, A.; Masters, M.N. *J. Natl. Cancer Inst.* **1990**, *82*, pp. 840-848.

69. Lee, M.J.; Waters, H.C. *Blood*, **1977**, *50*, pp. 949-951.

70. Ambinder, R.F. *Hematol. Oncol. Clin. North Am.*, **1990**, *4*, pp. 821-823.

71. Sobti, R.C.; Krishav, A.; Pfanffenberger, C.D. *Mutat. Res.* **1982**, *102*, pp. 89-102.

72. Kilgerman, A. D.; Cambell, J.A.; Erexson, J. L.; Allen, J. W.; Shelby, M. D., *Environ. Mutagen* **1987**, *9*, pp.29-36.

73.Hugget, R.J.; Kimerle, R.A.; Mehrle, P.M.; Bergman, H.L., *Biomarkers, Biochemical,Physiological and Histological Markers of Anthropogenic Stress.* Lewis Publishers, Chelsea, MI, 1992.

74. Nicholson, J. K.; Wilson, I.D., *Progress in NMR Spectroscopy*, **1989**, *21*, pp. 449-501.

75. Lipkowitz, S.; Garry, V.F.; Kirsch, I.R. *Proc. Natl. Acad. Sci., USA*, **1992**, 89, pp. 5301-5305.

76. Shields, P. G. *Environ. Health Perspectives*, **1994**, *102*, pp. 81-87.
77. OTA (US Congress Office of Technology Assessment). *The Role of Genetic Testing in the Prevention of Occupational Diseases*. OTA-RA-194. U.S. Government Printing Office, Washington, D.C., 1983.
78. Vinke, J; In *Enviromental Education, An Approach to Sustainable Development*; Schneider, H.; Vinke, J. Eds.; Organization for Economic Co-operation and Development, Paris, France,  1993, p.39.
79. Emmerij, L. In *Environmental Education, An Approach to Sustainable Development*; Schneider, H.;Vinke, J., Eds.; Organization for Economic  Co-operation and Development, Paris, France, 1993, p.9.

BIOMARKERS OF HUMAN EXPOSURE

# Chapter 5

# Blood Cholinesterase Activity: Inhibition as an Indicator of Adverse Effect

S. Padilla[1], L. Lassiter[2], K. Crofton[1], and V. C. Moser[1]

[1]Neurotoxicology Division, National Health and Environmental
Effects Research Laboratory, U.S. Environmental Protection Agency,
Research Triangle Park, NC 27711
[2]Toxicology Program, University of North Carolina,
Chapel Hill, NC 27514

Organophosphate and carbamate insecticides are designed to inhibit cholinesterase (ChE), the serine protease which hydrolyses the neuro-transmitter acetylcholine. Inhibition of ChE activity may lead to central and peripheral nervous system dysfunction in both humans and labora-tory animals. Although blood ChE does not participate in neurotransmis-sion, depression of blood ChE activity has been traditionally used as a convenient biomarker for **exposure** to ChE-inhibiting insecticides. Our laboratory has been engaged in numerous studies to explore the possibility that blood ChE inhibition might also be utilized as a biomarker of adverse effect (in addition to a biomarker of exposure). Our results to date indicate that the degree of depression of blood ChE activity usually correlates well with the amount of ChE inhibition in the target tissues or the degree of clinical impairment; the exact numerical relationship, however, is dependent upon time after dosing, choice of tissue or clinical endpoint, and the insecticide tested.

Acetylcholinesterase (AChE; EC 3.1.1.7), an enzyme present throughout the body, normally hydrolyses the neurotransmitter acetylcholine which is present in the central nervous system, the neuromuscular junction, the parasympathetic nervous system, the sympathetic synapses, and the sympathetic innervation of the adrenal and sweat glands. Exposure to cholinesterase-inhibiting insecticides such as organophosphate or carbamate insecticides retards the breakdown of acetylcholine, resulting in overstimulation of the target cell (1,2).

Although there has been considerable research surrounding the toxicology of cholinesterase-inhibiting insecticides, many basic questions remain. One of the most pertinent questions concerns the pattern of blood cholinesterase inhibition in insecticide-dosed animals, and how that inhibition correlates with inhibition in the target tissue or with changes in behavior. That is, how well does blood cholinesterase inhibition correlate with biochemical or behavioral adverse effects in animals treated with anticholinesterases? The prevailing opinion in the toxicology community is that blood cholinesterase is a useful gauge for detecting exposure to anticholinesterase insecticides (*3,4*), but the relationship between blood cholinesterase inhibition and adverse effect is tenuous, at best (e.g., *5-8*).

Because defining a relationship between blood cholinesterase levels and insecticide-induced adverse effect is pivotal for interpreting human blood cholinesterase measurements, we have been striving to answer the above questions in a laboratory context. In general, the experimental design is to observe insecticide-dosed laboratory animals for signs of adverse effects and correlate those behavioral changes with the levels of cholinesterase activity in blood (and its components, plasma and erythrocytes) and in target tissue (e.g. brain or muscle tissue). In this way we hope to define the variables which determine if and when blood cholinesterase inhibition correlates with the appearance of adverse effects in mammals.

### Does Blood Cholinesterase Inhibition Correlate with Cholinesterase Inhibition in the Target Tissues?

Although cholinesterase activity is present in the blood, these cholinesterases apparently do not participate in neurotransmission. Consequently, inhibition of blood cholinesterase activity is not commonly considered an adverse effect. The cholinesterase activity in target tissues such as the nervous system or muscle motor endplate, however, does have a defined and necessary role: to hydrolyze the extracellular acetylcholine, ending its actions, and preventing the over-stimulation of the target cell. Therefore, inhibition of target tissue cholinesterase activity is normally viewed as an adverse effect. The question to be considered here is whether the blood cholinesterase inhibition is related to the inhibition in the target tissue and therefore could be considered a surrogate biomarker for a biochemical adverse effect.

Our general approach to answering this question was to take measures of target tissue inhibition as well as blood cholinesterase inhibition in rats dosed with various anticholinesterase insecticides. We first established whether there was a correlation and then determined under what circumstances the correlation was optimal. We also wanted to determine the best blood component to use in the correlation and if there was a reasonable correlation during subchronic as well as acute dosing. Some of these studies has been previously published using chlorpyrifos (*9, 10*), and paraoxon (*11*).

One experimental variable emerged as a crucial consideration: the choice of time after dosing for assessing cholinesterase inhibition. An example of this is illustrated in Figure 1. These data are taken from a group of animals dosed with 30, 60 or 125 mg/kg (sc) chlorpyrifos and tissues collected for cholinesterase determinations at 1, 7, or 35 days after dosing. The data presented in this figure are hippocampal data, but almost all central nervous system areas showed the same relationship with blood cholinesterase activity. With this subcutaneous dosing regimen, the tissue cholinesterase is maximally inhibited from 4 to 21 days after dosing (9). Therefore, in Figure 1, the 1 and 35 day time points represent times when the inhibition was either still increasing (i.e., 1 day) or was recovering (i.e., 35 days). Only the 7 day time point was during the time of peak cholinesterase inhibition. As can be seen clearly from this figure, the best correlation between hippocampal cholinesterase activity and activity in the whole blood occurs at 7 days. At 1 day, the blood cholinesterase decreases at a faster rate than the hippocampal cholinesterase activity, whereas at 35 days the opposite is true. At 7 days, however, both have reached minimal activity (i.e.,maximal inhibition) and the levels of blood cholinesterase activity (expressed as % of control) closely predict the levels of activity in the hippocampus for each animal. This type of pattern is also apparent in repeated dosing with chlorpyrifos (10) where the best correlations were noted during the actual dosing period when inhibition had reached a "steady state" level as opposed to during the recovery period.

To sum up our results to date, blood cholinesterase levels do correlate with inhibition in target tissue, but the following caveats apply:

1. The best correlation is achieved during maximal cholinesterase inhibition (i.e., time of peak effect) either after an acute dose or during repeated dosing.

2. The degree of cholinesterase inhibition is tissue-dependent (see comparison between striatum and diaphragm in 9). Moreover, although there may be approximately a "1-to-1" relationship between blood and target tissue cholinesterase inhibition (for example, see Fig. 1, 7 day graph), that is not necessarily to be expected.

3. Usually during the maximal cholinesterase inhibition, all three blood components (i.e., whole blood, plasma, or erythrocytes) will exhibit good correlation with the target tissue inhibition. During the initial inhibition phase and also during the recovery phase, plasma cholinesterase tends to show the poorest correlation. Conversely, during the recovery phase, erythrocyte and whole blood cholinesterase activity may lag behind recovery in the target tissues.

Our data collected to date, along with data from other laboratories (12, 13) indicate that blood cholinesterase inhibition does predict inhibition in the target tissues and therefore should be considered a biomarker for the biochemical adverse effect in the target tissues.

## Does Blood Cholinesterase Inhibition Correlate with Changes in Behavior?

In other words, does the level of cholinesterase inhibition in the blood predict the presence of behavioral adverse effects? Many authors have expressed surprise that more research has not been done to investigate this question (e.g., 14-17). This is an extremely important question in the realm of human monitoring, and we are attempting to address it in a systematic manner.

In our studies the behavioral endpoint we selected was motor activity, because previous investigations have shown that changes in motor activity are a relatively sensitive and reliable measure of exposure to anticholinesterases (11, 18, 19, Moser, V.C., *Neurotox. Teratol.*, in press). Moreover, motor activity is measured on a continuous scale (i.e., not quantal) which allows easier correlation with cholinesterase activity. Therefore in the following groups of studies, we determined the degree of cholinesterase inhibition in blood from insecticide-dosed rats *immediately* after they were tested for approximately one hour in motor activity chambers. In this manner, we

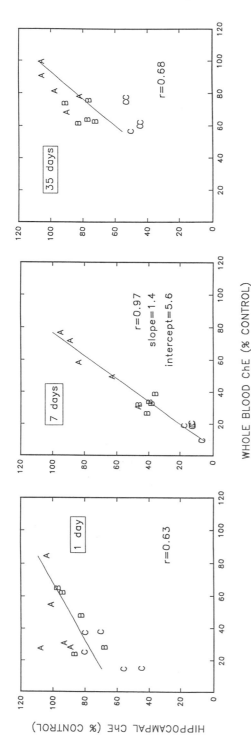

WHOLE BLOOD ChE (% CONTROL)

HIPPOCAMPAL ChE (% CONTROL)

*Figure 1: Correlation between whole blood cholinesterase and hippocampal cholinesterase activity in male, Long-Evans rats treated with one subcutaneous dose of chlorpyrifos [30(A), 60(B), or 125(C) mg/kg]. Animals were killed and tissues taken for analysis at 1, 7, or 35 days after dosing. Each point is an individual animal and each panel depicts the relationship between the cholinesterase inhibition in the two tissues at each time point. Cholinesterase activity was determined using the spectrophotometric method of Ellman and coworkers (34) with 1 mM final concentration of acetylthiocholine. This figure is a graphic representation of experimental data collected for (9). For easier comparison, all cholinesterase activity were converted to % control by dividing the experimental value by the mean of the respective control values.*

could correlate the degree of blood cholinesterase inhibition with the degree of behavioral impairment elicited by anticholinesterase dosing in each individual animal. Two organophosphates, chlorpyrifos and paraoxon, and two carbamates, aldicarb and carbaryl, were selected for evaluation. Toxicokinetic considerations are germane to the overall toxicity profile of these compounds. In general, animals treated with carbamates show rapid onset of signs and cholinesterase inhibition which dissipate within hours, as opposed to animals treated with organophosphates which show clinical signs and cholinesterase inhibition within 2 to 4 hours followed by a much slower recovery (for a review see *20*). We tried to minimize the impact of toxicokinetic differences among the compounds by performing all behavioral and biochemical assessments at the time of peak effect. By doing that, we hoped to just ask the question of whether the blood cholinesterase inhibition correlated with the changes in behavior when the animal was experiencing the more severe signs, independent of the time after dosing. Presumably the time of peak effect would be very dependent upon toxicokinetic considerations, but once the time of peak effect had been pinpointed, other endpoints might be considered more independent of toxicokinetic parameters.

All measurements were taken during the time of behavioral peak effect which was 3.5 hours for chlorpyrifos (single, oral dose), 1 hour for carbaryl (single, oral dose), 3 hours for paraoxon (single, sc dose) and 1.5 hours for aldicarb (single, oral dose). In order to determine the time of peak effect, preliminary studies were conducted to establish the most appropriate post-dosing time for testing. To accomplish this, a few rats were dosed with a high, but not lethal, level of the insecticide in question and then examined at 30 minute to 1 hour intervals for the next 6 to 7 hours, using a modified battery of tests (typically cholinergic signs, alertness, and gait changes) which is a subset of the Functional Observational Battery (*21*). The time of peak effect was designated as the time after dosing when the most signs were present and were the most severe. Interestingly, in our hands, this time of peak effect (defined behaviorally) is usually concomitant with the peak cholinesterase inhibition, although the cholinesterase inhibition often, but not always, persists longer than the behavioral changes (*19, 22*).

Figure 2 depicts the results from all four studies. The level of whole blood cholinesterase activity (expressed as a percentage of the mean control value) is plotted against the level of motor activity (also expressed as a percentage of the mean control value) for each animal dosed with each of the insecticides. Upon linear regression of these points for each compound, one finds that the slopes and x axis intercepts of the regression lines may be quite different. In other words, the relationship between the two endpoints (i.e., whole blood cholinesterase activity and motor activity) is not the same with each insecticide. For instance, the chlorpyrifos-treated animals show a linear relationship between blood cholinesterase and motor activity impairment, but the blood cholinesterase has to be depressed to almost 15% of control levels before decreases in motor activity are noted. With aldicarb and paraoxon dosed animals, however, a different relationship between the two endpoints was evident (Fig. 2). No changes in motor activity were observed until blood cholinesterase was decreased to a level of $\simeq 50\%$ of control. Further inhibition of blood cholinesterase was linearly correlated with depressed motor activity. In other words, with these insecticides, i.e., aldicarb and paraoxon, a different threshold existed: 50-60% inhibition of blood cholinesterase was required before the levels of blood cholinesterase correlated with the presence of an adverse effect. Finally, the carbaryl-dosed animals yielded yet another slope and

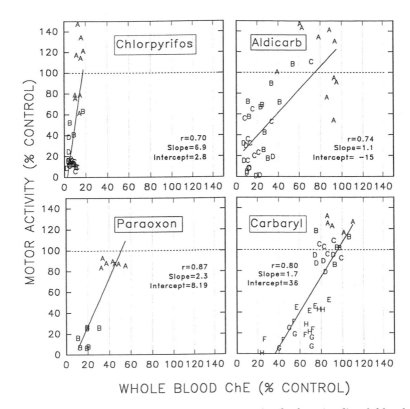

WHOLE BLOOD ChE (% CONTROL)

*Figure 2:* Graphic representation (by individual animal) of blood cholinesterase activity (% control) with motor activity levels (% control). Each point is an individual animal. Long-Evans male rats were dosed with one of the following: **Chlorpyrifos:** Animals (n=10/dosage group) were dosed with 0, 10(A), 30(B), 60(C) or 100(D) mg/kg chlorpyrifos in corn oil (1 ml/kg) by gavage. **Aldicarb:** Animals (n=10/dosage group) were dosed with 0, 0.065(A), 0.195(B), 0.390(C) or 0.650(D) mg/kg aldicarb in corn oil (1 ml/kg) by gavage. **Paraoxon:** Animals (n=7/dosage group) were dosed (sc) with 0, 0.17(A), or 0.34(B) mg/kg paraoxon in peanut oil (1 ml/kg). **Carbaryl:** Animals (n=5/dosage group) were dosed with 0, 1.5(A), 3.0(B), 6.0(C), 12.5(D), 25.0(E), 27.5(F), 50.0(G), or 75.0(H) mg/kg carbaryl in corn oil (1 ml/kg) by gavage. Cholinesterase activity was analyzed using the radiometric assay (carbaryl, aldicarb, and paraoxon) by Johnson and Russell (35) or the spectrophotometric assay (chlorpyrifos) by Ellman and coworkers (34); both assays were conducted with a substrate concentration of approximately 1 mM. Special care was taken during the cholinesterase measurements of the carbamate-inhibited blood to limit the degree of decarbamylation (36-38). For easier comparison, all cholinesterase activities and motor activity levels were converted to % control by dividing the experimental value by the mean of the respective control values. Lines were fitted to all of the data points for each insecticide, and the given correlation coefficient represents all of the data.

intercept configuration. In this case there was little to no threshold for blood cholinesterase inhibition. Small amounts of blood cholinesterase inhibition (i.e., > 20-15% of control) were related to depressed motor activity. When motor activity was maximally depressed, blood cholinesterase was still at least 40% of control.

How well does inhibition of blood cholinesterase activity correlate with behavioral adverse effect? The present data as well as data collected by others (23,24) indicate that there is a high probability that there is a linear relationship between blood cholinesterase inhibition and presence of clinical signs or changes in behavior. Sometimes there is a high threshold (e.g., chlorpyrifos); in some cases, there is a moderate threshold (e.g., aldicarb or paraoxon); and in other cases, the threshold is minimal (e.g., carbaryl). All the animals exposed to the insecticides showed a dose-dependent depression of cholinesterase activity in the blood, but the degree of behavioral impairment was dependent on the degree of blood cholinesterase inhibition as well as the insecticide the animal had been exposed to.

## Conclusions

The most pertinent question is what does the above information mean for human monitoring? The prevailing "rule of thumb" is that more than 50% blood cholinesterase inhibition must be noted before concern is warranted about "adverse effects" due to exposure to anticholinesterases (25,26). Our results show that this relationship (i.e. between blood cholinesterase and adverse effect) is time- and insecticide-dependent, and that there is no general rule which can be stated which applies to every anticholinesterase insecticide.

What could explain this compound specificity? The first consideration, of course, would be differing pharmacokinetic profiles, but there may be additional explanations. It is assumed that the primary mechanism of action of carbamate and organophosphate insecticides is the inhibition of cholinesterase activity, and if the degree of cholinesterase inhibition is known then one can predict effects. That may be partially true, but, given the above data, making predictions may not be so simple. Adverse effects may not just be dependent upon the degree of cholinesterase inhibition, but may also be related to tissue distribution of the inhibitor (pharmacokinetic profile), the rate of inhibition (pharmacodynamic profile), or it could be that these chemicals have effects in addition to cholinesterase inhibition (e.g., 27, 28) or that cholinesterase has functions other than hydrolysis of acetylcholine (e.g., 29-33). Whatever the explanation, one must keep in mind that it may be imprudent to make generalizations encompassing all anticholinesterases using knowledge garnered from only a few chemicals in the class. A judicious approach for both the scientist and the regulator would be to view these chemicals not in a generic sense, but as individuals.

## Acknowledgements

The research described in this article has been reviewed by the Health Effects Research Laboratory, U.S. Environmental Protection Agency, and approved for publication. Approval does not signify that the contents necessarily reflect the views and policies of the Agency, nor does mention of trade names and commercial products constitute endorsement or recommendation for use.

## Literature Cited

1.  Ecobichon, D.J., In *Casarett and Doull's Toxicology: The Basic Science of Poisons*; Amdur,M.O., Doull, J., and Klaassen C.D., Eds., 4th ed., Pergamon Press, New York, New York, 1991, pp. 580-592.
2.  Marrs, T.C. *Pharmac. Ther.* **1993**, *5*, pp. 51-66.
3.  Wilson, B.W.; Henderson, J.D. *Rev. of Environ. Contamin. and Toxicol.* **1992**, *128*, pp.55-69.
4.  Peakall, D.B. *Toxicol. and Ecotoxicol. News,* **1994**, *1*, pp. 55-60.
5.  Karnik, V.M.; Ichaporia, R.N.; Wadia, R.S. *J. Assoc. of Physicians of India.* **1970**, *18*, pp. 482-487.
6.  Endo, G.; Horiguchi, S.; Kiyota, I.; Kurosawa, K. *Proc. of the ICMR Seminar*, **1988**, *8*, pp. 561-564.
7.  Jimmerson, V.R.; Shih, T.-M.; Mailman, R.B. *Toxicol.* **1989**, *57*, pp. 241-254.
8.  Carr, R.L.; Chambers, J.E. *Pharmacol Biochem Behavior.* **1991**, *40,* pp. 929-936.
9.  Padilla, S.; Wilson, V.Z.; Bushnell, P.J. *Toxicol.* **1994**, *92*, pp.11-25.
10.  Padilla, S. *Toxicol.* **1995**, *102*, pp. 215-220.
11.  Padilla, S.; Moser, V.C.; Pope, C.N.; Brimijoin, W.S. *Toxicol. Appl. Pharmacol.* **1992**, *117*, pp. 110-115.
12.  Pope, C.N.;Chakraborti, T.K. *Toxicol.* **1992**, *73*, pp. 35-43.
13.  Pope, C.N.; Chakraborti, T.K.; Chapman, M.L.; Farrar, J.D.; Arthun, D. *Toxicol.* **1991**, *68*, pp. 51-61.
14.  Annau, Z. In *Organophosphates: Chemistry, Fate, and Effects.* Chambers, J.E., and Levi, P.E., Eds., Academic Press, New York, New York, 1992 pp. 419-432.
15.  D'Mello, G.D., In *Clinical and Experimental Toxicology of Organophosphates and Carbamates*, Ballantyne, B. and Marrs, T.C., Eds. Butterworth-Heinemann, Ltd., Oxford, 1992 pp. 61-74.
16.  D'Mello, G.D. *Human Experimen. Toxicol.* **1993**, *12*, pp. 3-7.
17.  U.S. Environmental Protection Agency, Report of the SAB/SAP Joint Study Group on Cholinesterase: Review of Cholinesterase Inhibition and Its Effects. EPA-SAB-EC-90-014, **1990**.
18.  Ruppert, P.H.; Cook, L.L.; Dean, K.F.; Reiter,L.W. Pharmacol. *Biochem. Behav.* **1983**, *18*, pp. 579-584
19.  Moser, V.; McDaniel,K.; Padilla, S.;Nostrandt, A. *The Toxicologist.* **1995**, *15*, p. 205.
20.  Fukuto, T.R. *Environ. Health Perspect.* **1990**, *87*, pp. 245-254.
21.  Moser, V.C. *J. Am. Coll. Toxicol.* **1989**, *8*, pp.85-93.
22.  Nostrandt, A.; Padilla, S.: Moser, V.; Willig, S. *The Toxicologist*, **1995**, *15,* p. 205.
23.  Roney, P.L.; Costa, L.G.; Murphy, S.D. *Pharmacol. Biochem. and Behav.* **1986**, *24*, pp. 737-742.24.
24.  Kurtz, P.J. *Pharmacol. Biochem. Behav.* **1977**, *6*, pp. 303-310.
25.  Namba, T. *Bull WHO*, **1971**, *44*, pp. 289-307.
26.  Ames, R.G.; Brown, S.K.; Mengle, D.C.; Kahn, E.; Stratton, J.W.; Jackson, R.J. *J. Soc. Occup. Med.* **1989**, *39*, pp. 85-92.

27. Eldefrawi, A.T.; Jett, D.; Eldefrawi, M.E. In *Organophosphates: Chemistry, Fate, and Effects,* Chambers, J.E. and P.E. Levi, Eds. Academic Press, New York, New York, 1992 pp. 257-270.

28. Chiappa, S.; Padilla, S.; Koenigsberger, C.; Moser, V.C.; Brimijoin, S. *Biochem. Pharmacol.* **1995**, *49*, pp.955-963.

29. Layer, P.G. In *Multidisciplinary Approaches to Cholinesterase Functions,* Shafferman, A., Velan, B. Eds., Plenum Press, New York, 1992 pp. 223-231.

30. Small, D.H. *Trends in Biological Sciences*, **1990**, *15*, pp. 213-216.

31. Greenfield, S.A.; Chubb, I.W.; Grünewald, R.A.; Henderson, Z.; May, J.; Portnoy, S.; Weston, J.; Wright, M.C. *Exp. Br. Res.* **1984**, *54*, pp. 513-520.

32. Zakut, H.; Lapidot-Lifson, Y.; Beeri, R.; Ballin, A.; Soreq, H. *Mutat. Res.* **1992**, *276*, pp.275-284.

33. Massoulié, J.; Pezzementi, L.; Bon, S.; Krejci, E.; Vallette,, F.-M. *Progress in Neurobiol.* **1993**, *41*, pp.31-91.

34. Ellman, G.L.; Courtney, K.D.; Andres, K.D., Jr.; Featherstone, R.M. *Biochem. Pharmacol.* **1961**, *7*, pp. 88-95.

35. Johnson, C.D.; Russell, R.L. *Anal. Biochem.* **1975**, *64*, pp. 229-238.

36. Pickering, C.E.; Pickering, R. G. *Arch. Toxikol.* **1971**, *27*, pp. 292-310.

37. Padilla, S.; Hooper; M.J. Proceedings of the U.S. EPA Workshop on Cholinesterase Methodologies. Office of Pesticides Programs. U.S. EPA, Washington, DC., 1992, pp. 63-81.

38. Nostrandt, A.C.; Duncan, J.A.; Padilla, S. *Fundamental. Appl. Toxicol.* **1993**, *21*, pp. 196-203.

## Chapter 6

# Neurotoxic Esterase Inhibition: Predictor of Potential for Organophosphorus-Induced Delayed Neuropathy

**Marion Ehrich**

**Laboratory for Neurotoxicity Studies, Virginia–Maryland Regional College of Veterinary Medicine, Virginia Polytechnic Institute and State University, Duckpond Drive, Blacksburg, VA 24061–0442**

Organophosphorus compounds (OPs) are used in agriculture and industry. Acute toxicity follows inhibition of acetylcholinesterase (AChE), but some OPs can induce a delayed neuropathy (OPIDN) that occurs weeks after a single exposure. Manifestations of OPIDN differ among species, with locomotor effects prominent in humans and hens but lacking in laboratory rats. Potential for development of this progressive and irreversible neuropathy is determined by capability of the OP to significantly and irreversibly inhibit neuropathy target esterase (NTE, neurotoxic esterase). NTE inhibition can be used to identify OPs that induced delayed neuropathy whether NTE is examined in neural tissue, lymphocytes, or neuroblastoma cell lines. Relative inhibition of NTE and AChE, determined shortly after exposure, can be used to distinguish the likelihood of causing delayed neuropathy or acute toxicity following exposure to OPs. Use of NTE as a biomarker for neuropathy-inducing OPs is, however, limited by a less than direct relationship between NTE inhibition and OPIDN.

### Organophosphorus Compounds and Esterase Inhibition

Organophosphorus compounds (OPs) are one of the many types of synthetic insecticides in use today. OPs are also used as plasticizers, lubricants, petroleum additives, and chemical warfare agents. It is their use as insecticides that receives the most attention, however, because they command the largest segment (more than 1/3) of the total $6.1 billion insecticide market. Over 89 million acres of the United States are sprayed annually with OPs. OP insecticides are widely used because they are effective and because they are biodegradable. The toxicity of OP insecticides has been studied for decades, and we know that they have capability to phosphorylate and, thereby, inhibit esterases. The reason they kill insects and cause risk of neurotoxicity to man, domestic animals, and wildlife is that they inhibit acetylcholinesterase (AChE), the enzyme that degrades acetylcholine, a neurotransmitter of the central and peripheral nervous system (1-5). In avians and mammals, the inhibition of AChE is relatively rapid and causes signs of cholinergic poisoning related to excessive acetylcholine stimulating receptors in smooth muscle, skeletal muscle, autonomic ganglia, and the central nervous system. The capability of OPs to inhibit AChE, which is found in mammalian erythrocytes as well as in the nervous system, has provided a biochemical marker for both exposure to and effects of OPs that are discussed elsewhere in this monograph (6,7).

AChE is not, however, the only esterase inhibited by OPs. OPs also inhibit other serine esterases less specific as to location and substrates than AChE. Carboxylesterases, for example, can be inhibited by OPs. These esterases are found in liver, kidney, plasma, lymphocytes, platetets, lung, skin, mucosa and gonads. Carboxylesterases include a number of isozymes which differ in tissue (including species) source, substrate specificity, enzyme kinetics, and molecular weight. These enzymes can hydrolyze carboxylesters, thioesters, and aromatic amides (8,9). Although inhibited by OPs, this alone is not sufficient to directly result in clinical evidence of OP-induced neurotoxicity in insects, man, or animals. Combinations of OPs with carboxylesterases will, however, decrease OPs available for inhibition of AChE and prior inhibition of carboxylesterases will potentiate cholinergic poisoning with subsequent OP exposure (1,8,9).

Among the carboxylesterases is one that has been known since the mid-1970's to have particular usefulness in the evaluation of OP-induced neurotoxicity. This is because OPs not only can cause acute neurotoxicity that could follow AChE inhibition, but some may cause a progressive neuropathy that only begins to appear weeks after exposure. Inhibition of this carboxylesterase, known as neuropathy target esterase (NTE, neurotoxic esterase) is useful in identifying which OPs have potential for causing delayed neuropathy. Extent of NTE inhibition, especially if determined < 48 hr after exposure, can also be used to predict whether or not organophosphate-induced delayed neuropathy (OPIDN) will appear in susceptible animal species (including man). Inhibition of NTE alone, however, is not sufficient to predict the development of OPIDN in exposed subjects, as non-OPs (including carbamate insecticides) can inhibit NTE. A strong bond, not easily reversible, must form between NTE and the OP. The process of forming such a bond, usually called "aging," occurs with OPs that have a chemical structure that permits formation of a negative charge on the OP-esterase complex (10-13). As with other carboxylesterases, the location of NTE is not restricted to brain, spinal cord, and peripheral nerves. This particular esterase, which is integrally associated with intracellular membranes (e.g., the smooth endoplasmic reticulum) in the nervous system, is also found in lymphocytes, platelets, spleen, small intestine, placenta and the adrenal medulla (12, 14-17). NTE is also one of the carboxylesterases found in neuroblastoma cell lines, a feature that is currently being exploited in our laboratory as we develop a means of detecting neuropathy-inducing OPs without exposure of animals (18-20). (Active neuropathy-inducing OPs are potent NTE inhibitors in these cells.) NTE activity is determined by differential inhibition of the carboxylesterase that hydrolyzes phenyl valerate in the presence of OPs (paraoxon and paraoxon + mipafox)(*13*), a procedure that can be done in microassay (*21*).

Acetylcholinesterase inhibition and OPIDN represent two very different types of OP-induced neurotoxicity.  The first occurs in virtually all animals (including insects and man) after sufficient systemic exposure to an OP insecticide.  The second only occurs following sufficient inhibition of NTE, and inhibition, with only an occasional exception (*23,23*), must be followed by "aging" of the OP-NTE complex (*10,12,13*).  Furthermore, clinical signs of OPIDN (locomotor deficits progressing to paralysis) and neuropathological lesions do not appear in all animal species.  Most OPs used as insecticides and/or human or veterinary drugs have little effect on NTE relative to their effect on acetylcholinesterase (*2,3,12*).  Those OPs that cause OPIDN are (a) derivatives of phosphoric acid, phosphonic acid, or phosphoramidic acid (including phosphorofluoridates), compounds which cause the characteristic clinical and morphological manifestations of OPIDN (Type I OPIDN), or (b) triphosphites, compounds which cause a neuropathy that is morphologically and clinically distinct (sometimes called Type II OPIDN)(*24,25*).  A third syndrome, called an "intermediate syndrome," has been reported (*26*) after some human patients who had recovered from muscarinic signs of cholinergic poisoning died from respiratory failure several days after OP exposure.  A recent report (*27*) suggests that this syndrome is due to lesions of the diaphragm that follow acute acetylcholinesterase inhibition at this nicotinic site.

**Neurotoxic Esterase - Relationship Between Inhibition and OPIDN**

Just what is NTE?  And what is its relationship to OPIDN?  These questions are relevant in a discussion of biomarkers for agrochemicals.  Inhibition of NTE within hours of exposure to OPs predicts potential for developing OPIDN in susceptible animal models.  This connection is sufficiently viable that registration of OPs under the Federal Insecticide, Fungicide, and Rodenticide Act (FIFRA) requires that data on NTE inhibition be obtained as a biochemical determinant of potential to cause OPIDN (*28*).  For these studies, the hen is the animal model of choice because it exhibits both obvious locomotor deficits and a distinct and unique histopathology.  Although early and significant inhibition of NTE is an excellent predictor of potential for developing OPIDN, the relationship between NTE inhibition and OPIDN itself is less clear.  The following call to question the NTE - OPIDN relationship:  (1) The temporal differences between NTE inhibition and OPIDN;  (2) The high specific activity and inhibition of NTE at sites that have little morphological evidence of OPIDN;  (3) The discordance between NTE inhibition and OPIDN when drugs and chemicals are used to interfere with OPIDN development;  (4) The inhibition of NTE in species and age groups that do not demonstrate clinical manifestations of OPIDN; and (5) The potentiation or promotion of OPIDN when non-OP, reversible NTE inhibitors are administered after OPs.

**The temporal relationship between NTE inhibition and OPIDN following OP exposure**. The relationship between NTE inhibition and OPIDN is not as clear as the relationship between acetylcholinesterase inhibition and cholinergic poisoning. NTE inhibition occurs early (within hours) of exposure, yet notable clinical and morphological manifestations of OPIDN do not occur until weeks later. By the time OPIDN appears, NTE activity has begun to recover and it may actually return to control levels while OPIDN continues to progress (*12,29,30*). Whether or not NTE is the specific target that initiates events that culminate in OPIDN has been debated for some time (*5,11-13,24,31,32*), as the precise mechanisms responsible for the development of the neuropathy remain undefined. Also yet undefined is the normal  physiological role for NTE, a role that could be less difficult to investigate if pure, active NTE were available. NTE is, however, integrally associated with intracellular membrane lipids, making purification of active enzyme difficult, although significant progress toward purification has recently been reported (*13,33-35*).

**The high specific activity and inhibition of NTE at sites that have little morphological evidence of OPIDN**. OP-induced NTE inhibition in the brain, spinal cord, and sciatic nerve of the hen, which is the accepted animal model for OPIDN (*28*), has been used to indicate potential for development of OPIDN (*12,36,37*). Whole tissue homogenates are used for NTE determinations, even though lesions are restricted to or more significant in certain areas in each of these tissues (e.g., the cerebellum and medulla of the brain, the cervical region of the spinal cord, and the muscle branches of the sciatic nerve)(*38-40*). In fact, the highest specific activity of NTE is in brain (*12,21*), yet lesions are least in this tissue (*38*). NTE activity was similar in all segments of the spinal cord (*21*), yet lesions are most prominent in the cervical region (*39*). NTE activities were reported similar in a series of sciatic nerve segments (*41*), although lesions of OPIDN are more notable in the distal muscular branches of the sciatic nerve (*40*) and in the biventer cervicis nerve (*42,43*), nerves too small to isolate and use for biochemical assays. Also, NTE activity in sciatic nerve is <10% of the activity in brain (*12,21*), and it is technically more difficult to obtain and prepare sufficient quantities even of this relatively large peripheral nerve for NTE assays than it is to obtain and prepare brain or spinal cord. Although it has been suggested that NTE in peripheral nerve may better predict OPIDN (*41,44*), relative OP-induced NTE inhibition has been demonstrated to be similar among regions of the brain, segments of the spinal cord, and the sciatic nerve (*21*). Furthermore, the relationship between inhibition of NTE in brain and development of OPIDN has been remarkably consistent (*12,26,27*). Registration of OPs under FIFRA (*28*) requires NTE determinations to be made both in hen brain and spinal cord; NTE inhibition may tend to be slightly less in spinal cord than in brain, but NTE inhibition in both tissues demonstrates a dose-response relationship that correlates with lesion development (*12,36*). No lesions have been reported following OP-induced NTE inhibition in extraneural tissues. Why the most notable lesions of OPIDN occur in restricted portions of the spinal cord and in the distal portions of peripheral nerves of the hen when OP-induced NTE inhibition is relatively equivalent throughout the nervous system remains undefined. These data indicate that the relationship between basal NTE levels, specific regions in which NTE is inhibited, and specific regions in which lesions develop is unclear.

Although dose-response studies have indicated a high correlation between the extent of early NTE inhibition and clinical and morphological manifestations of OPIDN in hens given single doses of neuropathy-inducing OPS (*37,45*), this relationship is somewhat different when NTE inhibition and OPIDN are examined in multiple dose studies, or when hens are treated with OPs whose isomeres have differential capability to induce OPIDN. For example, OPIDN in multiple dose studies requires cumulative dosages greater than acute dosages that cause OPIDN, and there appears need for NTE inhibition to reach a critical point before OPIDN appears (*46*). The extent of NTE inhibition for OPIDN, however, does not appear to have to be as high as that necessary to cause equivalent OPIDN in single dose studies (*12,47*). Chemical isomerism has also been demonstrated to make a difference in manifestations of OPIDN for at least one OP. Even though relatively high levels of NTE inhibition were reached with methamidophos, for example, OPIDN did not occur unless the D-(+) isomere was used (*23*), suggesting that NTE inhibition may not require the chemical specificity that is necessary for OPIDN.

**The discordance between NTE inhibition and OPIDN when drugs and chemicals interfere with OPIDN.** Frequency of dosing and the chemical isomere used are not the only features that confound the direct correlation of NTE inhibition to OPIDN for OPIDN can be ameliorated by use of therapeutic agents that do not affect NTE. It is possible that these agents antagonize an action or process that contributes to OPIDN, but these actions or processes differ at least temporally from NTE inhibition. Corticoids and calcium channel blockers, for example, decrease clinical, electrophysiological and morphological manifestations of OPIDN, but neither class of drug altered OP-induced NTE inhibition (*48-54*). Amelioration with calcium channel blockers supports the hypothesis that there is a role for calcium and calcium-mediated enzymes in OPIDN (*24,29*), but the interaction of OPs and calcium channel blockers on the vascular system cannot be discounted (*55*).

Although the studies cited above suggest that the correlation between extent of NTE inhibition and OPIDN is not always direct, the effects of the ameliorating agents may be on processes, as yet undefined, that occur following NTE inhibition but before obvious clinical, electrophysiological, or morphological manifestations of OPIDN. There are other compelling experiments that suggest NTE inhibition and aging are necessary predeterminants in the development of OPIDN. These experiments demonstrate that OPIDN can be prevented by pretreatment with reversible inhibitors of NTE (*12*), much as OP-induced cholinergic poisoning can be prevented by pretreatment with reversible cholinesterase inhibitors (*3,56*). The decreased availability of biochemical targets at which OPs can strongly bind due to the presence of the reversible inhibitors on the esterases (NTE or acetylcholinesterase) support the relationship between permanent inhibition of the esterase and subsequent OP-induced symptoms of toxicity. Experiments demonstrating protection from OPIDN by pretreatment with reversible inhibitors of NTE may, therefore, be used to support the suggestions that NTE is a target that initiates events leading to OPIDN. Demonstration that these same reversible inhibitors of NTE could potentiate or promote OPIDN when given after the OP [see below] has, however, called to question the suggestion that NTE is the specific and only target responsible for initiation of OPIDN.

**The inhibition of NTE in species and age groups that do not demonstrate clinical manifestations of OPIDN.** The relationship of NTE to OPIDN is also confounded by demonstration of OP-induced NTE inhibition without obvious clinical signs in young chicks, rats, and mice (*12*), and by the modest NTE inhibition that can be followed by clinical and morphological alterations in the ferret (*57*). NTE inhibition without clinical manifestations of OPIDN can also occur in quail (*25,58*) and ecothermic vertebrates (*59*). Although obvious behavioral signs restricted to OPs responsible for OPIDN have not been recognized in rats (*60*), NTE inhibition followed by neuronal degeneration is recognized to occur in this species (*30,45*). The neuronal degradation that occurs in rats exposed to neuropathy-inducing OPs is an exaggeration of normal background lesions that occur as rats age. The lesions are, however, distinguished by an increase in the number of damaged fibers and by the extent of swelling that occurs when compared to age-matched controls. Furthermore, lesions in the rat are restricted to rostral levels of the fasciculus gracilis (*36,44,61*). In the hen, the accepted animal model for OPIDN (*28*), the lesions of OPIDN are extensive and distinct, appearing in the rostral spinocerebellar and caudal medial pontine spinal tracts and in peripheral nerves, as well as in the rostral (medullary) levels of the dorsal funiculi containing the fasciculus gracilis in the spinal cord. Lesions in the hen are qualitatively different from those of the rat, and include swollen edematous or debris-laden myelinated axons and collapsed axons with associated myelin debris representing various stages in Wallerian-like degeneration (*28,36,61*). In addition, there is substantial morphological reconstitution in the rat fasiculus gracilis (the only site of OP-induced injury in this species), whereas lesions in the hen evolve to Wallerian-like degeneration (*61*).

Other species have also been examined for morphological indices of OPIDN in the absence of locomotor deficits that identify OPIDN in the adult hen model. Variable spinal cord damage in mice dosed once (*62*) and more extensive spinal cord damage in mice dosed for 9 months (*63*) have been reported. In addition, extensive morphological examination has been done of 2- and 10-week-old chicks dosed with the neuropathy-inducing OP, diisopropylfluorophosphate (DFP)(*64*), with demonstration of spinal cord lesions without neurological deficits in the 2-week-old birds. The clinical evidence of neurotoxicity in 2-week-old chicks induced by OPs that were inhibitors of NTE and/or acetylcholinesterase, if it appeared, was reversible and differed from that seen in 10-week-old chicks and adult hens given neuropathy-inducing OPs (*64-66*). Reasons why significant inhibition of NTE can appear without subsequent progressive and irreversible locomotor deficits considered typical of OPIDN, even when morphological alterations occur, have not been determined. Differences in repair have, however, been suggested based on the ability to recover from NTE inhibition and the reversibility of morphological damage (*61,65*). Although brain from a species relatively less susceptible to OPIDN (the rat) has less NTE activity than does brain from a susceptible species (the hen)(*21,61,67*), a lower basal NTE level cannot be correlated with lower susceptibility to OPIDN over a wide range of species (*12*).

**The potentiation or promotion of OPIDN.** Recent studies demonstrating that OP-induced locomotor deficits can be made to appear (promotion) in animal species and ages previously thought unsusceptible (e.g., 35-day-old chicks and 3.5 month-old rats (*68-70*) and made worse following administration of low doses of neuropathy-inducing OPs to adult hens (potentiation)(*71,72*) have again caused scientists to question the specificity of the relationship of NTE inhibition to OPIDN. These studies suggest that NTE is not the site for promotion or potentiation of OPIDN, primarily because clinical manifestations of OPIDN can be intensified by increasing the doses of a reversible inhibitor (e.g., phenylmethylsulfonyl fluoride, PMSF) beyond where such dosages, along with those of the OP, would inhibit NTE more than 90-100% (*13,32,70,73,74*). A recent report (*75*) also noted that median clinical scores for groups of hens given a dose of OPs that caused only mild clinical signs (0-4 on an 8-point scale) could be statistically increased by administration of doses of a phophorothioic acid that did not inhibit NTE (range of scores in hens given both compounds = 1-8). OPIDN was found to be accompanied by a reduction in retrograde axonal transport, as indicated by accumulation of radiolabeled iodine in the spinal cord of hens after injection of labeled tetanus toxin into the gastrocnemius muscle. This determinant of OPIDN, however, was not significantly altered when PMSF administration following OP exposure promoted ataxia in chicks given a dose of a neuropathy-inducing OP too low to cause ataxia when given alone (*76*). And although more axonal degeneration, as indicated by the Fink-Heimer silver impregnation method, has been seen in tracts of the cervical spinal cord of 35-70 day-old chicks given 2 mg/kg DFP followed by the reversible NTE inhibitor PMSF as a promotor than was noted in chicks given only DFP (*70*), studies using perfusion-fixation, plastic embedding, and ultrastructural examination to compare lesions in promoted OPIDN with characteristic lesions of OPIDN (*39,40,45*) have not yet appeared in the literature.

The relationship of NTE inhibition to neurotoxicity has been further challenged by reports of some increase in locomotor deficits in hens and rats given the reversible NTE inhibitor PMSF after some neurotoxicants that are not OPs. This would suggest that PMSF may be altering some general process that contributes to neurotoxicity that is unrelated to NTE inhibition. For example, mild ataxia (average less than 2 on an 8-point scale) was noted in hens given non-ataxia-inducing doses of 2,5-hexanedione followed by treatment with PMSF (*76*). No change, however, was seen when rats were the experimental subjects, as locomotor ability was the same whether they were given 2,5-hexanedione with or without PMSF (*77*). Preliminary studies in rats also indicated that PMSF did not alter locomotor scores if these rodents were given acrylamide or 3,3'-iminodiproprionitrile (*77,78*). A preliminary report, however, noted an increase in maximal clinical scores of rats treated with bromphenylacetylurea from 5.0 to 7.5

on a 10-point scale when administration of the neurotoxicant bromphenylacetylurea was followed by administration of the reversible NTE inhibitor PMSF (77).

## Neurotoxic Esterase, Biomarker

With this discordance between NTE inhibition and OPIDN, is there enough of a relevant relationship between NTE inhibition and OPIDN to make it possible to use NTE inhibition to predict the severity of OP-induced neuropathic effects in man and animals? Could NTE inhibition be used to monitor exposure to OPs inducing delayed neuropathy? These are important questions because OPIDN is progressive and irreversible, with no specific post-exposure treatments delineated that will completely halt the neuropathy, although early therapeutic intervention may slow its severity and rate of progression (11,29,49,54). Early significant OP-induced NTE inhibition that is not readily reversible, along with protection from OPIDN by reversible NTE inhibitors, provides substantial evidence for a relationship between NTE inhibition and OPIDN (12,32,36,37). The relationship between NTE inhibition and development of OPIDN has potential to be exploited as a biomarker.

Biomarkers have been defined as measurable biochemical, physiologic or other alterations within an organism that can indicate potential or established impairment (79). NTE inhibition can, indeed, indicate potential for impairment before the impairment (OPIDN) is established. This is why NTE activity, along with AChE activity, is determined following single and multiple dosing of the hen model if OPs are to be registered under FIFRA (28). The premarket testing that is done provides dose-related information on inhibition for both NTE and AChE in hen brain and spinal cord. These data can be used to predict the likelihood of acute cholinergic poisoning and/or OPIDN (10,36). NTE inhibition alone is not usually sufficient to keep an OP from being marketed. Doses that inhibit NTE would also have to be considerably less than dosages that cause cholinergic poisoning that follows AChE inhibition if there is to be concern about marketed OP insecticides causing OPIDN (10,36). Comparing inhibition of NTE and AChE after administration of similar dosages or after exposure *in vitro* to similar concentrations of OPs has been demonstrated to identify OPs causing OPIDN (20,36,80). Therefore, for purposes of predicting potential for OPIDN before OPs are marketed, NTE inhibition, measured in nervous tissue, is of great value.

NTE is not restricted to the nervous system, however, and there have been suggestions that inhibition of its activity in lymphocytes and/or platelets could be used to monitor exposure to and potential risk from neuropathy-inducing OPs in human and animal patients (15,16,81,83). Blood samples are easily taken, and they are routinely used to monitor and provide indication of worker exposure to cholinesterase-inhibiting insecticides because erythrocyte AChE activity is related to AChE activity in nervous tissue (3,84). Although suggested as useful, little has been done to verify advantages associated with collection and determination of NTE activities in platelets following OP exposures (81). Other reports have, however, noted problems with using lymphocytic NTE activity for monitoring OP exposure that do not occur with using erythrocyte AChE activity for monitoring OP exposure (*81,82*). The total mass of erythrocytes in the circulatory system is well regulated (*85*), allowing for a relatively narrow range of basal AChE levels within an individual, although interindividual variations may be substantial (*81*). In addition, erythrocytes continuously circulate during their life span of about 120 days. Neither lymphocyte concentration in the blood stream nor lymphocyte life span within an individual are as well regulated as they are for erythrocytes. Lymphocytes continually enter and are removed from the blood stream, with the interval that they remain at any one time only being a few hours long (*85*). Concentration can vary with hormonal and environmental influences. Although lymphocytes have a life span of months and even years, their inability to stay continually where access is easy (the blood stream) makes them a less consistent monitor for OP exposure than erythrocytes. This sequestering of lymphocytes may explain the variabilities encountered when lymphocyte NTE activities were measured at time points more than 24 hr after exposure to neuropathy-inducing OPs (*86,87*). The sequestering of lymphocytes could seriously limit the use of lymphocytic NTE as a biomarker for wildlife, as it would be exceedingly difficult to know when exposure to OPs occurred and some individuals within the exposed population may have normal NTE values whereas others would not.

Experimental and clinical studies support the contention that use of lymphocyte NTE as a biomarker for exposure to neuropathy-inducing OPs could be difficult. In the hen, correlation between lymphocyte NTE, nervous system NTE, and potential for OPIDN were good within 24 hr of exposure, and poor after that time (*86,87*); lymphocyte NTE values also did not correlate with nervous system NTE values more than 24 hr after a human exposure (*88*). If lymphocyte NTE activities were to be used as a biomarker of exposure to neuropathy-inducing OPs in humans, the variability in basal levels would need consideration. It could be difficult to interpret postexposure lymphocyte NTE activities in exposed individuals unless lymphocyte NTE activities had been determined in these individuals prior to OP exposure (*81,88, 89*). As this may not be feasible when monitoring human exposure to OPs, it is possible that lymphocyte NTE activity relatively early after exposure could be compared to NTE activity at a much later time (*82*). This is because NTE inhibition, if it occurs, occurs early after OP exposure, and activity is generally back to preexposure levels by the time OPIDN occurs (*12,29,30*). In one case of a human exposure, lymphocyte NTE activity was, indeed, lower earlier (at 30 days) than later (at 90 days) after OP exposure (*82*). This suggests that lymphocyte NTE activity could be a viable biomarker of human exposure under special circumstances (*15,81-83,89*). One must, however, still consider that sequestering of lymphocytes occurs and that isolation of lymphocytes is a time-consuming process that provides a relatively low yield of cells for assay (*89*). For small subjects, such as some wild birds, obtaining an appropriately sized sample could be difficult.

Concern for public health means that biomarkers for neurotoxicity are needed. Useful biomarkers could reliably predict potential for damage when intervention still could make some difference, if they are used to recognize subtle neurotoxic effects and/or monitor exposure to neurotoxicants (*79,90*). So where would it be useful to use NTE as a biomarker? Certainly, premarket testing of OPs should continue to use NTE inhibition and the relationship of this inhibition to AChE inhibition to indicate potential for development of OPIDN. Premarket testing should include insecticides, therapeutic agents (e.g., cholinesterase inhibitors used for treatment of glaucoma, such as echothiophate), and OP-containing lubricants (because there is possibility that some non-insecticidal OPs can inhibit NTE without AChE inhibition). Early screening could use *in vitro* methods (NTE and AChE inhibition in hen brain tissue or in human neuroblastoma cell lines (*20,80,91*)), as relative sensitivity of these enzymes to inhibition in such systems appears to indicate potential for causing delayed neuropathy or acute cholinergic poisoning. Confirmation of capability to inhibit NTE and cause OPIDN should be followed, as required under FIFRA, by testing in the hen model (*28*).

Although there may be argument for use of NTE as a biomarker, the question is, would this be useful in field applications? Because premarket testing means so few potent NTE-inhibiting OPs reach the market, especially for use as insecticides, and because collection of lymphocytes is a tedious process, it is probably unnecessary to routinely monitor NTE activity among individuals exposed to insecticidal OPs in field situations. Information on the relative capability of commonly-used insecticides to inhibit NTE and AChE, should however, be considered important and be made available. There are, for example, some NTE-inhibiting insecticides in common use today (e.g., dichlorvos, trichlorfon, chlorpyrifos). They are, however, far more likely to inhibit AChE than NTE (*10,91,92*). Therefore, it is much more likely that AChE inhibition and symptoms of cholinergic poisoning would be noted at exposures far lower than those necessary for sufficient inhibition of NTE to be followed by OPIDN. For these reasons, if one were to have interest in monitoring potential for OPIDN that would be applicable to field situations, the most efficient and sensitive biomarker would be erythrocyte AChE activity. Inhibition of the activity of this enzyme would not only indicate exposure and potential for acute toxicity (*3,84*), but this, along with information on the relationship of NTE inhibition to AChE inhibition, could be more predictive of OP-induced neurotoxicity than use of lymphocyte NTE alone (*10,36,80,89,92*).

## Summary

In summary, therefore, capability of an OP to cause NTE inhibition, especially when considered in relation to AChE inhibition, is important because it provides indication of potential for a particular OP to cause OPIDN in man and susceptible animal species. NTE inhibition, however, is unlikely to provide a useful field marker of exposure to OPs, since those OP insecticides in current use are far better inhibitors of AChE than of NTE, and, therefore, inhibition of AChE would provide a better indicator of exposure. NTE activity can be and has been measured in lymphocytes of intact animals, but monitoring exposure this way can be confounded by the sequestering and recirculation of lymphocytes and by the difficulty of deciding what is normal or abnormal due to the high variability of lymphocyte NTE activities among the general population.

## Abbreviations

OPs = organophosphorus compounds
AChE = acetylcholinesterase
OPIDN = organophosphorus-induced delayed neuropathy
NTE = neuropathy target esterase, neurotoxic esterase
FIFRA = Federal Insecticide, Fungicide and Rodenticide Act
PMSF = phenylmethylsulfonyl fluoride
DFP = diisopropylfluorophosphate, diisopropylphosphorofluoridate, isofluorophate

## Acknowledgments

The author would like to acknowledge the following colleagues, graduate students, and laboratory specialists who have contributed to the work of The Laboratory for Neurotoxicity Studies, Virginia-Maryland Regional College of Veterinary Medicine: B.S. Jortner, L. Shell, K. Dyer, S. Padilla (USEPA), B. Veronesi (USEPA), H.A.N. El-Fawal, A. Nostrandt, D. Carboni, W. McCain, L. Correll, K. Fuhrman, and S. Perkins.

## Literature Cited

1. Ecobichon, D. J. In *Toxicology, The Basic Science of Poisons, 4th ed.;* (Amdur, M.O., Doull, J., Klaassen, C.D., Eds.; Pergamon Press: New York, NY, 1991; pp 565-622.
2. Chambers, J.; Levi, P. *Organophosphates: Chemistry, Fate, and Effects*; Academic Press, Inc.: San Diego, CA, 1992.
3. Ballantyne, B.; Marrs, T.C. *Clinical and Experimental Toxicology of Organophosphates and Carbamates*; Butterworth-Heinemann Ltd.: Oxford, UK, 1992.
4. Smith, G. *Pesticide Use and Toxicology in Relation to Wildlife: Organophosphorus and Carbamate Compounds*; U.S. Dept. of the Interior Fish and Wildlife Service Resource Publication 170, Washington, DC, 1987.
5. Sultatos, L.G. *J. Toxicol. Environ. Hlth.* **1994**, *43*, pp 271-289.
6. Baile, J.J.; Henderson, J.D.; Sanborn, J.R.; Padilla, S.; Wilson, B.W. In *ACS Symposium on Field Application of Biomarkers for Agrochemicals and Toxic Substances;* Blancato, J., Brown, R., Dary, C., Saleh, M., Eds.; American Chemical Society: Washington, DC, 1995.

7.    Padilla, S. In *ACS Symposium on Field Application of Biomarkers for Agrochemicals and Toxic Substances*; Blancato, J., Brown, R., Dary, C., Saleh, M,. Eds.; American Chemical Society: Washington, DC, 1995.
8.    Maxwell, D.M. In *Organophosphates: Chemistry, Fate, and Effects;* Chambers, J., Levi, Eds.; Academic Press, Inc.: San Diego, CA, 1992; pp. 183-196.
9.    Satoh, T. In *Reviews in Biochemical Toxicology, Vol. 8;* Hodgson, E., Bend, J.R., Philpot, R., Eds.; Elsevier Press: New York, NY, 1987; pp. 155-181.
10.    Richardson, R.J. *J. Toxicol. Environ. Hlth.* **1995,** *44,* pp 135-165.
11.    Lotti, M. *CRC Crit. Rev. Toxicol.* **1992,** *21,* pp 465-488.
12.    Johnson, M.K. In *Reviews in Biochemical Toxicology, Vol. 8*; Hodgson, E., Bend, J.R., Philpot, R., Eds,; Elsevier Press: New York, NY, 1987; pp 141-212.
13.    Johnson, M.K. *Chem.-Biol. Interactions* **1993,** *87,* pp 339-346.
14.    Dudek, B. R,; Richardson, R.J. *Biochem. Pharmacol.* **1982,** *31,* pp 1117-1121.
15.    Maroni, M.; Bleeker, M.L. *J. Appl. Toxicol.* **1986,** *6,* pp 1-7.
16.    Gurba, P.E.; Richardson, R.J. *Toxicol. Lett.* **1983,** *15,* pp 13-17.
17.    Sogorb, M.A.; Viniegra, S.; Reig, J.A,; Vilanova, E. *J. Biochem. Toxicol.* **1994,** *9,* pp 145-152.
18.    Nostrandt, A.C.; Ehrich, M. *Toxicol. Appl. Pharmacol.* **1993,** *121,* pp 36-42.
19.    Veronesi,B.; Ehrich, M. *In Vitro Toxicol.* **1993,** *6,* pp 57-65.
20.    Ehrich, M.; Correll, L.; Veronesi, B. *NeuroToxicol.* **1994,** *15,* pp 309-314.
21.    Correll, L.; Ehrich, M. *Fund. Appl. Toxicol.* **1991,** *16,* pp 110-116.
22.    Jokanovic, M.; Johnson, M.K. *J. Biochem. Toxicol.* **1993,** *8,* pp 19-31.
23.    Johnson, M.K.; Vilanova, E.; Read, D.J. *Arch. Toxicol.* **1991,** *65,* pp 618-624.
24.    Abou-Donia, M.B.; Lapadula, D.M. *Ann. Rev. Pharmacol. Toxicol.* **1990,** *30,* pp 405-440.
25.    Varghese, R.G.; Bursian, S.J.; Tobias, C.; Takaka, D. Jr. *NeuroToxicol.* **1995,** *16,* pp 45-54.
26.    Senanayake, N.; Karalliedde, L. *N.E. J. Med.* **1987,** *316,* pp 761-763.
27.    Vannestra, Y.; Lison, D. *Human Expt. Toxicol.* **1993,** *12,* pp 365-370.
28.    *U.S. Environmental Protection Agency Pesticide Assessment Guidelines, subdivision E. Hazard Evaluation: Human and Domestic Animals. Addendum 10: Neurotoxicity, series 81, 82, and 83.* EPA 540/09-1-123, Office of Prevention, Pesticides and Toxic Substances: Washington, DC, 1991. (Available from NTIS, Springfield, VA, PB91-154617).
29.    El-Fawal, H.A.N.; Correll, L.; Gay, L.; Ehrich, M. *Toxicol. Appl. Pharmacol.* **1990,** *103,* pp 133-142.
30.    Padilla, S.; Veronesi, B. *Toxicol. Appl. Pharmacol.* **1985,** *78,* pp 78-87.
31.    Lotti, M.; Moretto, A.; Capodicasa, E.; Bertolazzi, M.; Peraica, M.; Scapellato, M.L. *Toxicol. Appl. Pharmacol.* **1993,** *122,* pp 165-171.
32.    Aldridge, N. *Chem. Biol. Interac.* **1993,** *87,* pp 463-466.

33.  Glynn, P.; Read, D.J.; Guo, R.; Wylies, S.; Johnson, M.K. *Biochem. J.* **1994**, *301*, pp 551-556.
34.  Seifert, J.; Wilson, B.W. *Comp. Biochem. Physiol.* **1994**, *108C*, pp 337-341.
35.  Pope, C.N.; Padilla, S. *Toxicol. Appl. Pharmacol.* **1989**, *97*, pp 272-278.
36.  Ehrich, M.; Jortner, B.S.; Padilla, S. *Fund. Appl. Toxicol.* **1995**, *24*, pp 94-101.
37.  Ehrich, M.; Jortner, B.S.; Padilla, S. *Chem. Biol. Interac.* **1993**, *87*, pp 431-437.
38.  Tanaka, D., Jr.; Bursian, S.J. *Brain Res.* **1989**, *484*, pp 240-256.
39.  Jortner, B.S.; Ehrich, M. *NeuroToxicol.* **1987**, *8*, pp 97-108.
40.  Krinke, G.; Ullmann, L.; Sachsse, K.; Hess, R. *Agents and Actions* **1979**, *9*, pp 227-231.
41.  Carrera, V.; Barril, J.; Mauricio, M.; Pellin, M.; Vilanova, E. *Toxicol. Appl. Pharmacol.* **1992**, *177*, pp 218-225.
42.  El-Fawal, H.A.N.; Jortner, B.S.; Ehrich, M. *Fund. Appl. Toxicol.* **1990**, *15*, pp 108-120.
43.  Dyer, K.R.; El-Fawal, H.A.N.; Ehrich, M.F. *NeuroToxicol.* **1991**, *12*, pp 687-696.
44.  Moretto, A.; Lotti; Spencer, P.S. *Arch. Toxicol.* **1989**, *63*, pp 469-473.
45.  Dyer, K.; Jortner, B.S.; Shell, L.; Ehrich, M. *NeuroToxicol.* **1992**, *13*, pp 745-756.
46.  Lotti, M.; Johnson, M.K. *Arch. Toxicol.* **1980**, *45*, pp 263-271.
47.  Olajos, E.L.; DeCaprio, A.P.; Rosenblum, I. *Ecotoxicol. Environ. Safety* **1978**, *2*, pp 383-399.
48.  Ehrich, M.; Jortner, B.S.; Gross, W.B. *Toxicol. Appl. Pharmacol.* **1986**, *83*, pp 250-260.
49.  Ehrich, M.; Jortner, B.S.; Gross, W.B. *Toxicol. Appl. Pharmacol.* **1988**, *92*, pp 214-223.
50.  Lidsky, T.; Manetto, C.; Ehrich, M. *J. Toxicol. Environ. Hlth.* **1990**, *29*, pp 65-75.
51.  El-Fawal, H.A.N.; Jortner, B.S.; Ehrich, M. *NeuroToxicol.* **1990**, *11*, pp 573-592.
52.  Drakontides, A.B.; Baker, T.; Riker, W.F., Jr. *NeuroToxicol.* **1982**, *3*, pp 165- 178.
53.  Baker, T.; Drakontides, A.B.; Riker, W.F., Jr. *Expt. Neurol.* **1982**, *78*, pp 397- 408.
54.  Baker, T.; Stanec, A. *Toxicol. Appl. Pharmacol.* **1985**, *79*, pp 348-352.
55.  McCain, W.C.; Wilcke, J.; Lee, J.C.; Ehrich, M. *J. Toxicol. Environ. Hlth.* **1995**, *44*, 167-187.
56.  Koplovitz, I.; Harris, L.W.; Anderson, D.R.; Lennox, W.J.; Steward, J.R. *Fund. Appl. Toxicol.* **1992**, *18*, pp 102-106.
57.  Stumpf, A.M.; Tanaka, D., Jr.; Aulerich, R.J.; Bursian, S.J. *J. Toxicol. Environ. Hlth.* **1989**, *26*, pp 61-73.
58.  Bursian, S.J.; Brewster, J.S.; Ringer, R.K. *J. Toxicol. Environ. Hlth.* **1983**, *11*, pp 907-916.

59. Fulton, M.H.; Chambers, J.E. *Pest. Biochem. Physiol.* **1985**, *23*, pp 282-288.
60. Ehrich, M.; Shell, L.; Rozum, M.; Jortner, B.S. *J. Am. Coll. Toxicol.* **1993**, *12*, pp 55-68.
61. Carboni, D.; Ehrich, M.; Dyer, K.; Jortner, B.S. *NeuroToxicol.* **1992**, *13*, pp 723- 734.
62. Veronesi, B.; Padilla, S.; Blackmon, K.; Pope, C. *Toxicol. Appl. Pharmacol.* **1991**, *107*, pp 311-324.
63. Lapadula, D.M.; Patton, S.E.; Campbell, G.A.; Abou-Donia, M.B. *Toxicol. Appl. Pharmacol.* **1985**, *79*, pp 83-90.
64. Funk, K.A.; Henderson, J.D.; Liu, C.H.; Higgins, R.J.; Wilson, B.W. *Arch. Toxicol.* **1994**, *68*, pp 308-316.
65. Moretto, A.; Capodisca, E.; Peraica, M.; Lotti, M. *Biochem. Pharmacol.* **1991**, *41*, pp 1497-1504.
66. Farage-Elawar, M.; Magnus Francis, B. *J. Toxicol. Environ. Hlth.* **1987**, *21*, pp 455-469.
67. Correll, L.; Ehrich, M. *Toxicol. Lett.* **1987**, *36*, pp 197-204.
68. Peraica, M.; Capodiscasa, E.; Moretto, A.; Lotti, M. *Biochem. Pharmacol.* **1993**, *45*, pp 131-135.
69. Moretto, A.; Capodiscasa, E.; Lotti, M. *Toxicol. Lett.* **1992**, *63*, pp 97-102.
70. Pope, C.N.; Chapman, M.L.; Tanaka, D., Jr.; Padilla, S. *NeuroToxicol.* **1992**, *13*, pp 355-364.
71. Pope, C.N.; Padilla, S. *J. Toxicol. Environ. Hlth.* **1990**, *31*, pp 261-273.
72. Lotti, M.; Caroldi, S.; Capodiscasa, E.; Moretto, A. *Toxicol. Appl. Pharmacol.* **1991**, *108*, pp 234-241.
73. Pope, C.N.; Tanaka, D., Jr.; Padilla, S. *Chem. Biol. Interac.* **1993**, *87*, pp 395- 406.
74. Moretto, A.; Lotti, M. *Toxicol. Indust. Hlth.* **1993**, *9*, pp 1037-1046.
75. Moretto, A.; Bertolazzi, M.; Lotti, M. *Toxicol. Appl. Pharmacol.* **1994**, *129*, pp 133-137.
76. Moretto, A.; Bertolazzi, M.; Capodiscasa, E.; Peraica, M.; Richardson, R.J.; Scapellato, M.L.; Lotti, M. *Arch. Toxicol.* **1992**, *66*, pp 67-72.
77. Ray, D.E.; Johnson, M.K. *Human Expt. Toxicol.* **1992**, *11*, pp 421-422.
78. Osman, K.A.; Moretto, A.; Lotti, M. *Toxicologist* **1995**, *15*, p 208.
79. Landrigan, P.J.; Graham, D.G.; Thomas, R.D. *Environ. Hlth. Perspect.* **1994**, *102S2*, pp 117-120.
80. Lotti, M.; Johnson, M.K. *Arch. Toxicol.* **1978**, *41*, pp 215-221.
81. Maroni, M.; Barbieri, F. *Neurotoxicol. Teratol.* **1988**, *16*, pp 479-484.
82. Lotti, M.; Moretto, A.; Zoppellari, R.; Dainese, R.; Rizzuot, N.; Barusco, G. *Arch. Toxicol.* **1986**, *59,* pp 176-179.
83. McConnell, R. *NIOSH Research and Demonstration Grants, Annual Report* **1992**, pp. 95-96.
84. Padilla, S.; Wilson, V.Z.; Bushnell, P.H. *Toxicology* **1994**, *92*, pp 11-25.
85. Guyton, A.C. *Textbook of Medical Physiology, 8th ed.* W.B. Saunders Inc.: Philadelphia, PA, 1991; Chapter 32, pp. 356-364; Chapter 33, pp. 365-373.

86. Schwab, B.W.; Richardson, R.J. *Toxicol. Appl. Pharmacol.* **1986**, *83*, pp 1-9.

87. Nostrandt, A.; Ehrich, M. *Toxicologist* **1990**, *10*, p 107.

88. Osterloh, J.; Lotti, M.; Pond, S.M. *J. Anal. Toxicol.* **1983**, *7*, pp 125-129.

89. Bertoncin, D.; Russolo, A.; Caroldi, S.; Lotti, M. *Arch. Environ. Hlth.* **1985**, *40*, pp 139-144.

90. Silbergeld, E.K. *Environ. Res.* **1993**, *63*, pp 274-286.

91. Ehrich, M.; Correll, L.; Veronesi, B. *Toxicologist* **1996**, *16*, in press.

92. Johnson, M.K. *Acta Pharmacol. et Toxicol.* **1981**, *49SV*, pp 87-98.

Chapter 7

# Increased Urinary Excretion of Xanthurenic Acid as a Biomarker of Exposure to Organophosphorus Insecticides

Josef Seifert

Department of Environmental Biochemistry, University of Hawaii, Honolulu, HI 96822

This paper considers increased urinary excretion of xanthurenic acid caused by organophosphorous insecticides (OPI's) as a possible biomarker of exposure to OPI's. Urinary xanthurenic acid is assayed by high-performance liquid chromatography. Pyrimidinyl phosphorothioates and crotonamide phosphates are the most potent OPI's, increasing xanthurenic acid urinary excretion up to six-fold. Elevation of xanthurenic acid occurs within 6 hr of OPI administration and persists for 48 hr following a single OPI dose. Age, dietary regimen and several drugs related to OPI action or L-tryptophan metabolism do not affect OPI-increased xanthurenic acid urinary excretion, while its onset and extent is gender-dependent. The magnitude of a single OPI dose for increasing xanthurenic acid excretion (5 mg of diazinon/kg) indicates importance of this phenomenon for occupational and accidental poisonings.

Organophosphorous insecticides (OPI's) are the major type of insecticides used in the U.S.A. and worldwide. The mechanism of their action is based on inhibition of acetylcholinesterase with toxic consequences due to accumulation of acetylcholine (1). The inhibition of acetylcholinesterase is routinely used as a biochemical marker of exposure to OPI's (2).

Some toxic effects of OPI's are caused by mechanisms other than inhibition of acetylcholinesterase (3-5) and have the potential for use as complementary biomarkers of exposure to OPI's. Recently, we discovered that pyrimidinyl phosphorothioate and crotonamide phosphate

0097–6156/96/0643–0094$15.00/0

insecticides increased urinary excretion of xanthurenic acid in mice (6,7). Inhibition of kynurenine formamidase (aryl formylamine hydrolase, EC 3.5.1.9), an enzyme in the L-kynurenine pathway of L-tryptophan metabolism (Figure 1), was proposed as the trigger for changes in L-tryptophan metabolism. This paper summarizes our findings on this novel phenomenon and provides the information needed for its use as a biomarker of exposure to OPI's.

## Experimental

**Animals and Their Treatment.** Mice (Swiss Webster; 8-12 week-old males except for the study of the effects of age and gender) were used throughout this project. OPI's and other tested drugs were administered intraperitoneally into mice in methoxytriglycol for the nonpolar compounds and in isotonic saline for the polar ones. Cholinergic toxicity of the potent acetylcholinesterase inhibitors (e.g., parathion) was alleviated by atropine. Diazinon [O,O-diethyl    O-(2-isopropyl-4-methyl-6-pyrimidinyl) phosphorothioate] was used as a model OPI since it is both a potent kynurenine formamidase inhibitor and an inducer of xanthurenic acid urinary excretion (6-8).

**Assay of Liver Kynurenine Formamidase Activity.** Kynurenine formamidase activity was assayed spectrophotometrically with N-formyl-L-kynurenine as a substrate (9). The enzyme was prepared from livers of control and/or OPI-treated mice by homogenization and subsequent centrifugations at 10,000 and 150,000xg. Increase in absorbancy due to enzymatically released L-kynurenine was monitored at 365 nm.

**Assay of Urinary Xanthurenic Acid.** Urine collected from 2-4 mice was preserved with the addition of hydrochloric acid. Acidified urine can be stored at -5°C without substantial changes in xanthurenic acid concentration for several days. Urine samples were cleared by centrifugation and evaporated to dryness on a vacuum rotary evaporator at 40°C. The residue was dissolved in 10 mM sodium hydroxide in 80% methanol and the precipitate formed after 2 hr-standing at 0°C was removed by centrifugation. The resulting supernatant was filtered through a microfilter prior to high-performance liquid chromatography (HPLC). The sample was separated on a reversed-phase $C_{18}$ column with 1 mM monosodium phosphate buffer, pH 2.3 as a mobile phase. Xanthurenic acid was detected at 254 nm (Figure 2).

## Results

**OPI Effects on Liver Kynurenine Formamidase.** Liver kynurenine formamidase was almost completely inhibited by a single dose of diazinon within 30 min after its administration (Figure 3). In vivo spontaneous reactivation of the inhibited enzyme was slow. Full

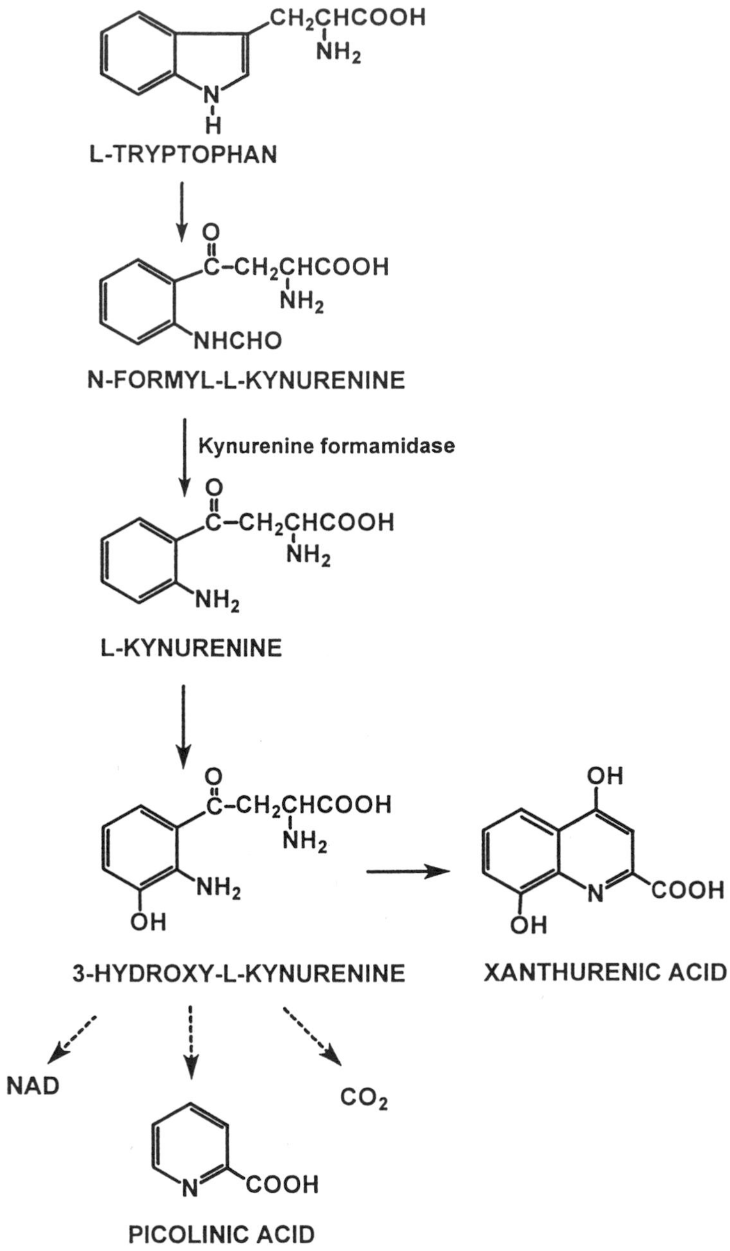

**Figure 1. The L-Kynurenine Pathway of L-Tryptophan Metabolism.** Decreased formation of liver L-kynurenine due to inhibition of kynurenine formamidase by OPI's is compensated by increase in L-kynurenine plasma pool (6, 7).

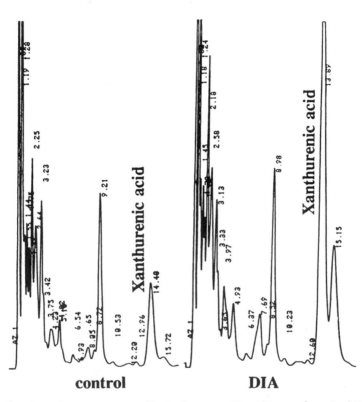

control                              DIA

**Figure 2. Assay of Urinary Xanthurenic Acid by HPLC.** Mice were administered 20 mg of diazinon/kg. Xanthurenic acid in the urine collected for 20 hr was separated on a reversed-phase $C_{18}$ column (4.6x30 mm, 3 um CR Pecosphere, Perkin-Elmer) and detected at 254 nm. DIA - diazinon-treated. Adapted from ref. 6.

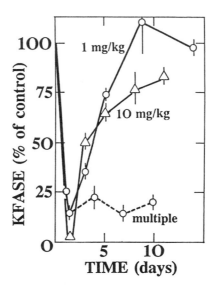

**Figure 3. Liver Kynurenine Formamidase in Mice Treated with Diazinon.** Mice were administered a single (1 or 10 mg/kg) or multiple doses of diazinon (1 mg/kg on days 0, 3, 6 and 9). Liver kynurenine formamidase was assayed spectrophotometrically with N-formyl-L-kynurenine. Values are the means of two assays with the range indicated by bars. Activity at time 0 - $424 \pm 75$ (N=14) $nmol.(mg\ protein)^{-1}.min^{-1}$. Adapted from ref. *6*.

recovery of the enzyme activity in mice treated with 10 mg of diazinon/kg has not occured even after 10 days.

A single dose of pirimiphos-ethyl, diazinon, pirimiphos-methyl, etrimfos, dicrotophos and monocrotophos, cited in descending order of their potency, inhibited liver kynurenine formamidase by more than 80% (Table I). Majority of OPI's were intermediate kynurenine formamidase inhibitors that reduced the enzyme activity by 28-78%. Ronnel, trichlorfon and leptophos did not inhibit liver kynurenine formamidase.

**Table I. Potency of Organophosphorous Insecticides to Inhibit Mice Liver Kynurenine Formamidase and Increase Xanthurenic Acid Urinary Excretion**

| OPI | Kynurenine formamidase inhibition, % | Xanthurenic acid[a], % of control (N) |
|---|---|---|
| *Potent inhibitors/increased xanthurenic acid* | | |
| Pirimiphos-ethyl | 99 | 576±195 (2) |
| Diazinon | 97±3 | 445±213 (14) |
| Pirimiphos-methyl | 95 | 387 |
| Etrimfos | 89 | 184 |
| Dicrotophos | 89 | 294±22 (2) |
| Monocrotophos | 86 | 183±22 (3) |
| *Intermediate inhibitors/no increase in xanthurenic acid* | | |
| Diazoxon | 78 | 50±11 (2) |
| Parathion-methyl | 76 | 104 |
| Profenophos | 73 | 69 |
| Parathion-ethyl | 69 | 72 |
| Malathion | 58 | 95±2 (2) |
| Chlorpyriphos-methyl | 28 | 56 |
| *Noninhibitors/no increase in xanthurenic acid* | | |
| Leptophos | 11 | 100±12 (2) |
| Trichlorfon | 8 | 72 |
| Ronnel | 1 | 113 |

Liver kynurenine formamidase was assayed 12-18 hr after administering 20 mg/kg of a test compound (10 mg of diazoxon/kg) to mice. Xanthurenic acid was determined in urine collected for 15-24 hr. Values are means±S.D. for $\underline{N}$ independent experiments (a range for two experiments). Basal values for liver kynurenine formamidase activity and urinary xanthurenic acid were 424±75 (N=14) nmol L-kynurenine.min$^{-1}$.(mg protein)$^{-1}$ and 737 ± 256 (N=27) nmol.(24 hr)$^{-1}$,respectively.
[a]A similar values were obtained when xanthurenic acid was expressed in nmol.(ml of urine)$^{-1}$. Adapted from ref. 7.

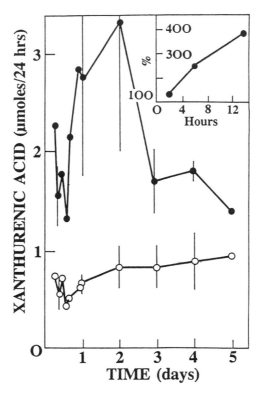

**Figure 4. Urinary Excretion of Xanthurenic Acid in Control and Diazinon-Treated Mice.** Mice were administered 40 mg of diazinon/kg. Urinary xanthurenic acid was assayed by HPLC. Standard deviations indicated by bars were calculated from 3-4 independent experiments. The inset shows xanthurenic acid excretion for the initial period. Key: (o) control; (●) diazinon-treated. Reproduced from ref. *6* with kind permission from Elsevier Science Ltd, The Boulevard, Langford Lane, Kidlington OX5 1GB, UK.

**OPI Effects on Xanthurenic Acid Urinary Excretion.**
Xanthurenic acid increased two to six-fold in the urine
of mice treated with a single dose (20 mg OPI/kg) of the
six most potent liver kynurenine formamidase inhibitors
(Table I). The highest increase was obtained with
pirimiphos-ethyl and diazinon. Elevated xanthurenic acid
excretion was detected as early as six hours after a
diazinon treatment, reaching a maximum in 24- and 48-
urine collections (Figure 4). The dose-dependence study
established that 5 mg of diazinon/kg was a threshold for
increasing xanthurenic acid urinary excretion (Table II).
Diazinon at doses higher than 10 mg/kg did not further
increase xanthurenic acid.

## Table II. Effect of Varying Diazinon Doses on Xanthurenic Acid Urinary Excretion

| Diazinon, $mg.kg^{-1}$ | Xanthurenic acid, $nmol.(24\ hr)^{-1}$ (N) |
|---|---|
| 0 | 737±256 (27) |
| 1 | 855±199 (3) |
| 5 | 1754 |
| 10 | 3913±479 (2) |
| 20 | 2417 |
| 40 | 3670±752 (13) |
| 80 | 3169 |

Urine was collected for 24 hr after diazinon
administration. Values are means±S.D. for $\underline{N}$
independent experiments (a range for two
experiments). Adapted from ref. *6*.

**Effects of Endogenous and Exogenous Factors on Diazinon-
Induced Increase in Xanthurenic Acid Urinary Excretion.**
Increases in xanthurenic acid urinary excretion, caused
by diazinon, were not altered by the mice age (Table
III). The increase was similar in prepubertal, adult and
"aging" mice. The dietary regimen was not critical for
elevation of urinary xanthurenic acid (Table III).
Diazinon-induced increases were proportional to the
changes in basal levels of xanthurenic acid that were
caused by dietary deprivation. Gender was the only factor
that affected xanthurenic acid urinary excretion after
diazinon treatment (Table III). The increase in
xanthurenic acid occured earlier and at higher rates in
female mice, $\underline{i}.\underline{e}.$ a maximal 11-fold increase compared to
4-fold increase in males, relative to control levels.
    The increase in xanthurenic acid urinary excretion
caused by diazinon was not affected by any of the tested
drugs (Table III). Replacement of sulfur by oxygen ($\underline{i}.\underline{e}.$,
P=S in diazinon to P=O in diazoxon) reduced <u>in vivo</u> liver

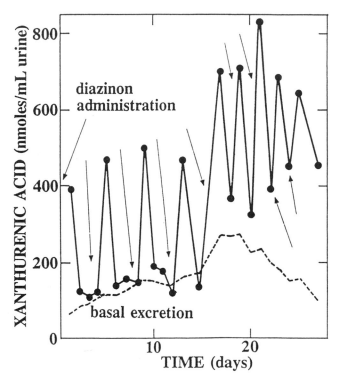

**Figure 5. Effect of a Multiple Administration of Diazinon on Urinary Excretion of Xanthurenic Acid.** Mice were administered 20 mg of diazinon/kg at indicated times. Urine was collected for 24 hr and xanthurenic acid assayed by HPLC.

kynurenine formamidase inhibition and abolished the OPI-enhanced xanthurenic acid urinary excretion (Table I).

**Table III.  Effects of Age, Gender, Diet and Drugs on Diazinon-Increased Urinary Excretion of Xanthurenic Acid**

| Factor | Xanthurenic acid % of control | | |
|---|---|---|---|
| Age (weeks) | | | |
| 3-4 | 513±85 | | |
| 8-12 | 445±213 | | |
| 55-60 | 411±135 | | |
| Gender - urine collected after diazinon administration for: | | | |
| | 0-2 | 2-6 | 6-13 [hr] |
| Male | 137 | 255 | 385 |
| Female | 604±154 | 1104±73 | 200 |
| Diet | | | |
| Normal | 445±213 | | |
| Restricted | 402±39 | | |
| Drugs[a] | | | |
| Atropine | 519 | | |
| Cycloheximide | 464±32 | | |
| Pyridine-2-aldoxime | 496±48 | | |
| Phenylmethyl- sulfonyl fluride | 408 | | |

Urine was collected for 24 hr except for the gender examination. Values are means±range for two independent experiments (mean±SD, N=27, for 8-12 week-old mice maintained on a normal diet). [a]Adapted from ref. 7.

**Liver Kynurenine Formamidase and Urinary Xanthurenic Acid in Multiple and Prolonged Exposures of Mice to Diazinon.** Multiple administration of diazinon maintained the inhibition of liver kynurenine formamidase for the entire period of treatment (Figure 3). Urinary xanthurenic acid excretion increased rapidly following mice treatment and returned to basal values in 24 to 48 hours after each diazinon administration (Figure 5).

**Discussion**

Pyrimidinyl phosphorothioates and crotonamide phosphates are the two groups of OPI that increased urinary xanthurenic acid excretion with a single dose of 20 mg/kg. Urinary xanthurenic acid is elevated only when inhibition of liver kynurenine formamidase exceeds 80%.

Since the majority of OPI's inhibit kynurenine formamidase to some extent [Table I, reference (8)], it is likely that even less potent OPI inhibitors, at higher doses, have the potential to inhibit kynurenine formamidase to the critical level with the consequences of increased urinary excretion of xanthurenic acid.

Changes in urinary xanthurenic acid caused by a single dose of OPI can be monitored from several hours up to two days after exposure to OPI. Response to a multiple diazinon administration follows the pattern of a single dose and allows one to monitor exposure to the OPI for the entire period of treatment. The magnitude of the diazinon dose that is needed for elevation of xanthurenic acid excretion implies the importance of this phenomenon in accidental and occupational poisonings.

The L-tryptophan metabolism and its biochemical components can be altered by numerous endogenous and exogenous factors (10). Gender was the only observed factor that affected OPI-induced enhancement of xanthurenic acid urinary excretion and therefore needs to be considered when using urinary xanthurenic acid for the assessment of OPI exposure. Abolishment of the diazinon effect on xanthurenic acid excretion after a replacement of sulfur by oxygen is probably due to reduction of diazoxon metabolic stability and consequently lower inhibition of kynurenine formamidase [Table I, reference (7)].

Increase in xanthurenic acid urinary excretion is a unique consequence of OPI interference with the L-tryptophan metabolism. However, there are other conditions and environmental factors under which xanthurenic acid urinary excretion is elevated. For instance, urinary excretion of xanthurenic acid is raised in several mental disorders (11) and during pregnancy (12) in humans, and in poisoning by carbon disulfide in rats (13). Thus, when the medical history of subjects tested is unknown, use of urinary xanthurenic acid concentrations will also require measurement of cholinesterase activities for a comprehensive assessment of OPI exposures.

Two lines of biochemical evidence support extrapolation of OPI-increased xanthurenic acid urinary excretion from mice to humans. First, the mouse L-kynurenine pathway of the L-tryptophan metabolism is similar to that in man (14). Second, the susceptibility of liver kynurenine formamidases from various mammals and avians to OPI's is an established phenomenon (15). It is reasonable to speculate that the basic assumption for OPI-increased xanthurenic acid excretion can be met, i.e. that the human liver enzyme is inhibited by OPI to the critical level. Use of urinary xanthurenic acid as a biomarker of exposure to OPI's needs to be validated by examining workers exposed to higher amounts of OPI's, such as pesticide formulators and applicators, farm workers and survivors of accidental or suicidal poisonings.

**Acknowledgments**

This work was supported in part by the USDA Special Grants Program for Tropical and Subtropical Agriculture Research. It is contributed as Journal Series no. 4104 by the Hawaiian Institute of Tropical Agriculture and Human Resources. I thank Drs. B.M. Brennan, R. Dashwood and J.W. Hylin for their reading and criticism of this manuscript.

**Literature Cited**

1. Holmstedt, B. *Pharmacol. Rev.* **1959**, *11*, 567–688.
2. Wilson, B.W.; Henderson, J.D. *Rev. Environ. Contam. Toxicol.* **1992**, *128,* 55–69.
3. O'Neill, J.J. *Fund. Appl. Toxicol.* **1981**, *1*, 154–160.
4. Clement, J.G. *Fund. Appl. Toxicol.* **1985**, *5*, S61–S77.
5. Karczmar, A.G. *Fund. Appl. Toxicol.* **1984**, *4*, S1–S17.
6. Seifert, J.; Pewnim, T. *Biochem. Pharmacol.* **1992**, *44*, 2243–2250.
7. Pewnim, T.; Seifert, J. *Eur. J. Pharmacol.* **1993**, *248*, 237–241.
8. Eto, M.; Seifert, J.; Engel, J.L.; Casida, J.E. *Toxicol. Appl. Pharmacol.* **1980**, *54*, 20–30.
9. Seifert, J.; Casida, J.E. *Pest. Biochem. Physiol.* **1979**, *12*, 273–279.
10. *Kynurenine and Serotonin Pathways. Progress in Tryptophan Research*; Schwarcz, R.; Young, S.N.; Brown, R.R., Eds.; Adv. Exp. Med. Biol. *294*; Plenum Press: New York, **1991**.
11. Hoes, M.J.A.J.M. In *Quinolinic Acid and the Kynurenines*; Stone, T.W., Ed.; CRC Press, Inc.: Boca Raton, Florida, **1989**; pp 229–239.
12. Wachstein, M.; Gudaitis, A. *J. Lab. Clin. Med.* **1953**, *42*, 98–107.
13. Okayama, A.; Ogawa, Y.; Goto, S.; Yamatodani, A.; Wada, H., Okuno, E.; Takikawa, O.; Kido, R. *Toxicol. Appl. Pharmacol.* **1988**, *94*, 356–361.
14. Takikawa, O.; Yoshida, R.; Kido, R.; Hayaishi, O. *J. Biol. Chem.* **1986**, *261*, 3648–3653.
15. Seifert, J.; Casida, J.E. *Comp. Biochem. Physiol.* **1979**, *63C*, 123–127.

# Chapter 8

# Serum Protein Profile: A Possible Biomarker for Exposure to Insecticides

Mahmoud Abbas Saleh, Mohamed Abou Zeid,
Zaher A. Mohamed[1], and Fawzia Abdel Rahman

Environmental Chemistry and Toxicology Laboratory,
Department of Chemistry, Texas Southern University,
3100 Cleburne Avenue, Houston, TX 77004

Serum protein profile as a possible biomarker for detecting exposure risks to pesticides was assessed using lindane, endrin, endosulfan, fenvalerate, toxaphene, heptachlor, and the two industrial pollutants trichlorophenol and polychlorinated biphenyls in an *in-vitro* study. Rat serum was incubated individually with the sublethal dose of each tested compound at $37^0C$ for 24 hours. The change in serum protein profile was monitored using Sodium Dodecylsulfate Polyacrylamide Gel Electrophoresis (SDS-PAGE) and Fast Protein Liquid Chromatography (FPLC) techniques. The impact of each tested compound on the protein profile was expressed as changes in the number of bands/peaks and areas and/or formation of new bands/peaks. From the obtained results, it seems that the molecular effect of the tested compounds is unique and specific to the type of each compound. Thus, serum protein profile may be used as a biomarker for single exposure to an individual insecticide.

The exposure biomarkers most commonly monitored, including the parent compound or its metabolites in blood or urine, typically reflect only those exposures occurring during the last 24-48 hours. Less recent exposures may be detectable as reaction products of xenobiotics (or their metabolites) with macromolecules such as DNA and protein. The indirect analysis of protein adducts has long been utilized to monitor the exposure to organophosphates via the assay of plasma or serum pseudoacetylcholinesterase and erythrocyte acetylcholinesterase activities. Recently other classes of pesticides were reported to form protein adducts, including serum proteins. A survey of the literature indicates that more than 30 different pesticides may form adducts with hemoglobin, and that many of these adducts have been suggested as biomarkers of exposure in animals and/or humans *(1)*. Stable protein adducts integrate exposure from all sources via all portals of entry over the life span of the protein and reflect inter-individual variation in absorption, metabolism, and in some cases, susceptibility. Blood is known to be the major carrier of most essential substances as well as the endogenous defender of the body

[1]Current Address: Department of Biochemistry, Faculty of Agriculture,
Cairo University, Fayoum, Egypt

against foreign organisms and xenobiotics *(2)*. Thus, blood/serum is a rich pool for biomarkers *(3)*. It was suggested that insecticides may have more affinity to bind to serum and/or plasma proteins than to other blood constituents. The binding of these pesticides to any protein may cause some changes in the chemical and physiological characteristics of such proteins *(4)*. Binding of some insecticides to hepatic microsomal proteins caused a conversion of cytochrome P-450 to cytochrome P-420 in rat liver *(5)*. The binding of DDT, dieldrin, and parathion to rat and cockroach blood proteins was studied. The results showed that the insecticides bind mostly to high molecular weight proteins and that this binding is non-specific and hydrophobic *(6)*. It was also reported that malathion has induced changes in catfish serum proteins as well as in hematological parameters *(7)*. The number of protein bands in the SDS-PAGE electrophoresis were indicative of such changes. The changes in these bands could be explained by the physiological reaction to the insecticide or as a breakdown of some relatively high molecular weight proteins. Previous studies from this laboratory showed that *in-vivo* dermal exposure of rats to sublethal doses of malathion has a similar effect on the protein profile as that observed after *in-vitro* incubation using the same concentration *(8)*. Other studies showed that most DNA, hemoglobin, and albumin adducts are selective biomarkers for the effective dose of xenobiotics *(9)*. It was also shown that $Al^{3+}$ cations bind preferentially to transferrin more than albumin in blood serum *(10)*.

The biological monitoring of pesticide residues and metabolites is becoming important in the surveillance of occupationally and environmentally exposed individuals. Detection of these compounds in the body indicates that an exposure has occurred; that the pesticide is bioavailable, having been absorbed; and that a dose to critical tissues may have been incurred. Methods are becoming more sensitive as advances are made in analytical instrumentation systems *(11)*. Recently, macromolecular adducts such as proteins and DNA have received attention as possible internal dosimeters (biomonitors or biomarkers) of exposure to pesticides *(3)*.

The present work was conducted to evaluate the potentiality of using serum protein profile as a possible biomarker of exposure to insecticides and other toxic chemicals through an *in-vitro* study using SDS-PAGE and FPLC.

## Materials and Methods

**Chemicals.** Endrin, lindane, toxaphene, heptachlor, fenvalerate, trichlorophenol (TCP), and polychlorinated biphenyls (PCB's 1221) were obtained from the US EPA (CAS 8001-35-2/QAT 5/83) either as neat samples or as standard pure solutions. Other chemicals and reagents were of high purity or HPLC grade, obtained from Sigma Chemical Co.

**Incubation of Serum with the Tested Compounds.** Stock solutions of each of the *in-vitro* tested compounds (endrin, lindane, toxaphene, endosulfan, heptachlor, fenvalerate, TCP and PCB's) were prepared by dissolving 1 mg of the technical compound in 1 ml methanol in clean dry screw capped tubes. Aliquots of 20 μl of each stock solution was added to 200 μl serum taken from untreated rats and 780 μl of Tris-HCL buffer (pH 7.4) to bring the final concentration up to 20 ppm. Untreated control serum (200 μl) was added to 800 μl buffer and served as control. All serum/buffer tubes were incubated in a water bath at 37°C for 24 hours. The incubated serum samples were then used for the study of protein profile using SDS-PAGE and FPLC. The 20 ppm concentration of insecticides was selected as an exaggerated dose to maximize the interaction with the blood serum proteins. However, a further study will be carried out to verify the dose response interaction at the practically monitored levels in human blood.

**Sodium Dodecyl Sulfate Polyacrylamide Gel Electrophoresis (SDS-PAGE).** The electrophoretic separation of serum proteins was carried out using 5% and 15% polyacrylamide in Tris-Glycine buffer at pH 8.3 with 0.5% SDS according to the method of Laemmli 1970 *(12)*. Serum/buffer samples (50µl each) were treated with 4% SDS and 2% β-mercaptoethanol and heated at 100 °C for 3 minutes. After separation, the gel was stained with 0.1 % Coomassie brilliant blue R-250 stain in 10% acetic acid and 20% aqueous methanolic solution. The stained gels were destained and stored in 7% acetic acid solution for scanning. The gels were scanned using a LKB Ultroscan XL laser scanner, while the data integration was done using Gelscan XL computer software (Pharmacia version 2.0).

**Fast Protein Liquid Chromatography (FPLC).** Aliquots of 50 µl of serum/test compound/buffer were filtered through 0.25µm filter and injected into a Pharmacia/ LKB Fast Protein Liquid Chromatography (FPLC) instrument equipped with two P-500 pumps and a VWM 2141 ultraviolet detector and attached to a Mono Q HR 5/5 cation exchange column (Pharmacia Inc.). Separation was carried out using 0.02M piperazine at pH 6 (buffer A) and 0.02M piperazine containing 0.3 M sodium chloride at pH 6 (buffer B) with a linear gradient from 0% to 100% B in a total time of 30 minutes. The flow rate was 0.5 ml/min and the obtained peaks were monitored simultaneously at 280 and 254 nm. The instrument was controlled with a LCC-501 plus controller, while the quantitative analysis and data processing were carried out using the FPLC manager software (ver. 2.1).

**Results and Discussion**

The electrophoretic separation of serum proteins using SDS-PAGE following incubation of serum with different chlorinated insecticides (endrin, lindane, toxaphene, endosulfan, heptachlor, and the synthetic pyrethroid fenvalerate), the industrial pollutants TCP and PCB's 1221 as well as the untreated control serum, is shown in Figure (1) and Table I. A comparison between the protein profile of the treated and untreated samples reveals variation in both the number and intensity of the major protein bands transferrin, albumin, and pre-albumin. A marked decrease in the percentage of pre-albumin was found as a result of the effect of toxaphene, heptachlor, and PCB's 1221. The change in low molecular weight proteins (15,000-16,000 dalton) was accompanied with an increase in the percentage of higher molecular weight proteins. In the case of fenvalerate, the formation of a new protein band with a molecular weight of 18,000-22,000 daltons was accompanied by the disappearance of smaller molecular weight bands (15,000-16,000 daltons) as shown in Fig. 1. The behavior of the industrial pollutants TCP and PCB's 1221 was almost the same.

Table I. Average Percentage Distribution of Serum Proteins Incubated *in-vitro* with Chlorinated Insecticides and Industrial pollutants for the Scanned SDS-PAGE

| M.wt. (dalton) | Control | Endrin | Lindane | Toxaphene | Endo-sulphan | Hepta-chlor | Trichloro-phenol | Fenvalerate | PCB's |
|---|---|---|---|---|---|---|---|---|---|
| 15,000 | 10.0 | 8.4 | 3.6 | 3.3 | 2.6 | 2.4 | 0 | 0 | 2.5 |
| 18,000 | 0 | 0 | 6.4 | 2.0 | 3.0 | 3.8 | 5.1 | 9.3 | 4.0 |
| 20,000 | 5.3 | 3.1 | 5.6 | 6.1 | 5.9 | 4.9 | 3.2 | 3.7 | 4.9 |
| 22,000 | 0 | 0 | 0 | 0 | 0 | 2.4 | 0 | 2.3 | 1.7 |
| 29,000 | 0 | 0 | 0 | 0 | 0 | 0 | 0 | 2.0 | 2.6 |
| 33,000 | 6.4 | 6.8 | 10.2 | 4.1 | 3.6 | 6.2 | 4.4 | 3.3 | 5.9 |
| 38,000 | 2.0 | 2.0 | 2.7 | 1.7 | 2.1 | 2.7 | 2.1 | 4.0 | 2.0 |
| Pre-albumin | 8.1 | 8.4 | 4.1 | 2.0 | 4.1 | 2.7 | 6.6 | 4.3 | 1.9 |
| Albumin | 25.1 | 26.7 | 27.9 | 29.8 | 27.6 | 26.5 | 25.4 | 22.6 | 26.6 |
| Transferrin | 3.7 | 7.2 | 2.4 | 3.8 | 8.6 | 1.7 | 8.0 | 6.8 | 3.6 |
| 95,000 | 0 | 0 | 3.3 | 2.5 | 0 | 2.6 | 0 | 0 | 5.3 |
| 100,000 | 11.6 | 8.9 | 6.7 | 6.6 | 7.8 | 7.8 | 8.2 | 8.2 | 8.0 |
| 160,000 | 27.8 | 28.5 | 27.1 | 38.1 | 34.7 | 36.3 | 37.0 | 33.5 | 31.0 |

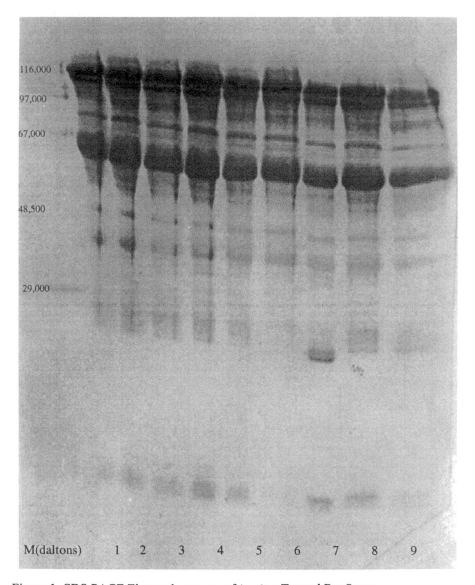

Figure 1. SDS-PAGE Electropherogram of *in-vitro* Treated Rat Serum.
(1) Endrin, (2) Lindane, (3) Toxaphene, (4) Endosulfane, (5) Heptachlor, (6) Trichlorophenol, (7) Fenvalerate, (8) PCB's, (9) Serum Control, (M) Protein Marker.

Figure (2) illustrates the FPLC chromatograms of normal serum and serum incubated *in-vitro* with different insecticides and the industrial pollutants TCP and PCB's. Generally, the relative percentage distribution of serum protein fractions differed in all treatments as compared with the normal serum pattern. The changes in height and area of the peaks of protein fractions indicate the interference of the tested compounds which might conjugate with some specific protein(s). The appearance of new bands (6a and 6b) and (9a and 9b) was the result of the interference of lindane, endosulfan, and fenvalerate with serum proteins. The splitting of peak 9 into two new peaks (9a and 9b) was the result of interference of heptachlor and TCP with serum proteins. These results agree with those previously reported in which new protein bands were detected after incubation of serum with $^{14}$C-aldrin, $^{14}$C-telodrin and $^{14}$C-DDT *(4, 6, 13,14)*. The results suggested that these new proteins are fractions of the high molecular weight lipoprotein. The new bands formed in the electrophoretic profile may be due to the breakdown of high molecular weight proteins *(7)*. The number of fractions formed due to the adduct formation may also vary according to the structure and concentration of the pollutant *(15,16)*.

Biomarkers provide an important tool to the toxicologist/risk analyst in incorporating biological insights relevant to mechanisms of action into risk assessments. The purpose of this study was to provide accurate and reliable tools for assessing the impact of pesticide exposure. Serum has been the medium of choice in this course of study as it is the pool for proteins and other biological compounds. Serum protein was a prominent biomarker for detecting the impact of either organophosphates or organochlorine insecticides on the profile of serum proteins. The use of FPLC technique in the assessment of protein profile proved to be a promising tool for preliminary identification of the adducts between proteins and insecticides. In addition, this technique may provide a rapid and accurate method for detection. Further studies are needed to use the FPLC as a reliable tool for assessing the hazards of exposure to pesticides through separation and identification of each protein-insecticide adduct. However, such a technique will be limited for detection of exposure to a single pollutant at a time. Further development of the methodology, by identification of the molecular characteristics of the specific adduct for each pollutant, might lead to the ability of detecting exposure to a mixture of toxicants simultaneously.

Thus, for any given toxicant, there may be more than a single biomarker of effect which can be assessed with various specificities in different body fluids and tissues. Hence, different types of biomarkers need to be considered in conjunction whenever possible and appropriate using sensitive and advanced instrumentation techniques.

## Acknowledgment

This work was funded by the U.S. EPA Grant # CR 818220-02-5. Although the research described in this article has been supported by the U.S. EPA, it has not been subjected to Agency review and, therefore does not necessarily reflect the view of the Agency, and no official endorsment should be inferred.

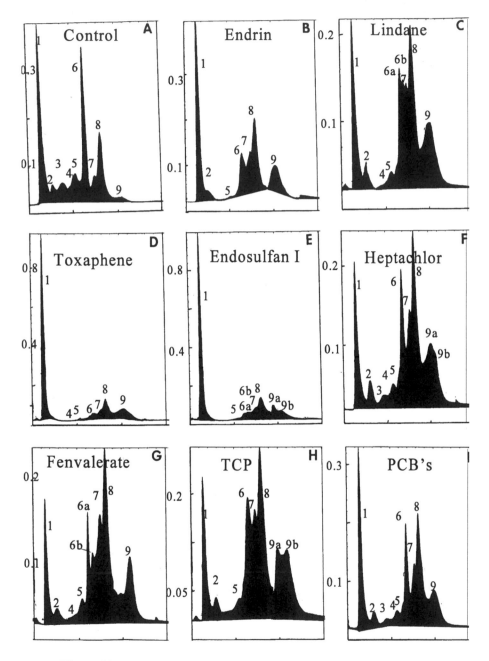

Figure (2). FPLC Chromatograms of Rat Serum Protein Profiles of
Untreated Control as well as *In-Vitro* incubation with
20 ppm of the tested compounds.
1= Transferrin          7,8= Albumin

## Literature Cited

1. Schnell, F.C. *Rev. Environ. Toxicol.* **1993**, *5*, 51-160.
2. Luzio, J.P.; Thompson, R.J., *Molecular Medical Biochemistry*; 1990, Cambridge Univ. Press; Cambridge, pp. 264.
3. Stevens, D.K.; Eyre, R.J., *Fund. Appl. Toxicol.*, **1992**, *19*, 336.
4. Moss, J.A.; Hathway, D.E., *Biochem.*, **1964**, *91*, 384-389.
5. James,T.S., *LifeScience*, **1974**, *14*, 2215-2229.
6. Skalsky, H.L.; Guthrie, *Pesticide, Biochem. Physiol*, **1975**, *7*, 289-296.
7. Dutta, H.M.; Dogra, J.V.; Singh, N.K.; Roy, P.K.; Nasar, S.T., Adhikara, S.; Munshi, J.S.; Richmonds, C., *Bull. Environ. Contam. Toxicol.*, **1992**, *49*, 91-97.
8. Abou-Zeid, M.M.; El-Baroty, G.; Abdel-Reheim, E.; Blancato, J.; Dary, C.; El-Sebae, A.H.; Saleh, M.A., *J. Environ. Sci. Health*, **1993**, *B 28*, 413-430.
9. Timbrell, J.A.; Draper, R.; Waterfield, C.J., *Toxicol. And Ecotoxicol. News*, **1992**, *1(1)*, 4-14.
10. El-Sebae, A.H.; Abou-Zeid, M.M., Abdel-Rahman, F.H.; Saleh, M.A., *J.Environ. Sci. Health,* **1994**, *B 29(2)*, 303-321
11. Nauman, C.H.; Santolucito, J.A.; Dary, C.C. In *Biomarkers of Human Exposure to Pesticides;* Saleh, M.A.; Blancato, J.N.; Nauman, C.H., Eds.; 1994, ACS Symposium Series 542, Washington D.C., pp. 1-2.
12. Laemmli, U.K. *Nature*, **1970**, 277-280.
13. El-Sebae, A.H.; Abdel-Ghany, A.E.; Shalloway, D.; Abou-Zeid, M.M.; Blancato, J.; Saleh, M., *J. Environ. Sci. Health*, **1993**, *B 28(6)*, 763.
14. Eliason, B.C.; Posner, H.S., *Amer. J. Obestet. Gynecol*, **1971**, *III*, 925-927.
15. Rai, R. *J.Environ.Biol.*, **1987**, *8(2)*, 225-228.
16. Larcen, G.L.; Davison. K.L.; Bakke, J.E.; Pass, N.M. In *Biomarkers of Human Exposure to Pesticides;* Saleh, M.A.; Blancato, J.N.; Nauman, C.H., Eds.; 1994, ACS Symposium Series 542, Washington D.C., pp. 166-177.

# Chapter 9

# Breast Milk as a Biomarker for Monitoring Human Exposure to Environmental Pollutants

**Mahmoud Abbas Saleh, Abdel Moneim Afify, Awad Ragab, Gamal El-Baroty, Alaa Kamel, and Abdel Khalek H. El-Sebae[1]**

**Environmental Chemistry and Toxicology Laboratory, Department of Chemistry, Texas Southern University, 3100 Cleburne Avenue, Houston, TX 77004**

Residues of lead and chlorinated insecticides in human breast milk of 120 Egyptian women representing 20 different governorates (provinces) throughout Egypt were determined. The mean values of lead in rural areas and small cities were significantly lower than those in industrial and highly populated cities. Higher levels of lead in mother's human milk from Cairo, Alexandria and Assiut may be attributed to heavy automobile traffic and the use of leaded gasoline in addition to the use of lead water pipe lines in these areas. On the other hand, the most abundant residues of organochlorine pesticides and their metabolits found in mother's milk were $p,p'$-DDE and lindane. Endosulfan I, and $p,p'$-DDT residues were detected in some samples, but aldrin and endrin were not detected in most of the samples. The mean values for $p,p'$- DDE, lindane, endosulfan I and $p,p'$-DDT levels in the milk samples of the 20 governorates studied were 21.37 ppb, 8.42 ppb, 4.84 ppb and 2.93 ppb respectively, which are lower than the levels previously reported in some of the developed countries. Relatively higher levels of organochlorine insecticide residues were recorded in the samples from intensive agricultural activity regions.The results showed that population living in large cities have higher levels of lead but lower levels of pesticide residues while those living in rural areas have lower levels of lead and higher levels of pesticides suggesting that levels of pollutants in human milk may serve as a good biomarker for body burden exposure.

The 60 million Egyptian inhabitants can be grouped into three main different community types. The urban population, living in the capital city of Cairo (14 million) and other big cities, is generally exposed to air pollutants, especially lead evolving with vehicle exhaust, petroleum and gasoline vapors, carbon monoxide, and mineral dusts. In addition, the urban communities may also be exposed to toxic residues in food and

---

[1]Current Address: Department of Pesticides Chemistry and Toxicology, Faculty of Agriculture, Alexandria University, Chatby, Alexandria, Egypt

0097–6156/96/0643–0114$15.00/0

drinking water which may include pesticides and other toxicants. The second large community consists of those living in rural villages who are more likely to be exposed to pesticides and other agrochemicals. The third community includes those living in remote desert and mountain areas in the western and eastern deserts, the Sinai peninsula, and northern and eastern sea coasts. The population in these areas is still engaged in few agricultural activities. There are also some Bedouin tribes in the southern part of Egypt such as Bassharia who live mainly on raising herds of camels and sheep in the desert areas. In such locations, the exposure to man made chemicals and pollutants is a minimum. Therefore, it is reasonable to consider such inhabitants as a real unexposed reference group compared to the urban or rural communities. This Egyptian model structure, where clear differentiation can be drawn between rural, urban and remote desert inhabitants, is expected to be successful in reaching significant correlations between types of human activities and levels of exposure to hazardous chemicals. Since most organochlorine insecticides are environmentally persistent and fat soluble, they may accumulate in food and be stored in high concentrations in tissues and lipid rich organs such as the adipose tissues, liver, meat, and milk *(1)*. Human milk is the most important and indispensable food for the newborn. During lactation, fat mobilization could take place from the adipose tissue, and therefore, organochlorine compounds such as DDT, and HCHs and their metabolites are mobilized and released mainly by breast milk. Recent epidemiological studies indicated that breast fat and serum lipids of women with breast cancer contain significantly elevated levels of some chlorinated organic compounds compared to non-cancer controls. Among the chemicals found in this investigation were several lipophilic, persistant compounds such as *o,p'*-DDT, an isomer of the DDT insecticide, heptachlor, PCB's, and other chlorinated pesticides and industrial chemicals *(2)*. Thus, monitoring of such residues in mother's milk will be a good criterion to measure their impact on general population, not only on the existing population, but also on the next generation of the newly born children. In Egypt, a few surveys have been carried out in this regard, mainly in two governorates (Beny Sweif and Fayoum). Although the environmental contamination by chlorinated hydrocarbon insecticides has been reduced since 1970, when these pesticides were banned, some amounts of these chemicals were recently found in human breast milk. However, these amounts were within the range of acceptable daily intakes according to the FAO/WHO guidelines *(3,4)*.

The main objectives of this research was to evaluate the Egyptian mother's milk contents of organochlorine insecticides and the heavy metal lead (Pb) as criterion for measuring the body burden with environmental pollutants due to long term exposure.

## Material and Methods

A total of 120 Egyptian women (20 to 40 years of age, 6 from each governorate) participated in the study voluntarily and agreed with its aims. The women were giving birth to their first or second child and all had lived in the indicated areas during the past ten years. Apart from these selection criteria, they were chosen at random. Collection sites are shown in Figure 1. Milk samples, 10-15 ml each were obtained from the out-patient/pediatric unit of each governorate's central hospital in May and June of 1993 from lactating women. The samples were collected by manual expression into sterilized vials (ca. 20 ml), transported in a solid $CO_2$ ice box, and stored in a deep freezer at -80°C until

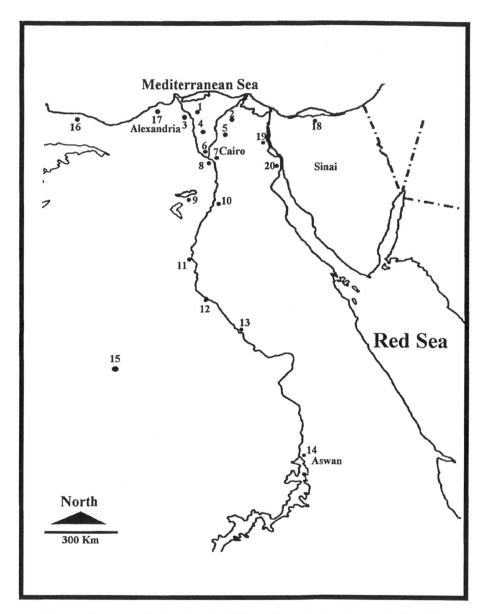

Figure 1. Sample collection sites in different Egyptian locations as indicated in Table I.

analyzed. Care was taken to avoid contamination during and after sample collection. The milk was re-homogenized with an ultrasonic homogenizer prior to analysis.

**Preparation of Samples for Lead Analysis.** All samples were analyzed in triplicate, using a graphite furnace Perkin Elmer 4100 ZL Atomic Absorption spectrometer, after digestion in a nitric perchloric acid mixture as described by Vahter and Slorach *(5)*.

**Preparation of the Samples for GC Analysis.** The organochlorine insecticides in human milk were extracted according to the method of Brevik *(6)* with slight modification,  The method includes the extraction of 1 ml of milk sample with acetone:hexane (3:4, v/v) by ultrasonic disintegration for 10 sec, followed by a second extraction with acetone:hexane (1:2, v/v). The organic layers were combined together, evaporated to dryness, redissolved in 1.5 ml of hexane and cleaned with concentrated $H_2SO_4$ (1.5 ml). Duplicate subsamples were analyzed for each gross sample. The hexane layers were made up to 1 ml and analyzed by gas chromatography/electron capture and gas chromatography/mass spectrometry.

**Gas Chromatography and Gas Chromatography/Mass Spectrometry.** Gas chromatography (GC) was carried out using a Hewlett Packard 5890 gas chromatograph series II controlled with chemstation software and equipped with [63]Ni electron capture detector.  The column used was a DB-608 megabore column (30m, 0.53 mm i.d.), and the flow rate was 3.5 ml/min of helium.  The oven temperature was programmed at 4°C/min from 190°C to 240°C and held at 240°C for 10 min.  The detector and injector temperatures were adjusted at 220°C and 390°C respectively.  The injections were split with the split ratio (1:10).
The GC/MS was carried out using a 5972 MSD mass selective detector attached to 5890 plus gas chromatograph equipped with electronic pressure control. The column installed was an HP-5MS capillary column (30m, 0.25 mm i.d., 0.25 mm film), and the flow rate was 1 ml/min of helium.   Injections were splitless and purged after 1.5 min with an injector temperature of 250°C.  The oven's initial temperature was 85°C, held for 1.5 min., then programmed to 190°C at 30°C/min., then to 240°C at 3°C/min. The mass spectrometer was adjusted to the selective ion monitoring mode for the ions at *m/z* 181, 108, and 246. The dwell time was 40 msec/ion. Peak identification was performed by comparing the retention times of each peak with those of known standard compounds and by the characteristic ion peak in the SIM-GC/MS.

## RESULTS AND DISCUSSION

**Lead Concentration in Mother's Milk and its Hazardous Impact.** Table I  presents the mean lead concentrations  in milk samples collected from the 20 different locations shown in Figure 1.   According to the daily permissible level established by the WHO *(5)* of 5.0 $\mu$g/kg body weight/day, which is equivalent to about 15.5 ppb in mother's milk, it can be seen from Table I that the mean values of lead levels were below the permissible level  in the governorates of Fayoum, Matruh, Minia and Suez. Lead levels in Aswan, Beheira, Beny Sweif, Dakahlia, Gharbia, Giza, Ismailia, Kaliobia, Menoufia,

New Valley, North Sinai, Sharkia and Sohag were slightly higher than the WHO permissible value. Mean lead levels in mother's milk from Alexandria, Assiut, and Cairo were significantly higher than the permissible value. The higher levels of lead in mother's milk from Cairo, Alexandria and Assiut may be attributed to heavy automobile traffic and the use of leaded gasoline (the only type of gasoline available in Egypt), in addition to contamination of drinking water from the lead drinking water pipe lines. There is very little information regarding the transfer of lead via the milk from mothers who are at high risk, such as those living in big cities i.e. Cairo, Alexandria and Assiut. Chronic adult exposure to lead occurs mainly through the inhalation of dust and fumes, and incidental ingestion of polluted food and drinks, and the inhalation of cigarette smoke. The threshold limit value-time weighted average (TLV-TWA) for lead, dust and fumes is $0.15$ mg/m$^3$. An average of 66 $\mu$g/l in mother's milk from women in Cairo is equivalent to $10.54$ $\mu$g/kg/day while a child weighing 5.5 kg may receive at least 58 $\mu$g of lead per day. According to Mahaffey *(7)*, the tolerable or maximum daily intake of lead from all sources for infants between birth and the age of 6 months should be as low as possible and should be less than 100$\mu$g/day. In Table (II) the lead levels in human milk recorded in this study were compared with those of other countries from 1971-1994 *(8-21)*. By comparison with the published data from other countries, the lead levels in Egypt are between moderate lead levels found in Japan, Germany, Sweden, and the United States, and the high values found in Mexico, Indonesia, and Thailand . The current lead level in Egypt, although higher in some cases than the recommended WHO levels still appears not to be at the level of a serious risk except in those cases where the lead level is higher than 100 ppb in Assiut, followed by Cairo and Alexandria.

Due to the high risk of lead exposure between urban population in the heavily populated cities of Cairo, Alexandria, and Assiut, it is recommended that the leaded-gasoline and the leaded pipelines for drinking water should be banned as early as possible because it is the main source of lead pollution in these crowded urban communities. This ban will reduce the burden on both the human adults and their children who are actually more vulnerable for maternal and childhood exposure to a lead polluted diet.

**Chlorinated Insecticides Levels in Human Milk.** The data in Table III shows that the main detected organochlorine insecticides and their metabolites were DDE and lindane. DDT and endosulfan I residues were also detected in some samples. Endrin was only detected in one of the samples in New Valley, while aldrin was not detected in any of the samples.

**DDE (*p,p'*- DDE):**    DDE was identified by GC and GC/MS in 100% of the samples. The lowest mean levels, 5.80-8.95 ppb, were found in New Valley, North Sinai, and Aswan, while the highest mean levels, 41.7-54.5 ppb, were found in Sharkia, Behera and Dakahlia. The mean value of DDE levels in the 20 governorates studied was 21.37 ppb which is lower than the levels reported in many developed countries as shown in Table IV *(23-37)*. Since DDE levels reflect previous long term exposure to DDT, these results prove the effectiveness of the restrictions imposed on the use of DDT in agriculture, vector control, and households since the early 70's.

**DDT (*p,p'*- DDT):**    DDT is the organochlorine compound most extensively studied throughout the world and has been shown to be present universally in the milk

Table I. Lead Concentration in Human's Breast Milk from Egypt $\mu g/l$. Error % are shown in parenthesis

| Location | Subject | | | | | | Average | Range |
|---|---|---|---|---|---|---|---|---|
| | 1 | 2 | 3 | 4 | 5 | 6 | | |
| Alexandria [17]ᵃ | 40 (1.9) | 40 (1.7) | 66 (3.4) | 42 (1.8) | 68 (0.4) | 40 (3.8) | 49.2 | 40-68 |
| Assiut [12] | 48 (8.1) | 40 (3.6) | 158 (14.3) | 128 (1.1) | 116 (3.5) | 118 (3.3) | 101.4 | 40-158 |
| Aswan [14] | 74 (0.9) | 8 (2.9) | 18 (1.9) | 20 (0.6) | 8 (2.5) | 16 (1.3) | 24.0 | 8-74 |
| Beheira [3] | 52 (7.6) | 40 (6.2) | 28 (1.6) | 32 (5.6) | 24 (7.6) | 40 (7.8) | 36.0 | 24-52 |
| Beny Sweif [10] | 10 (4.7) | 24 (6.8) | 26 (6.3) | 20 (4.3) | 82 (4.0) | 30 (4.5) | 32.0 | 10-82 |
| Cairo [7] | 64 (3.9) | 48 (5.9) | 82 (5.2) | 68 (3.6) | 46 (5.7) | 88 (2.8) | 66.0 | 46-88 |
| Dakahlia [2] | 36 (3.9) | 10 (9.4) | 16 (3.0) | 18 (3.4) | 10 (8.0) | 16 (4.6) | 17.6 | 10-36 |
| Fayoum [9] | 10 (2.0) | 18 (1.7) | 4 (4.2) | 16 (1.4) | 10 (1.1) | 12 (1.6) | 11.6 | 4-18 |
| Gharbia [1] | 28 (8.0) | 22 (7.8) | 24 (2.1) | 26 (7.5) | 28 (1.9) | 36 (5.2) | 27.4 | 22-36 |
| Giza [8] | 28 (7.8) | 12 (5.3) | 24 (5.2) | 28 (3.1) | 12 (1.4) | 20 (1.8) | 20.6 | 12-28 |
| Ismailia [19] | 82 (6.2) | 24 (4.9) | 20 (1.6) | 8 (1.2) | 20 (5.8) | 12 (1.6) | 27.6 | 8-82 |
| Kaliobia [6] | 28 (1.8) | 36 (2.7) | 24 (4.6) | 36 (7.1) | 28 (8.5) | 36 (6.5) | 30.6 | 24-36 |
| Matrouh [16] | 12 (2.9) | 6 (1.3) | 8 (2.6) | 8 (3.1) | 12 (1.2) | 8 (0.3) | 9.0 | 6-12 |
| Minia [11] | 16 (4.3) | 16 (8.7) | 8 (4.5) | 6 (4.1) | 0 (1.9) | 8 (1.8) | 9.0 | 0-16 |
| Menoufia [4] | 36 (4.2) | 24 (1.4) | 26 (7.3) | 36 (2.6) | 86 (4.6) | 8 (4.8) | 36 | 8-86 |
| New Valley [15] | 16 (3.3) | 22 (5.1) | 20 (3.8) | 14 (5.6) | 24 (0.5) | 20 (2.4) | 19.4 | 14-24 |
| North Sinai [18] | 40 (5.1) | 20 (7.5) | 46 (8.6) | 20 (7.2) | 18 (4.9) | 28 (2.9) | 28.6 | 18-46 |
| Sharkia [5] | 22 (3.0) | 26 (4.0) | 42 (5.3) | 24 (6.6) | 32 (7.3) | 36 (2.1) | 30.4 | 22-42 |
| Sohag [13] | 14 (8.9) | 24 (1.6) | 20 (3.7) | 36 (2.9) | 10 (2.1) | 26 (3.4) | 21.6 | 10-36 |
| Suez [20] | 0 | 72 (1.4) | 0 | 0 | 0 | 0 | 15.2 | 0-72 |

ᵃ Sample collection sites as indicated in Figure (1).

**Table II.  Lead Concentration ($\mu$g/l) in Breast Milk Reported Internationally**

| Location | Number of Subjects | Average | Range | Reference |
|---|---|---|---|---|
| Cincinnati (USA) | 22 | 0.012 | - | (8) |
| Connecticut (USA) | 7 | 20.00 | 0-70.0 | (9) |
| New York (USA) | 10 | 5.00 | - | (10) |
| Iowa (USA) | 4 | 26.00 | 15.0-64.0 | (11) |
| Washington, D.C. (USA) | 20 | 20.00 | 0-50.0 | (12) |
| Boston (USA) | 100 | 17.00 | 0-72.0 | (13) |
| Arizona (USA) | 39 | 2.80 | 0.90-10.0 | (14) |
| Bangkok (Thailand) | 164 | 85.00 | 136-220 | (15) |
| Uppsala (Sweden) | 41 | 20.00 | 5.0-90.0 | (16) |
| London (England) | 28 | 2.40 | 1.9-8.6 | (17) |
| Kuala Lumpur (Indonesia) | 114 | 47.8 | 24.9-106.1 | (18) |
| Hamburg (Germany) | 10 | 13.2 | 9.1-15.5 | (19) |
| Mexico City (Mexico) | 35 | 61.8 | 9.2-351 | (20) |
| Japan | 22 | 27.0 | 0-56.0 | (21) |
| Styria (Austria) | 64 | 3.40 | 0-20.4 | (22) |
| Egypt (20 Cities) | 120 | 30.6 | 0-158 | Present Study |

Table (III). Distribution of the Main Organochlorine Insectiside Residues
in Egyptian Mother's Milk

| Governorate* | Lindane | | Endosulfan I | | 4,4'- DDE | | 4,4'- DDT | |
|---|---|---|---|---|---|---|---|---|
| | Average | Range | Average | Range | Average | Range | Average | Range |
| **Greater Cairo** | | | | | | | | |
| Cairo [7] | 5.05 | 2.72 - 8.98 | 0.00 | 0.00 - 0.00 | 11.30 | 1.40 - 19.70 | 0.95 | 0.00 - 2.85 |
| Giza [8] | 4.96 | 2.69 - 6.59 | 0.69 | 0.00 - 2.08 | 12.02 | 1.96 - 19.70 | 0.00 | 0.00 - 0.00 |
| Kaliobia [6] | 13.87 | 1.21 - 33.20 | 0.00 | 0.00 - 0.00 | 20.62 | 8.95 - 27.30 | 2.04 | 0.00 - 3.87 |
| **Delta Region** | | | | | | | | |
| Sharkia [5] | 6.57 | 4.15 - 9.78 | 0.00 | 0.00 - 0.00 | 54.50 | 19.70 - 83.30 | 2.43 | 1.70 - 3.64 |
| Gharbia [1] | 4.63 | 3.47 - 6.49 | 10.50 | 0.00 - 18.90 | 25.82 | 4.06 - 58.30 | 1.68 | 0.00 - 2.53 |
| Behera [3] | 12.82 | 4.47 - 22.80 | 28.98 | 7.34 - 57.90 | 50.95 | 5.74 - 117.00 | 5.51 | 0.00 - 10.60 |
| Dakahlia [2] | 19.53 | 13.40 - 24.70 | 6.00 | 0.00 - 18.00 | 41.70 | 7.30 - 67.20 | 4.12 | 0.00 - 9.38 |
| Menoufia [4] | 1.52 | 0.00 - 3.44 | 0.00 | 0.00 - 0.00 | 7.93 | 2.40 - 10.90 | 2.59 | 0.00 - 7.76 |
| **Upper Egypt** | | | | | | | | |
| Fayoum [9] | 3.77 | 0.68 - 7.82 | 0.00 | 0.00 - 0.00 | 20.06 | 4.59 - 37.40 | 1.84 | 0.00 - 4.61 |
| Beny Sweif [10] | 2.04 | 0.95 - 2.90 | 3.25 | 0.00 - 5.81 | 27.26 | 9.17 - 46.70 | 1.59 | 0.00 - 3.41 |
| Minia [11] | 3.11 | 0.81 - 6.30 | 8.03 | 0.00 - 20.30 | 16.36 | 6.88 - 21.70 | 4.83 | 0.00 - 14.50 |
| Assiut [12] | 10.95 | 0.78 - 31.00 | 1.63 | 0.00 - 4.90 | 30.89 | 3.47 - 71.30 | 5.12 | 3.51 - 7.02 |
| Sohag [13] | 12.88 | 6.75 - 21.30 | 4.77 | 0.00 - 14.30 | 29.42 | 2.97 - 77 50 | 5.47 | 0.00 - 13.60 |
| Aswan [14] | 12.84 | 4.47 - 28.20 | 7.02 | 2.41 - 12.70 | 8.95 | 3.80 - 18.50 | 2.46 | 0.00 - 7.38 |
| **Coastal Areas** | | | | | | | | |
| Alexandria [17] | 12.46 | 2.37 - 20.50 | 0.00 | 0.00 - 0.00 | 9.35 | 7.32 - 12.10 | 0.00 | 0.00 - 0.00 |
| Matrouh [16] | 11.30 | 6.41 - 15.00 | 25.83 | 0.00 - 61.80 | 15.11 | 7.04 - 23.00 | 0.00 | 0.00 - 0.00 |
| North Sinai [18] | 11.81 | 9.92 - 12.90 | 0.00 | 0.00 - 0.00 | 8.84 | 5.17 - 13.70 | 14.52 | 4.77 - 32.90 |
| Ismailia [19] | 3.71 | 1.36 - 5.43 | 0.00 | 0.00 - 0.00 | 19.17 | 14.20 - 25.40 | 0.00 | 0.00 - 0.00 |
| Suez [20] | 1.44 | 0.00 - 2.21 | 0.00 | 0.00 - 0.00 | 11.35 | 5.78 - 20.30 | 2.39 | 0.00 - 7.16 |
| **Desert** | | | | | | | | |
| New Valley [15] | 13.08 | 8.93 - 18.20 | 0.00 | 0.00 - 0.00 | 5.83 | 4.13 - 8.54 | 1.04 | 0.00 - 3.13 |
| **Average in Egypt** | 8.42 | 0.00 - 31.00 | 4.84 | 0.00 - 61.80 | 21.37 | 1.40 - 117.00 | 2.93 | 0.00 - 32.90 |

*Number of collection site as indicated in Figure (1).

**Table (IV) Average Organochlorine Insecticide Residues in Human Milk from Various Countries  (μg/kg whole milk).**

| Country (City) | Year | Lindane | DDE | DDT | DDE/DDT | Reference |
|---|---|---|---|---|---|---|
| Brazil[#] (São Paulo) | 1985 | traces | 30 | 119 | 0.25 | (23) |
| Canada* | 1987 | 0.21 | 29.22 | 2.45 | 11.92 | (24) |
| Egypt (20 Governorates) | 1993 | 8.42 | 21.37 | 2.93 | 7.30 | Present study. |
| Finland* | 1985 | - | 21.35 | 2.1 | 10.16 | (25) |
| Finland* | 1982 | N.A. | 29.75 | 1.26 | 23.61 | (26) |
| Germany (Westfalia) | 1987 | 0.77 | 26.18 | 2.13 | 12.29 | (27) |
| Greece* | 1983 | 0.39 | N.A. | 1.22 | N.A. | (28) |
| Hong Kong | 1989 | 1.71 | 332.59 | 61.84 | 5.38 | (29) |
| India* | 1986 | N.A. | 254.8 | 55.89 | 4.55 | (30) |
| India* | 1988 | 1.65 | 44.27 | 8.75 | 5.06 | (31) |
| Israel (Jerusalem) | 1985 | 1.14 | 79.0 | 8.46 | 9.33 | (32) |
| Italy* | 1985 | N.D. | 1.4 | 0.25 | 5.6 | (33) |
| Jordan (Amman) | 1992 | 6.78 | 60.18 | 13.27 | 4.53 | (34) |
| Poland* | 1987 | 1.75 | 173.25 | 16.1 | 10.76 | (35) |
| Spain (Madrid) | 1991 | 0.3 | 18.7 | 0.4 | 46.75 | (1) |
| Spain | 1982[@] | 19 | 170 | 83 | 2.05 | (36) |
| Thailand (Bangkok) | 1987 | 0.11 | 126.35 | 25.58 | 4.94 | (27) |
| Turkey* | 1987[@] | 0.35 | 161.21 | 22.8 | 7.07 | (37) |
| USA (New York) | 1987 | 0.07 | 18.39 | 0.78 | 23.57 | (27) |
| Vietnam (Ho Chi Minh) | 1987 | 0.99 | 288.1 | 202.1 | 1.43 | (27) |

*Calculations based on 3.5% fat content. [@]Year of publication.
[#]Milk samples of occupationally exposed donors.

of lactating mothers *(38)*. However, from the 60 human milk samples, 51% of the samples were free from any detectable DDT levels, a fact which may suggest that there were no recent sources of pollution by intact DDT.  DDT levels from 0.00-5.00 ppb were detected in 75% of the samples, while the higher levels (5.00 ppb and above) were recorded only in 25% of the milk samples from the 20 Egyptian governorates.  The highest DDT level (32.90 ppb) was detected in the Dakahlia governorate.  The fact that the highest DDT residue levels were found in human milk from the Delta area may be attributed to the higher agricultural activity in the Delta and the fact that mothers in these areas mainly consume fresh fruits and vegetables.  New DDT residue levels reflect previous and present exposure of the mothers to this particular pollutant *(39)*.

**Hexachlorocyclohexanes (HCH isomers):**  $\gamma$-HCH (lindane) was detected in 95% of the analyzed human milk samples. The lowest levels were found in governorates between Cairo and Assiut, and in Suez (0.00-10.00 ppb), while the higher levels (10.00-33.00 ppb) were found in the Delta area and in Alexandria. The higher levels could be a reflection of the use of lindane in agriculture and in the control of cattle ecto-parasites. Also, this might be due to the human consumption of large quantities of polluted fatty fish.  Several studies *(40)* have pointed out the presence of organochlorine residues including lindane in different food stuffs (meat, dairy products, grain, and drinks). These contaminated foods, which represent the basic staples for the donor mothers, may explain the source of lindane in their milk.

Higher organochlorine insecticide residues were identified in breast milk samples from areas around the extensive agricultural regions.  The WHO (1986) has proposed an acceptable daily intake (ADI) of 20 µg/kg body weight for total DDT compounds, and 10 µg/kg body for $\gamma$-HCH *(41)*. If these acceptable ADI values are applied assuming a mean intake of 0.8 L human milk per day by infants with the mean body weight of 5 kg, the mean concentration will not exceed 125 µg/kg whole milk for the total DDT and metabolites, and 62.5 µg/kg whole milk for lindane ($\gamma$-HCH) *(42,43)*. None of the mean levels of the organochlorine insecticides recorded in this study exceeded the WHO recommended limits.  Although the averages were below the international ADI's, however, single values were recorded which actually exceeded that average and also exceeded the WHO ADI's.  Such values are alarming that there are epidemiological exposures which might be hazardous to the new born.  A more extensive survey where more samples are analyzed especially in the highly exposed governorates is needed for precise description of the status of body burden of both organochlorines and heavy metals.

In Table (II) the level of organochlorine components in human milk recorded in this study was compared with those of other countries from 1987- 1994.  It is noteworthy to mention that the reported amount of organochlorine insecticide residues in human milk worldwide have shown a steady decline and that the amounts for 1992 and 1993 are much less than those previously reported in 1985 and 1986.  This could be attributed to the effectiveness of governmental restrictions on the use of these insecticides regions.

The results showed that population living in large cities have higher levels of lead but lower levels of pesticide residues while those living in rural areas have lower levels of lead and higher levels of pesticides suggesting that levels of pollutants in human milk may serve as a good biomarker for body burden exposure.

**Acknowledgment**
This work was funded by the EPA Grant # CR 818220-02-5.

**Literature Cited**

1. Hernández, L.M.; Fernández, M.A.; Hoyas, E.; González, M.J. and Garcia, J.F., *Bull. Environ. Contam. Toxicol.* **1993**, *50*, pp. 308-315,
2. Davis, M. J. and Grant, D. L. In: *Similarities and Differences Between Children and Adults: Implication for Risk Assessment*; Guzelian, S.; Henry, C. and Olin S.S. Eds.; Washington D.C., ILSI Press.1986, pp.
3. Dogheim, S.M.; Almaz, M.M.; Kostandi, S.N; and Hegazy, M.E. *J. Assoc. Off. Anal. Chem.* **1988**, *71*, pp. 872-874.
4. Dogheim, S.M; El-Shafeey, M.; Afifi, A.M.; and Abdel-Aleem, F.E, *J. Assoc. Off. Anal.Chem.* **1991**, *74*, pp. 89-91.
5. Vahter, M. and Slorach, S. *Exposure Monitoring of Lead and Cadmium.* An International pilot study within the WHO/UNEP Human exposure Assessment Locations (HEAL) programme, World Health Organization and the United Nations Environment Programme, Nairobi, (1990) p.82.
6. Brevik, E. *Bull. Environm. Contam. Toxicol.* **1978**, *19*, pp.281-286.
7. Mahaffey, K.R. *Pediatrics* **1977**, *59* pp.448-456.
8. Murthy, G.K. and Rhea, S.U. *J. Dairy Sci.* **1971**, *54*, pp.1001-1005.
9. Lamm, S.; Cole, B.; Gly, K.; and Ullman, W. *N.Engl. J. Med.* **1973**, *289*, pp. 574-575.
10. Lamm, S. and Rosen, *Pediatrics* **1974**, *54*, pp.137-141.
11.Ryu, J. E.; Ziegler, E.E.; Nelson, S.E.; and Formon, S. J. *J. Pediatr.* **1978**, *93*, pp.476-478.
12. Walker, B. *J. Food Protect.* **1980**, *43*, pp.178-179.
13. Rabinowitz, M.; Leviton, A. and Needlemann, H. *Arch.Environm. Health* **1985**, *40*, pp.283-286.
14. Rockway, W.S.; Weber, W.; Lei, Y.K. and Kemberling, R.S. *Int. Arch. Occup. Environ. Health* **1984**, *53*, pp.181-187.
15. Chatranon, W.; Chavalittamrong, B.; Kritalugsana, S. and Pringsulaka, P. *Southeast. Asian. J. Trop. Public Health* **1978**, *9*, pp.420-422.
16. Larsson, B.; Slorach, A.; Hagman, U.; and Hofwander, Y. *Acta Paediatr.Scand.* **1981**, *70*, pp.281-284.
17. Kovar, Z.; Strelow, D.C.; Richmond, J. and Thompson, G.M. *Arch. Dis. Child.* **1984**, *59*, pp.36-39.
18. Ong, N.C.; Phoon, O.W.; Law, Y.H.; Tye, Y. and Lim, H.H. *Arch. Dis. Child.* **1985**, *60*, pp.756-759.
19. Sternowsky, J.H. and Wessolowski, R. *Arch.Toxicol.* **1985**, *57*, pp.41-45.
20. Namihira, D.; Saldivar, L.; Pustilnik, N.; Carreon, G.J.; Salinas, M.E. *J. Toxicol. Environ. Health* **1993**, *38*, pp.225-232.
21. Ding, H.C.; Lui, H.J.; Sheu, Y.L. and Chang, T.C.. *J. Food Drug Anal.* **1993**, *1*, pp.265-271.
22. Tiran, B.; Rossipal, E.; Tiran, A.; Karpf, E. and Lorenz, O. *Trace Elem. Electrolytes* **1994**, *11*, pp.42-45.

23.Matuo, Y.K.; Lopes, J.N.C.; Casanova, I.C.; Matuo, T. and Lopes, J.L.C. *Arch. Environ. Contam. Toxicol.*1992, *22*, pp.167-175.

24. Dewailly, E.; Nantel, A.; Weber, J.P. and Meyer, F. *Bull. Environ. Contam. Toxicol.* 1989, *43*, pp.641-646.

25.Mussalo-Rauhamaa, H.; Pyysalo, H.; Antervo, K. *J. Toxicol. Environ. Health* 1988, *25*, pp.1-19.

26.Wickstrom, K.; Pyysalo, H. and Srimes, M.A., *Bull. Environ. Contam. Toxicol.* 1983, *31*, pp.251-256.

27.Schecter A.; Fuerst, P.; Krueger, C.; Meemken, H.-A.; Groebel, W. and Constable,J. *Chemosphere* 1989, *18*, pp.445-454.

28.Fytianos, K.; Vasilikiotis, G.; Weil, L.; Kaviendis, E. and Laskaridis, N. *Bull. Environ. Contam. Toxicol.* 1985, *34*, pp.504-508.

29. Ip, H.M.H. and Phillips, D.J.H., *Arch. Environ. Contam. Toxicol.* 1989, *18*, pp. 490-495.

30.Zaidi, S.S.A.; Bhatnagar, V.K.; Banerjee, B.D.; Balakrishman, G. and Shah, M.P., *Bull. Environ. Contam.Toxicol.*1989, *42*, pp.427-430.

31.Tanabe, S.; Gondaira, F.; Subramanian, A.; Ramesh, A.; Mohan, D.; Kumaran, P., Venugopalan, V.K. and Tatsukawa, R., *J. Agric. Food. Chem.* 1990, *38*, pp. 899-903.

32.Weisenberg , E.; Arad, I.; Grauer, F.; Sahm, Z. *Arch. Environ. Contam. Toxicol.* 1985, *14*, pp.517-521.

33.Dommarco, R.; Di Muccio, A.; Camoni, I. and Gigli, B. *Bull. Environ. Contam. Toxicol.* 1987, *39*, pp.919-925.

34. Alawi, M.A.; Ammari, N. and Shuraiki, Y. *Arch. Environ. Contam. Toxicol.* 1992, *23*, pp.235-239.

35.Sitarska, E.; Gorski, T.; Ludwicki, J.K. *Bull. Environ. Contam. Toxicol.* 1987, *39*, pp.756-761.

36. Baluja, G.; Hernandez, L.M. and Rico, M.C. *Bull. Environ. Contam. Toxicol.* 1982, *28*, pp.573-577.

37.Karakaya, A.E.; Burgaz, S.; Kanzik, I. *Bull. Environm. Contam. Toxicol.* 1987, *39*, pp.506-510.

38.Spicer, P.E.and Kereu, R.K. *Bull. Environm. Contam. Toxicol.* 1993, *50*, pp.540-546.

39.Kanja, L.; Skåre, J.U; Nafstad, I.; Maitai, C. K.and Løkkken, P.,*J. Toxicol. Environm. Health* 1986, *19*, pp.449-464.

40.Kucinski, B. *Ciência Hoje* 1986, *4*, pp.58-62.

41.WHO *Env. Health Criteria* 1986 *59*, p.73.

42.WHO, *The Quantity and Quality of Breast Milk*. World Health Organization, Geneva, 1985.

43. Cordle F. and Kolbye, A., *Cancer* 1979, *43*, pp.2143-2150.

# Chapter 10

# Carnitine and Its Esters as Potential Biomarkers of Environmental−Toxicological Exposure to Nongenotoxic Tumorigens

John E. Garst

3008 Del Prado, Alamogordo, NM 88310

Recently it has become clear that *l*-carnitine (Cn) and/or its esters can protect against a wide variety of toxic substances. This protection arises both from their critical roles in normal and abnormal metabolism and as signal and regulatory molecules. Peroxisomal proliferating agents (PPA) are structurally diverse and include widely-used drugs, industrial-, and agrochemicals. PPA cause hepatomegaly by increasing both the number and size of liver cells. PPA also produce hepatic tumors in rodents, but PPA are termed non-genotoxic tumorigens because they do not directly affect DNA. Enhanced gene-expression and reduced natural cell death (apoptosis) are increasingly suspected of involvement in non-genotoxic tumor production. PPA-mediated liver tumors may arise because of the ability of PPA to induce conversion of free Cn mostly to acetylOCn. Because perturbations diminishing free Cn and increasing acetylOCn have the combined effect of fostering gene expression and of impeding apoptosis, respectively, they offer a direct explanation of PPA-induced hepatic tumors. Hence, perturbations in Cn are postulated to define in molecular terms the events underlying *non-genotoxic tumor promotion.*

The name carnitine (Cn), which is derived from carnivore, reflects its isolation in 1906 from animal flesh. Twenty years later, Cn was fortuitously assigned the correct structure, 3-hydroxy-4-(trimethylammonium)-butanoate. But, for nearly fifty years after its isolation, Cn was a substance without a function ( *1* ). Cn is a vitamin for some insects ( *1* ). Although omnivorous humans may obtain up to 7/8 of their Cn from dietary meat and meat products, technically Cn is not a vit-amin for humans, because considerable Cn can be provided by endogenous bio-synthesis ( *2* ). Cn is biosynthesized from lysine and methionine by protein-lysine N-trimethylation, protein hydrolysis, hydroxylation at the 3 position of ε-N-trimethyllysine, a chain-splitting aldol cleavage, an oxidation, and a hydrox-ylation at the β-position of γ-butyrobetaine ( *2,3* ). Ascorbic acid (vitamin C) and iron are critical cofactors for Cn biosynthesis ( *ref 12 in citation 2* ). Serum concentrations of free Cn are significantly lower in lactovegetarians and strict vegetarians, but the functional significance of this reduction is unclear ( *2* ).

NOTE: Please see Legend of Symbols on page 136 for abbreviations.

0097−6156/96/0643−0126$15.00/0

## Cn and Cn Esters Can Protect Against Diverse Toxic Substances

Recent evidence suggests that Cn and/or particularly short-chain Cn esters can moderate or even ameliorate adverse effects of diverse chemicals and drugs ( *4* ). These chemicals and drugs include halothane, 2,4,5-trichlorophenoxyacetic acid (2,4,5-T), 1-methyl-4-phenyl-1,2,3,6-tetrahydropyridine, dipropylacetic acid (i.e. valproic acid), adriamycin, aflatoxin $B_1$ , and even hyperbaric oxygen. Moreover, the toxicities of ammonia, carbon monoxide, thioacetamide, heavy metals like cadmium and mercury, and even the antibiotic cephaloridine seem mediated, at least in part, by actions that somehow affect Cn. Citations for the toxic effects of these compounds, the protection afforded by Cn and/or Cn esters, and other relevant actions of Cn and Cn esters were discussed in a 1995 Society of Toxicology symposium ( *4* ) and a manuscript on the subject will appear in a forthcoming issue of *Toxicology and Applied Pharmacology*.

Since 1990 several monographs ( *5-6* ) and articles ( *7-11* ) have reviewed the biology of Cn and its esters. Understanding how they can moderate or protect against toxic agents requires knowledge both of the fundamental roles of Cn and Cn esters 1) in buffering free coenzyme A (HSCoA), 2) in normal mitochondrial and peroxisomal metabolism, and 3) in detoxification, as well as recognition that Cn and Cn esters can have powerful signaling and regulatory actions.

## Fundamental Roles of Cn and Cn Esters

**Cn and Cn Esters Buffer the AcylSCoA/Free HSCoA Ratio.** Most carboxylic acids are readily converted to the corresponding acylcoenzyme A esters in the peroxisomes, the outer mitochondrial membrane, or the mitochondrial matrix. This multi-step transformation consumes ATP and free coenzyme A (HSCoA) and utilizes a chain-length specific acylSCoA synthase ( *9,12-15* ). An overload of acylSCoA esters, such as can occur after eating, would deplete all cytosolic free HSCoA ( *9,14,16* ), were it not for several mechanisms, including the fact that many acylSCoA can be readily converted to acylOCn esters. Optically active R-(L)-Cn has biological importance mostly, but not solely, because the Cn alcohol group can transesterify acylcoenzyme A esters (RCOSCoA) (equation 1) and facilitate their transport into the mitochondria as acylOCn esters for energy-producing β-oxidation ( *16,17* ). These acylOCn esters retain the ATP-derived

$$
\begin{array}{ccccc}
\text{Acylcoenzyme A} & & & \text{Coenzyme A} & \\
& & & & \\
\text{COO}^- & & & \text{COO}^- & \\
\text{C} & & & \text{C} & \\
\text{HOC} & + & \text{RCOSCoA} \; \rightleftarrows \; & \text{RCOOC} & + \quad \text{HSCoA} \qquad (1) \\
\text{C} & & & \text{C} & \\
\text{+N(Me)}_3 & & & \text{+N(Me)}_3 & \\
\text{R-(L)-Carnitine} & & & \text{R-(L)-Acylcarnitine} &
\end{array}
$$

high energy of the acylSCoA ester yet can be converted back to the acylSCoA ester as metabolism utilizes and reduces foodstuff supplies. Functionally like a pH buffer that sequesters or releases protons to maintain a relatively constant hydrogen ion concentration, formation of acylOCn esters from acylSCoA esters creates a buffering action that ensures a supply of free HSCoA, which is essential for many biochemical transformations ( *9,14,18* ). Besides being directed toward oxidation, the acylOCn ester can be transferred as an energy store to other organs. Cn esters can also be excreted, generally in the urine.

**Cn and Cn Esters in Mitochondrial and Peroxisomal Metabolism.** Cn is essential for synthesis of a variety of esters of which the acetyl- (AcOCn); propionyl- (PrOCn), isovaleryl- (IvOCn), octanoyl- (OcOCn) and palmitoyl- (PmOCn) esters seem to be the most important. These acylOCn esters are formed by an array of chain-length specific acylOCn transferases that catalyze the Cn-mediated transesterification of acylSCoA in the left to right forward reaction of equation 1, or the coenzyme A (HSCoA)-mediated thiolysis of Cn esters, producing the acylSCoA ester as the reverse reaction of equation 1 ( 16 ). Various acylcarnitine transferases and disposition enzymes, for which Cn and/or Cn esters are substrates, along with Cn and its esters, constitute what is referred to herein as the *Cn system.*

The various acylcarnitine transferases have a diverse localization within the cell. Inside their outer membrane, mitochondria contain a Cn palmitoyl transferase I (CPT-1) that acts on long-chain acylSCoA esters (Figure 1). Malonyl-SCoA, derived from food, down-regulates CPT-I activity and mitochondrial fatty acid β-oxidation, fostering fat storage ( 16 ). Cn palmitoyl transferase II (CPT-II), located on the inside of the inner mitochondrial membrane, performs essentially the reverse reaction of CPT-I, reforming the acylSCoA using the matrix supply of mitochondrial impermeable HSCoA (Figure 1). Lastly, mitochondria contain a soluble matrix Cn acetyltransferase (CAT). CAT processes short- and medium-chain carboxylic acids that can directly partition into the mitochondria (Figure 1). Other organelles also contain Cn enzymes. Very long-chain fatty acids undergo β-oxidation as the corresponding acylSCoA ester in the peroxisomes, reducing the chain-length by two carbon (AcSCoA) increments. Oxidation generally stops with octanoylSCoA (OcSCoA; 19 ). Hence, peroxisomes contain both CAT and Cn octanoyltransferase (COT) activities to convert peroxisomally-produced AcSCoA and OcSCoA to the corresponding acylOCn ( 16, 19 ). The endoplasmic reticulum also contains a Cn medium/long-chain transferase activity ( 20 ), the purpose of which is not well understood.

Various other enzymes also regulate the disposition of Cn or Cn esters. Many cells contain a sodium-dependent system for free Cn uptake located on the outer membrane (not shown in Figure 1) ( 6 ). Impairment or deficiency of this uptake system from the renal brush border can lead to reduced systemic Cn uptake and adverse systemic Cn deficiency (CnD; 21 ).

Regardless of their acyl chain-length, acylOCn esters enter the mitochondria only via a concentration-dependent Cn translocase (CnT), located in the inner mitochondrial membrane (Figure 1). Normal concentrations of acylOCn esters allow CnT to move cytosolic/peroxisomal Cn esters through the otherwise impenetrable inner mitochondrial membrane into the matrix while simultaneously exporting free Cn from the mitochondria to the cytosol/peroxisomes (Figure 1). Available evidence suggests there is but one form of CnT ( 22 ). In view of the importance of CnT for fatty acid β-oxidation and energy production, the three known cases of human CnT deficiency (< 1% of normal CnT) have been understandably fatal ( 22 ). Signs varied from hypothermia to seizures and neurological disorders.

**Cn and Cn Esters in Detoxifying Metabolism.** Some acylOCn esters, whether originating as xenobiotic- or as endogenous carboxylic acids, are directly excreted ( 2,16,23 ). This occurs because the compound itself cannot be further metabolized or because metabolism of the compound is prevented, often by genetic enzyme deficiencies. For example, unmetabolizable trimethylacetic (e.g. pivalic) acid is excreted in the urine as pivaloylOCn ( 18 ). Bieber *et al.* ( 23 ) discuss Cn as a detoxicant of other non-metabolizable acylSCoA esters and provide some information on Cn assays elsewhere in this volume. People with genetic deficiencies of medium-chain acylSCoA dehydrogenase, which normally

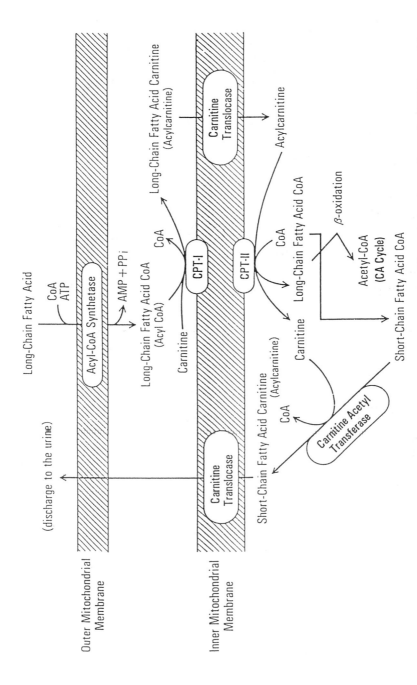

Figure 1. Schematic showing the mitochondrial Cn enzymes. CPT-I, carnitine palmitoyl transferase I; CPT-II, carnitine palmitoyl transferase II; CA, citric acid cycle (Adapted from ref. 31).

oxidizes OcSCoA, or of isovaleroylSCoA dehydrogenase, which normally oxidizes IvSCoA, can excrete large quantities of OcOCn and IvOCn, respectively ( *24,25* ). Missing enzymes are clearly the underlying cause. But excretion of Cn esters reflects the accumulation of acylSCoA esters and/or a deficiency of free HSCoA ( *14,18,25* ). Either process can cause various toxic effects and are exemplified by the toxicities of valproic acid and hypoglycin ( *18* ). But, because free Cn can transesterify some acylSCoA esters and because it can shuttle acyl residues between intracellular-, intercellular-, and even interorgan compartments ( *16,17* ), large-scale excretion of Cn esters can precipitate a systemic CnD ( *23,26-29* ). CnD can cause many effects, including muscle weakness, fatigue, and malaise ( *7-10,30-31* ).

### Signal and Regulatory Roles of Cn and Cn Esters

Signal potential for Cn and particularly Cn esters is substantial, but little recognized. Stryer discusses similarities between a number of biochemical signal molecules ( *15* ). He notes that many signal molecules are formed in a single step from a major metabolic intermediate (see Table I). Stryer also stresses the irreversibility of the synthesis and degradation of these signal molecules as being critical to controlling their concentrations.

Table I.  Signal Molecules Formed in One-step from a Major Intermediate

| Precursor Molecule | Signal Molecule |
| --- | --- |
| ATP | Cyclic AMP |
| GTP | Cyclic GMP |
| Glutamate | GABA[a] |
| Tyrosine | DOPA[b] |
| Choline | Acetylcholine |
| Arachidonate | Prostaglandin $PGE_2$ |

[a] $\gamma$-aminobutyric acid
[b] 3,4-dihydroxyphenylalanine
SOURCE: Adapted from ref 15.

   Cn esters meet one, but not both of Stryer's criteria. Like the other signal molecules, Cn esters are formed and cleaved in a single step involving relatively abundant free Cn by a set of chain-length specific acylcarnitine transferases. Cn and Cn esters have been suggested to act as integrators and regulators of the cellular response to xenobiotics ( *32* ). Accordingly Stryer's second point, namely the irreversibility of signal synthesis and degradation, cannot be retained. Since the effects of Cn esters are based on their concentration, to serve a integrative, regulatory role Cn and Cn esters must be free to traverse intracellular, intercellular, and interorgan compartments to affect equilibrium. Their signal or message can be diminished or even retracted if target compartments can reverse their formation or facilitate their excretion.
   That excretion may be an important aspect of Cn ester signaling is suggested because after years of study no process has been identified by which the valuable N-methyl groups from Cn and Cn esters are recovered. Like choline, Cn

and Cn esters contain three N-methyl groups. But choline undergoes stepwise degradation to betaine, N,N-dimethylglycine, and N-methylglycine, recycling each methyl group to S-adenosylmethionine ( *33* ). In contrast there is no evidence for recycling of Cn and/or Cn ester N-methyl groups. This suggests that recycling of these N-methyl groups might interfere with some inate function. If that function were detoxification of acylSCoA esters by excretion (as noted in the preceding section), the absence of excretion would accrue toxic, even lethal acylSCoA esters as occurs with certain genetic enzyme deficiencies. This might explain why evolution has not selected to recycle these N-methyl groups.

For Cn and Cn esters to act as integrators and regulators of the cellular response to xenobiotics as proposed ( *32* ), these substances must have biological actions. The actual signal and regulatory properties of Cn and Cn esters are diverse and include antioxidant and antiapoptotic actions, effects on known signal and regulatory pathways, as well as protein translation and nucleic acid transcription as well as other aspects of gene expression.

**Antioxidant Properties of Cn and Cn Esters.** The unusual "antioxidant" action of AcOCn has been discussed ( *34-35* ). In another example AcOCn can raise glutathione (GSH) concentrations and reduce lipofuscin in brain ( *36-37* ). But exactly how these "antioxidant" actions are mediated is unclear. They could stem from the ability of acetylOCn (AcOCn) to act as an energy reservoir in sperm ( *38* ), in flight muscles and heart ( *16,39* ), and possibly in peripheral mononuclear cells ( *40* ). Through catalysis by CAT AcOCn, serving as an energy reservoir, may maintain an AcSCoA pressure for citric acid cycle oxidation forming NADH and for NADH electron transport chain oxidation forming ATP. Via the malate shuttle, NADH and ATP can be converted to NADPH and provide reducing equivalents for maintenance of GSH ( *41* ). Hence, facilitating the mitochondrial citric acid cycle and the malate shuttle could explain the "antioxidant" action of AcOCn. Besides increasing mitochondrial efficiency, decreasing mitochondrial oxidant production has also been suggested for AcOCn ( *35* ).

Alternatively, the indirect "antioxidant" action of AcOCn could involve another effect of massive importance. The AcOCn "antioxidant" action might stem from its ability to be converted to free Cn and for Cn to prevent the mitochondrial permeability transition (MPT). The MPT involves the opening of large pores in the mitochondria that allow matrix contents to spill into the cell cytosol. Opening of these pores can be associated with mitochondrial swelling, but swelling may not always occur, even when the pores have opened ( *42* ). With or without swelling, development of the MPT causes release of GSH (without conversion to its disulfide), rapid oxidation of escaping pyridine nucleotides, and consequently depletion of ATP and ADP. Hence, occurrence of the MPT can dramatically reduce mitochondrial antioxidant protection in several ways ( *42* ). Cyclosporin A (at 1μM) is a very potent inhibitor of the MPT; it prevents loss of mitochondrial GSH, oxidation of mitochondrial pyridine nucleotides, and depletion of ATP and ADP. However, the less potent, but naturally abundant Cn (at 1 mM) may be the more relevant regulator ( *43* ). And MPT inhibition may underlie protection by Cn against the ATP-depleting, Parkinson-like disease-inducing agent 1-methyl-4-phenyl-1,2,3,6-tetrahydropyridine ( *43* ).

AcOCn is not the only Cn ester antioxidant. The propionyl ester (PrOCn), formed from PrSCoA generated by β-oxidation of odd-length fatty acids, also has marked antioxidant effects. They arise from the specific ability of PrOCn, but not AcOCn, to chelate iron and prevent Fenton-like reactions that can foster lipid peroxidation ( *44* ). While the short-chain acylOCn work differently and are more potent, studies using red blood cells, devoid of mitochondria, revealed the long-chain ester (PmOCn) to be a secondary antioxidant, critical to rebuilding oxidant-damaged cell membranes that can restrain oxidative processes ( *45* ).

**Antiapoptotic Properties of Cn and Cn Esters.** Apoptotic cell death was investigated in a teratocarcinoma cell line ( *46* ). Addition of fetal calf serum to the teratocarcinoma cell line reduced DNA fragmentation to less than 5% of the extent without fetal calf serum. Addition of AcOCn alone enhanced cell survival and retarded both DNA fragmentation and nuclear condensation by half its extent without the fetal calf serum. But AcOCn can only delay, not prevent apoptosis ( *46* ). The authors suggested that apoptosis in these cells does not seem to involve an oxidant mechanism, because addition of FCS does not change GSH concentrations ( *46* ). The ability of AcOCn to delay apoptosis is also interesting because various other work confirms that increasing AcOCn can reduce DNA fragmentation ( *47-49* ). Perhaps the apoptosis retarding action of AcOCn arises because it can sustain the citric acid cycle and the electron transport chain or, alternatively, because AcOCn can be readily converted to Cn, which can prevent the mitochondrial permeability transition ( *43* ). In this regard a structure-activity relationship study of various Cn derivatives found that only those Cn esters that could be converted to free Cn could prevent hepatic apoptosis ( *50* ). But the facile interconversion between AcOCn and free Cn and ambiguity about the antiapoptotic mechanisms were not addresssed in this study of derivatives.

**Other Signaling and Regulatory Properties of Specific Cn Esters.** Protein kinase C (PKC) is a critical enzyme regulating receptor function, cellular differentiation, and carcinogenesis. Some acylSCoA esters can potentiate the activity of PKC ( *12* ). If the acylOCn ester were without PKC activity, the buffering action of Cn that converts PKC-potentiating acylSCoA esters to the corresponding acyl-OCn ester would moderate any increase in PKC activity by reducing acylSCoA ester concentrations. But large, bulky acylSCoA, such as those derived from peroxisome proliferating agents that often have PKC activity ( *12* ), seem precluded from forming acylOCn esters.

Other specific Cn esters also can also serve as cellular signals. For example, IvOCn is a controlling factor in limiting lysosomal proteolysis-mediated muscle degradation ( *51* ). IvOCn is also associated with a specific form of proteolysis; IvOCn can activate a form of the calcium-activated neutral proteases, called the calpains-II ( *52* ). The calpains-II target for destruction those proteins involved in the cytoskeleton that connect the mitochondria and other organelles ( *53* ).

In contrast, the long-chain C-16 PmOCn plays integral roles in the cell as a product of CPT-1 and as substrates for CnT and CPT-II. But PmOCn itself also has clear signaling and regulatory properties. While conversion of acylSCoA to acylOCn might moderate any acylSCoA-linked increases in PKC, if the acylOCn ester is devoid of PKC action, one specific acylOCn, namely the long-chain Pm-OCn, is also a powerful inhibitor of PKC ( *54-55* ). Because PmOCn can accrue under a variety of physiological conditions, its antiPKC actions could reflect evolutionary protection. But PmOCn also has detrimental properties, which include the inhibition of CAT ( *23* ), increased ion flux ( *56-58* ), uncoupled cells ( *59* ), inhibited gap junction transmission ( *60* ), and a strong arrhythmic effect on heart ( *61* ).

**Cn and Cn Esters Can Affect Protein Translation and Nucleic Acid Transcription.** The ability of Cn and its esters to affect gene expression (GE) and protein production affords these substances additional regulatory potential. Several instances have been identified where Cn and/or its esters control protein translation ( *62* ), protein function ( *63-64* ), key cellular receptors ( *65* ), and even GE. Association between Cn and Cn esters and GE became most evident with the recent discovery of a genetically CnD mouse strain ( *66* ); these juvenile visceral steatosis (*jvs*) mice show hepatic fat deposition, hepatomegaly, and impaired urea cycle enzymes. Administration of Cn alleviates these problems; moreover, it actually restores to normal the reduced concentrations of mRNA's for the affected urea cycle enzymes. CnD *jvs* mice also produce high concentrations of the mRNA's for the tumor-linked proto-oncogenes (POG) c-*jun* and c-*fos* and also for $\alpha$-fetoprotein and aldolase A, which are typically expressed in undifferentiated or de-differentiated hepatocytes ( *67* ). Production of the AP-1 transcription factor (a tetradecanoylphorbol acetate-responsive binding element) might underlie the problems encountered by *jvs* mice ( *66* ).

CnD and Cn ester formation could increase protein products by increasing mRNA transcription or by reducing mRNA degradation OR they could increase protein translation or decrease protein degradation. And CnD and/or Cn esters seem to act in many ways to affect protein concentrations. Not only does CnD increase mRNA for c-*jun* and c-*fos* POG, but AcOCn, for example, enhances the integrity of cytochrome C oxidase by increasing subunit I mRNA ( *64* ). The specific mechanism underlying either increase in mRNA is unclear, however. A reduction in antioxidant, short-chain acylOCn esters may also be relevant to transcription, since oxidative conditions can activate transcription factors that produce c-*jun,* c-*fos* and other POG protein products ( *68-69* ).

Both Cn and AcOCn have been shown to prevent the degradation of specific proteins. In this regard feeding Cn with an ethanol (EtOH) diet increased microsomal cytochrome P-450 CYP2E1 isozyme content significantly higher than that afforded by EtOH alone ( *70* ). In this case Cn had no effect alone and it did not affect mRNA concentrations. Exactly how Cn affords this increase in CYP2E1 protein is not known. However, a recent discovery is pertinent to this issue. AcOCn acts as an antioxidant to prevent the degradation of certain post cardiac arrest brain proteins ( *71* ). This discovery suggests that the increase of CYP2E1 P-450 protein by EtOH + Cn might involve addition of Cn to the EtOH metabolite acetic acid to form AcOCn. Then, AcOCn could act as an antioxidant to protect the P-450 isozyme against oxidative processes.

**Cn and Cn Esters Can Affect GE.** Recent evidence suggests that Cn esters have the potential to affect DNA transcription directly. DNA is wound about a multi-protein nucleosome core that is composed of histones, containing many strongly basic amino acids. These basic amino acids bind the strongly acidic phosphate anions of the phosphate groups that connect the DNA bases, which wrap around the nucleosome. In addition to binding the protein core, these histones contain tails with more basic amino acids. These tails reach around to cover the top surface of the DNA. N-Acetylation of the basic amino acids in this tail removes their positive charge, releasing the acetylated tails from their grasp on the phosphate anions of DNA, enabling nucleosomal conformational changes and/or faciliting transcription factor binding. Histone tail release must happen for DNA transcription to occur and consequently, acetylation of histones is strongly correlated with actively transcribed DNA ( *72-73* ).

In principle, deacetylation of these tails restores histone-to-DNA binding and would terminate the active transcription process. Hence, inhibition of deacetylation should foster the transcription process. Much evidence indicates that 10 mM sodium butyrate inhibits deacetylation of histones. And, prevention of deacetylation by butyrate causes hyperacetylation of histone H4 and produces strong, but uncoordinated gene activity resulting in a dominant lethal mutation in the fruit fly *Drosophila melanogaster* ( *74* ). Recently PrOCn, and to a lesser extent AcOCn have been found to inhibit histone deacetylation nearly as well as butyrate ( *74* ). Hence, formation of large amounts of short-chain acylOCn should, in principle, foster not only CnD but GE, both of which may enhance DNA accessibility to transcription factors. This possible new role for short-chain acylOCn in transcription also raises another question. Just as PrOCn was found to inhibit histone deacetylation, might butyrylOCn mediate butyrate actions and might longer Cn esters, like peroxisomally-produced OcOCn, have similar effects on transcription?

## Cn and Cn Esters as Biomarkers of Environmental/Toxicological Exposure

Free HSCoA and acylSCoA esters have many signal and regulatory actions ( *12, 14,18* ), but their actions are constrained to cytosolic/peroxisomal or to mitochondrial compartments by impermeable membranes. By virtue of its buffering function, the Cn system reflects the status of the free and acylSCoA in those compartments. But unlike the thioesters, Cn and Cn esters can use the CnT to freely cross the membranes. Hence, it is not unreasonable that Cn and Cn esters have the potential to signal and regulate, as the above discussion illustrates. Following is but one example, albeit an example with hundreds of causative substances, where perturbations in the steady-state balance between free Cn and its esters, termed Cn perturbations (CnP), illustrates the potential of CnP as a biomarker of non-genotoxic tumorigenicity.

**Cn and Cn Ester Perturbations May Explain Peroxisomal Proliferating Agent(s) (PPA)-Induced Tumorigenicity.** Structurally-diverse PPA, which include the important drugs aspirin, ibuprofen, and gemfibrozil and the herbicides bifonazole, dicamba, and the phenoxyacetic acids, cause hepatomegaly by increasing both the number and size of liver cells. PPA generally produce hepatic tumors in rats and mice, which do not involve alteration of DNA ( *75-76* ). Hence, understanding the mechanism by which these non-genotoxic agents produce their effects is of great importance to risk-assessment for humans ( *75-76* ).

**PPA and POG Expression.** PPA uniformly increase expression of POG *c*-Myc, but they also increase expression of H-*Ras* (77-78 ). Induction of c-*myc* protein often follows c-*jun* and c-*fos* proteins ( *68,79* ). Induction of c-*jun* and c-*fos* also occurs in CnD *jvs* mice ( *67* ). PPA-induced expression of *c*-myc may be connected to the transient CnD and acylSCoA mediated-increases in calcium and PKC or to oxidative reactions ( *68* ) that increase formation of the oxidation-derived pigment lipofuscin after PPA exposure ( *76* ). Interestingly, CnD shows considerable pathological parallel to peroxisomal proliferation (PP); animal and human CnD shows a pleiotropic response involving hepatomegaly and liver PP ( *30* ), and other common effects ( *7,14,25,65* ).

**PPA Reduce Hepatic Apoptosis.** Besides effecting changes that increase POG expression, PPA, like nafenopin for example, also decrease hepatic apoptosis ( *80* ). Other potent PPA, including Wyeth 14,643, have a similar effect *in vitro* (Roberts, R. A., Zeneca Central Toxicology Laboratory, personal communication, 1995). The immediate and prolonged increase in AcOCn caused by PPA may be a major factor relevant to tumor development, because AcOCn itself can retard normal hepatic cell apoptosis.

**Tumor Development Often Seems Linked to Increased POG Expression and Reduced Apoptosis.** Increasingly tumor development seems correlated with both increased POG expression and reduced apoptosis. Three studies illustrate this point ( *81-83* ). Immortalized rat lung fibroblasts were transfected with plasmids containing three genomes or combinations thereof. Cell proliferation rates of thirteen transformed genomes were found to vary relatively little, but apoptosis rates differed considerably and were the biggest determinant of cell population increases. These genomes were injected into immune-suppressed mice to produce tumors. Tumors derived from fibroblast cell lines having high rates of apoptosis grew slowly. Tumors derived from fibroblast cell lines with low rates of apoptosis grew rapidly. Hence, the authors concluded that the specific POGs being expressed seem to determine the apoptosis rate ( *81* ).

In neuroblastomas *bcl-2* POG promotes cell growth by preventing apoptosis. In humans with this type of tumor, *bcl-2* antibodies were most strongly associated with "unfavorable histology," according to the Shimada classification system, and with Neuroblastoma N-*myc* gene amplification suggesting that expression of this POG may be relevant to tumor development and/or progression of malignancy ( *82* ).

PC3 human prostate cells contain a variant called PC3-(R) that is round and antineoplastic drug-resistant. While PC3 cells undergo apoptosis in response to such drugs, PC3(R) cells do not. PC3(R) cells do, however, express 2-3 times more *bcl-2* than the parental cell line and also overexpress POG c-*jun*, c-*myc*, and H-*ras*. Hence, expression of *bcl-2* and/or the other POG would seem to contribute to the differential apoptosis and chemosensitivity of the round versus parental cell line ( *83* ).

**PPA Cause CnP.** After weeks of exposure, PPA substantially increase both free Cn and AcOCn and dramatically raise hepatic CAT activity. Major increases in CAT activity is one of several definitional biomarkers of a PPA ( *76* ). Other studies of PPA have found decreased free HSCoA and increased PPA-derived acylSCoA, as well as decreased free Cn and increased AcOCn immediately after PPA exposure ( *26-28,84* ). This suggests a short period of hepatic CnD, followed by a substantial long-term increase in free Cn. Hence, PPA uniformly affect changes composed of an early CnP (transient CnD with increased AcOCn) as well as long-term CnP (increases in both Cn and AcOCn).

**CnP May Mediate PPA-POG Expression and PPA-Apoptosis.** GE, apoptosis, and CnP occur in response to PPA administration. Since early CnP (transient CnD with increased AcOCn) can have the same effect on GE (increased) and apoptosis (decreased) as does PPA, CnP may underlie PPA-mediated tumors. Hence, it is possible that increased POG/GE and decreased apoptosis may be better defined in terms of these molecular CnP and therefore CnP could better explain PPA tumorigenicity than other proposed theories ( *32* ). Moreover, such CnP support the premise that the Cn system may be an integrator/regulator of the cellular response by the organism to its environment ( *32* ).

**CnP Reflect Important and Subtle, but not Sweeping Changes.** Although more details could be presented, sufficient basis has been provided to make clear this hypothesis that CnP have the potential to be highly relevant biomarkers of exposure at least for the non-genotoxic PPA class of hepatic tumorigens.

In the effort to elucidate and to utilize biomarkers, it is often tempting to overestimate their importance. In this regard it is interesting that not all toxic substances cause dramatic perturbations of Cn and Cn esters. For example, a toxic dose of carbon tetrachloride, which exerts its effect after P-450 activation to covalent binding intermediates, raises free Cn and short- and long-chain acylOCn esters but does not greatly preturb the acylOCn to free Cn ratio in liver or serum ( *85* ). Addition of supplemental free Cn does, however, increase the acylOCn/free Cn ratio in serum, although it is without effect in liver ( *85* ). Hence, it seems reasonable that the CnP mentioned herein may have interesting utility as a possibly selective biomarker of subtle, non-genotoxic changes. Substantial research is necessary, however, to determine the full potential and scope of CnP in non-genotoxic tumorigenicity.

## Legend of Symbols

CAT, carnitine acetyl transferase; CnD, carnitine deficiency; CnP, carnitine perturbations; CnT, carnitine translocase; COT, carnitine octanoyltransferase; CPT-I or CPT-II, carnitine palmitoyl transferase I or II; GE, gene expression; GSH, glutathione; MPT, mitochondrial permeability transition; PKC, protein kinase C; POG, proto-oncogene; PPA, peroxisome proliferating agents.

Cn, carnitine; acylOCn, acylcarnitine; AcOCn, acetylcarnitine; PrOCn, propionylcarnitine; IvOCn, isovaleroylcarnitine; OcOCn, octanoylcarnitine; PmOCn, palmitoylcarnitine.

HSCoA, coenzyme A; acylOSCoA, acylcoenzyme A; AcSCoA, acetylcoenzyme A; PrSCoA, propionylcoenzyme A; IvSCoA, isovaleroylcoenzyme A; OcSCoA, octanoylcoenzyme A; PmSCoA, palmitoylcoenzyme A.

## Acknowledgments

The author wishes to thank 1) the Editors for their invitation to present the work in this review both orally and in written form, 2) Dr. Timothy McCurdy and Sigma Tau Pharmaceuticals for their interest and support making this work possible, 3) Dr. Loren Bieber of Michigan State University for his patient and friendly assistance, and 4) Drs. G. Stanley Smith of New Mexico State University and C. J. Rebouche of the University of Iowa for their helpful comments on various intermediate manuscripts.

## Literature Cited

1   Fraenkel, G. In *Recent Research on Carnitine. Its Relation to Lipid Metabolism* (Wolf, G., Ed.), MIT Press: Cambridge, MA. 1965, pp. 1-3.
2   Rebouche, C. J. *FASEB J.* 1992, *6*, 3379-3386.
3   Rebouche, C. J. *Toxicologist* (Society of Toxicology Abstract Issue of *Fundam. Appl. Toxicol.)* 1995, *15*, 2.
4   Garst, J. E. *Toxicologist* (Society of Toxicology Abstract Issue of *Fundam. Appl. Toxicol.)* 1995, *15*, 1.
5   Carter, A. L., Ed. *Current Concepts in Carnitine Research,* CRC Press: Boca Raton, FL, 1992.
6   Ferrari, R.; DiMauro, S.; Sherwood, G. *L-Carnitine and its Role in Medicine: From Function to Therapy,* Academic Press: San Diego, CA.
7   Anonymous 1990, *Lancet.* 1992, *335*, 631-633.
8   Scholte, H. R.; Rodrigues-Pereira, R.; de Jonge, P. C.; Luyt-Houwen, I. E. M.; Verduin, M. H. M.; Ross, J. D. *J. Clin. Chem. Clin. Biochem.* 1990, *28*, 351-357.
9   Ramsay, R. R.; Arduini, A. *Arch. Biochem. Biophys.* 1993, *302*, 307-314.
10   Tanphaichitr, V.; Leelahagul, P. *Nutrition.* 1993, *9*, 246-254.
11   Broquist, H. P. In *Modern Nutrition in Health and Disease,* 8th edn. (Shils, M. E.; Olson, J. A.; Shike, M., Eds.), Lea and Febiger: New York, NY, 1994; Vol. 1, pp. 459-465.
12   Bronfman, M.; Morales, M. N.; Amigo, L.; Orellana, A.; Nuñez, L.; Cárdenas, L.; Hidalgo, P. C. *Biochem. J.* 1992, *284*, 289-295.
13   Caldwell, J. *Biochem. Soc. Transact.* 1985, *13*, 852-854.
14   Stumpf, D. A.; Parker, W. D. Jr.; Angelini, C. *Neurology.* 1985, *35*, 1041-1045.
15   Stryer, L. *Biochemistry,* 3rd edn. Freeman: New York, NY, 1988, pp. 472-473,1027.
16   Bieber, L. L. *Annu. Rev. Biochem.* 1988, *57*, 261-283.

17 Bieber, L. L. *Toxicologist* (Society of Toxicology Abstract Issue of *Fundam. Appl. Toxicol.)* **1995**, *15*, 2.
18 Brass, E. P. *Chem. Biol. Interact.* **1994**, *90*, 203-214.
19 Osmundsen, H.; Hovik, R.; Bartlett, K.; Pourfazam, M. *Biochem. Soc. Transact.* **1994**, *22*, 436-441.
20 Chung, C. D.; Bieber, L. L. *J. Biol. Chem.* **1993**, *268*, 4519-4524.
21 Horiuchi, M.; Kobayashi, K.; Yamaguichi, S.; Shimizu, N.; Koizumi, T.; Nikaido, H.; Hayakawa, J.; Kuwajima, M.; Saheki, T. *Biochim. Biophys. Acta* **1994**, *1226*, 25-30.
22 Pande, S. V.; Murthy, M. S. R. *Biochim. Biophys. Acta* **1994**, *1226*, 269-276.
23 Bieber, L. L.; Huang, Z. H.; Gage, D. A. In: *Field Application of Biomarkers for Agrochemicals and Toxic Substances.* (Blancato, J.; Brown, R.; Dary, C.; Saleh, M., Eds.), ACS Book Symposium Series, American Chemical Society: Washington, DC. (this volume).
24 Van Hove, J. L. K.; Kahler, S. G.; Millington, D. S.; Roe, D. S, Chace, D. H.; Heales, S. J. R.; Roe, C. R. *Pediatr. Res.* **1993**, *35*, 96-101.
25 Angelini, C.; Vergani, L.; Martinuzzi, A. *Crit. Rev. Clin. Lab. Sci.* **1992**, *29*, 217-242.
26 Bhuiyan, A. K.; Bartlett, K.; Sherratt, H. S.; Agius, L. *Biochem. J.* **1988**, *253*, 337-343.
27 Deschamps, D.; Fisch, C.; Fromenty, B.; Berson, A.; Degott, C.; Pessayre, D. *J. Pharmacol. Exp. Ther.* **1991**, *259*, 894-904.
28 Thurston, J. H.; Hauhart, R. E. *Pediatr. Res.* **1992**, *31*, 419-423.
29 Guilbert, C. C.; Chung, A. E. *J. Biol. Chem.* **1974**, *249*, 1026-1030.
30 Gilbert, E. F. *Pathology.* **1985**, *17*, 161-171.
31 Kuratsune, H.; Yamaguti, K.; Takahashi, M.; Misaki, H.; Tagawa, S.; Kitani, T. *Clin. Infect. Dis.* **1994**, *18 (Suppl 1)*, S62-S67.
32 Garst, J. E. *Toxicologist* (Society of Toxicology Abstract Issue of *Fundam. Appl. Toxicol.)* **1995**, *15*, 2.
33 Zeisel S. H. *Modern Nutrition in Health and Disease,* 8th edn. (Shils, M. E.; Olson, J. A.; Shike, M., Eds.), Lea and Febiger: Philadelphia, PA, **1994**, Vol. 1, pp. 449-458.
34 Shigenaga, M. K.; Hagen, T. M.; Ames, B. N. *Proc. Natl. Acad. Sci. U. S. A.* **1994**, *91*, 10771-10778.
35 Shigenaga, M. K.; Ames, B. N. In *Natural Antioxidants in Human Health and Disease* (Frei, B., Ed.), Academic Press: San Diego, CA, **1994**, pp. 63-106.
36 Fariello, R. G.; Ferraro, T. N.; Golden, G. T.; DeMattei, M. *Life. Sci.* **1988**, *43*, 289-292.
37 Amenta, F.; Ferrante, F.; Lucreziotti, R.; Ricci, A.; Ramacci, M. T. *Arch. Gerontol. Geriatr.* **1989**, *9*, 147-153.
38 Bruns, K. A.; Casillas, E. R. *Biol. Reprod.* **1989**, *41*, 218-226.
39 Childress, C. C.; Sacktor, B.; Traynor, D. R. *J. Biol. Chem.* **1966**, *242*, 752-760.
40 Kurth, L.; Fraker, P.; Bieber, L. *Biochim. Biophys. Acta* **1994**, *1201*, 321-327.
41 Kehrer, J. P.; Lund, L. G. *Free Radic. Biol. Med.* **1994**, *17*, 65-75.
42 Reed, D. J.; Savage, M. K. *Biochim. Biophys. Acta* **1995**, *1271*, 43-50.
43 Pastorino, J. G.; Snyder, J. W.; Serroni, A.; Hoek, J. B.; Farber, J. L. *J. Biol. Chem.* **1993**, *268*, 13791-13798.
44 Reznick, A. Z.; Kagan, V. E.; Ramsey, R.; Tsuchiya, M.; Khwaja, S.; Serbinova, E. A.; Packer, L. *Arch. Biochem. Biophys.* **1992**, *296*, 394-401.
45 Arduini, A. *Am. Heart. J.* **1992** *123*, 1726-1727.
46 Galli, G.; Fratelli, M. *Exp. Cell Res.* **1993**, *204*, 54-60.

47  Boerrigter, M. E. *Toxicologist* (Society of Toxicology Abstract Issue of *Fund. Appl. Toxicol.)* 1995, *15,* 2.
48  Boerrigter, M. E.; Franceschi, C.; Arrigoni-Martelli, E.; Wei, J. Y.; Vijg, J. *Carcinogenesis* 1993, *14,* 2131-2136.
49  Di Giacomo, C.; Latteri, F.; Fichera, C.; Sorrenti, V.; Campisi, A.; Castorina, C.; Russo, A.; Pinturo, R.; Vanella, A. *Neurochem. Res.* 1993, *18,* 1157-1162.
50  Revoltella, R. P.; Dal Canto, B.; Caracciolo, L.; D'Urso, C. M. *Biochim. Biophys. Acta* 1994, *1224,* 333-341.
51  Miotto, G.; Venerando, R.; Khurana, K. K.; Siliprandi, N.; Mortimore, G. E. *J. Biol. Chem.* 1992, *267,* 22066-22072.
52  Pontremoli, S.; Melloni, E.; Viotti, P. L.; Michetti, M.; Di Lisa, F.; Siliprandi, N. *Biochem. Biophys. Res. Commun.* 1990, *167,* 373-380.
53  Di Lisa, F.; Menabò, R.; Barbato, R.; Miotto, G.; Siliprandi, N. In *Current Concepts in Carnitine Research* (Carter, A. L., Ed.), CRC Press: Boca Raton, FL, 1992, pp. 27-36.
54  Nakadate, T.; Yamamoto, S.; Aizu, E.; Kato, R. *Cancer Res.* 1986, *46,* 1589-1593.
55  Nakadate, T.; Blumberg, P. M. *Cancer Res.* 1987, *47,* 6537-6542.
56  Baydoun, A. R.; Markham, A.; Morgan, R. M.; Sweetman, A. J. *Biochem. Pharmacol.* 1988, *37,* 3103-3107.
57  Dainty, I. A.; Bigaud, M.; McGrath, J. C.; Spedding, M. *Br. J. Pharmacol.* 1990, *100,* 241-246.
58  Wu, J.; Corr, P. B. *Am. J. Physiol.* 1994, *266,* H1034-H1046.
59  Yamada, K. A.; McHowat, J.; Yan, G. X.; Donahue, K.; Peirick, J.; Kléber, A. G.; Corr, P. B. *Circ. Res.* 1994, *74,* 83-95.
60  Wu, J.; McHowat, J.; Saffitz, J. E.; Yamada, K. A.; Corr, P. B. *Circ. Res.* 1993, *72,* 879-889.
61  McHowat, J.; Yamada, K. A.; Wu, J.; Yan, G. X.; Corr, P. B. *J. Cardiovasc. Electrophysiol.* 1993, *4,* 288-310.
62  Roncero, C.; Goodridge, A. G. *Arch. Biochem. Biophys.* 1992, *295,* 258-267.
63  Gadaleta, M. N.; Petruzzella, V.; Renis, M.; Fracasso, F.; Cantatore, P. *Eur. J. Biochem.* 1990, *187,* 501-506.
64  Gadaleta, M. N.; Petruzzella, V.; Fracasso, F.; Fernandez-Silva, P.; Cantatore, P. *FEBS. Lett.* 1990, *277,* 191-193.
65  Heathers, G. P.; Yamada, K. A.; Kanter, E. M.; Corr, P. B. *Circ. Res.* 1987, *61,* 735-746.
66  Saheki, T. *Toxicologist* (Society of Toxicology Abstract Issue of *Fundam. Appl. Toxicol.)* 1995, *15,* 2.
67  Tomomura, M.; Nakagawa, K.; Saheki, T. *FEBS. Lett.* 1992, *311,* 63-66.
68  Maki, A.; Berezesky, I. K.; Fargnoli, J.; Holbrook, N. J.; Trump, B. F. *FASEB J.* 1992, *6,* 919-924.
69  Abate, C.; Patel, L.; Rauscher III, F. J.; Curran, T. *Science* 1990, *249,* 1157-1160.
70  Tainaka, H.; Naito, T.; Murayama, N.; Yamazoe, Y.; Kato, R. *Biol. Pharm. Bull.* 1993, *16,* 1240-1243.
71  Liu, Y.; Rosenthal, R. E.; Starke-Reed, P.; Fiskum, G. *Free Radic. Biol. Med.* 1993, *15,* 667-670.
72  Lee, D. Y.; Hayes, J. J.; Pruss, D. and Wolffe, A. P. *Cell.* 1993, *72,* 73-84.
73  Turner, B. *Cell.* 1993, *75,* 5-8.
74  Fanti, L.; Berloco, M.; Pimpinelli, S. *Mol. Gen. Genet.* 1994, *244,* 588-595.
75  Citron, M. *Environ. Health Persp.* 1995, *103,* 232-235.

76  Lake, B. G. *Annu. Rev. Pharmacol. Toxicol.* **1995**, *35*, 483-507.
77  Cherkaoui-Malki, M.; Lone, Y. C.; Corral-Debrinski, M.; Latruffe, N. *Biochem. Biophys. Res. Commun.* **1990**, *173*, 855-861.
78  Hsieh, L. L.; Shinozuka, H.; Weinstein, I. B. *Br. J. Cancer* **1991**, *64*, 815-820.
79  Müller, R.; Bravo, R.; Burckhardt, J.; Curran, T. *Nature* **1984**, *312*, 716-720.
80  Bayly, A. C.; Roberts, R. A.; Dive, C. *J. Cell. Biol.* **1994**, *125*, 197-203.
81  Arends, M. J.; McGregor, A. H.; Wyllie, A. H. *Am. J. Pathol.* **1994**, *144*, 1045-1057.
82  Castle, V. P.; Heidelberger, K. P.; Bromberg, J.; Ou, X.; Dole, M.; Nuñez, G. *Am. J. Pathol.* **1993**, *143*, 1543-1550.
83  Sinha, B. K.; Yamazaki, H.; Eliot, H. M.; Schneider, E.; Borner, M. M.; O'Connor, P. M. *Biochim. Biophys. Acta* **1995**, *1270*, 12-18.
84  Kähönen, M. T. *Biochem. Pharmacol.* **1979**, *28*, 3674-3677.
85  Sachan, D. S. and Dodson, W. L. *J. Environ. Pathol. Toxicol. Oncol.* **1992**, *11*, 125-129.

# Chapter 11

# L-Carnitine as a Detoxicant of Nonmetabolizable Acyl Coenzyme A

**L. L. Bieber, Z. H. Huang, and D. A. Gage**

**Department of Biochemistry, Michigan State University, East Lansing, MI 48824–1319**

Liver contains short-chain, medium-chain, and/or long-chain carnitine acyltransferases associated with mitochondria, peroxisomes, and microsomes. These data, plus the demonstration that elevated levels of acylcarnitines can occur in human urine due to impaired metabolism of specific acyl-CoAs, strongly support roles for L-carnitine in both modulating the acyl-CoA:CoA ratio and being an acceptor of specific nonmetabolizable acyl-CoAs. Heretofore, an overlooked aspect of this metabolic problem has been the impact and implications of the inhibitory effects of acyl-CoAs, such as palmitoyl-CoA, on the short-chain carnitine acyltransferase activity. Herein, we show that palmitoyl-CoA is a potent inhibitor of carnitine acetyltransferase and that a secondary carnitine deficiency associated with a renal loss of carnitine can result in urinary excretion of specific $\alpha$-keto acids that are substrates for the branched-chain $\alpha$-keto acid dehydrogenase, pyruvate dehydrogenase, and $\alpha$-ketoglutarate dehydrogenase. These latter data indicate mitochondrial carnitine deficiency can result in inhibition of mitochondrial $\alpha$-keto acid dehydrogenase activity.

Although the pioneering studies of Bremer (*1*) and Fritz (*2*) concomitantly established a role for L-carnitine in the mitochondrial $\beta$-oxidation of long-chain fatty acids, multiple roles for L-carnitine in intermediary metabolism are strongly supported by data showing: 1) tissues such as liver contain a family of carnitine acyltransferases with overlapping acyl-CoA specificities, 2) peroxisomes, endoplasmic reticulum, and mitochondria all contain short-chain, medium-chain, and long-chain carnitine acyltransferase activity, and 3) the medium-chain/long-chain enzyme is a different protein in each organelle (*3,4*). Each of the above-mentioned organelles contain at least two enzymes with broad acyl-CoA specificities capable of reversibly catalyzing the conversion of $C_2$ to greater than $C_{20}$ acyl-CoAs to acylcarnitines as shown below.

Although three different enzyme types, CAT, COT, and CPT, are shown, it should be recognized that COT and CPT use the same substrates. The medium-chain/long-chain carnitine acyltransferase was designated a COT because of its kinetic preference for medium-chain acyl-CoAs, while others refer to the enzymes as CPT on the basis of functional considerations.

0097–6156/96/0643–0140$15.00/0

## Reactions Catalyzed

1.   CAT (Carnitine Acyltransferase)

Short-Chain Acyl-CoA
        + ⟷⟶ Short-Chain Acylcarnitine
     L-carnitine                +
                            CoASH

2.   COT (Carnitine Octanoyltransferase)

Medium-Chain/Long-Chain Acyl-CoA
        + ⟷⟶ Medium-Chain/Long-Chain Acylcarnitine
     L-carnitine                +
                            CoASH

3.   CPT (Carnitine Palmitoyltransferase)

Long-Chain/Medium-Chain Acyl-CoA
        + ⟷⟶ Medium-Chain/Long-Chain Acylcarnitine
     L-carnitine                +
                            CoASH

The majority of our studies with CAT have focused on its roles 1) in catalyzing the reaction in which L-carnitine serves as an acceptor for non-metabolizable acyl-CoA derivatives and 2) in modulating the acyl-CoA:CoA ratios intracellularly. Support for such roles has come from studies on human disease states which produce secondary carnitine deficiency (5-8) and from mitochondrial studies documenting a direct effect of carnitine on specific acyl-CoA:CoA ratios (9,10). However, the potential inhibitory effect of specific acyl-CoAs on the activity of carnitine acyltransferases that do not use the specific acyl-CoA as a substrate has received little experimental attention. Herein, we show that palmitoyl-CoA strongly inhibits CAT and that a secondary carnitine deficiency due to renal carnitine loss can promote excessive excretion of specific $\alpha$-keto acids.

## Assays

CAT was assayed spectrally at room temperature in a 260 $\mu$l reaction mixture containing 50 mM potassium phosphate buffer, pH 7.4, 200 $\mu$M dithiopyridine, 98 $\mu$M acetyl-CoA, and 1 $\mu$l of dialyzed commercial pigeon breast muscle CAT, diluted 1:10 in potassium phosphate buffer. Reactions were normally started by addition of 19.5 mM L-carnitine and the change in optical density at 324 nm monitored (full-scale = 1 OD unit, charge speed 2 cm/minute). For figure 1, 10 $\mu$l of palmitoyl CoA, final concentration 98 $\mu$M, was added where indicated. For figures 2 and 3, the enzyme was preincubated with palmitoyl-CoA and acetyl-CoA for 1 minute and the reaction initiated with L-carnitine. Palmitoyl-CoA concentration was varied as shown in the graphs. Assays were performed at room temperature, 25°C. A minimum of three separate runs were done for each curve. Extracts for determination of free and total carnitine were prepared as described (11), and free carnitine was quantitated by the radiochemical method of Cederblad and Lindstedt (12). Total carnitine was assayed as described above, after alkaline hydrolysis.

## Effect of Carnitine on Short-chain and $\alpha$-Keto Acid Metabolism

It is well established, from investigations of secondary carnitine deficiency produced by some organic acid acidurias, that L-carnitine can serve as an acyl acceptor for some non-metabolizable acyl moieties (5,6); in humans, the urinary acylcarnitine profile can be used

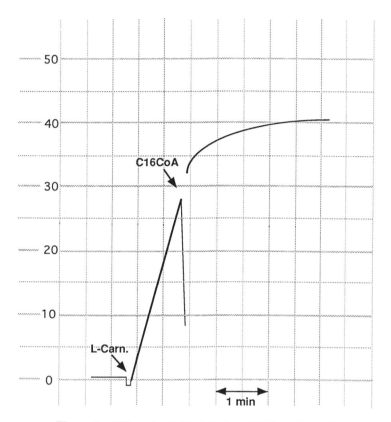

**Figure 1.** Effect of palmitoyl-CoA on CAT reaction rate.

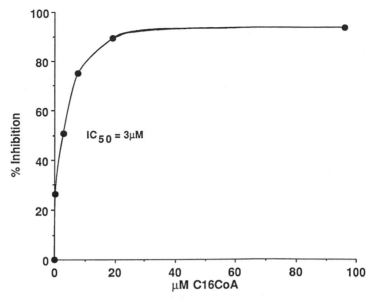

**Figure 2.** Effect of increasing palmitoyl-CoA concentration on CAT.

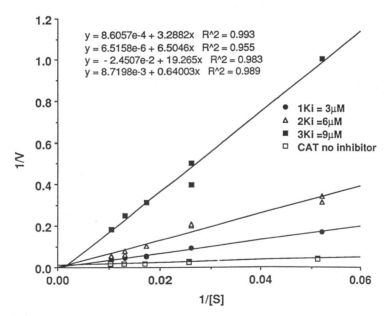

**Figure 3.** Kinetic analyses of the inhibitory effect of palmitoyl-CoA on CAT. The *y* values shown on the graph are in the same order as the symbols.

for diagnostic purposes. Table I shows the urinary acylcarnitine profile of an individual with impaired propionyl-CoA metabolism (sample provided by J. M. Saudubray). For the data in Table I, pivaloyl-L-carnitine is the internal standard and the DMCPE (N-demethylcarnitine propyl esters) derivatives were determined by GC-MS analyses as described (*13*). Note that the amount of propionyl-L-carnitine was 8- to 10-fold higher than the next most abundant acylcarnitine, (i-5:0) isovalerylcarnitine. The excreted acyl conjugated to carnitine in some disease states can greatly reduce tissue and blood carnitine levels which, in severe cases, may be life-threatening.

**TABLE I.  Urinary Acylcarnitine Profile of an Individual
with Compromised Propionyl-CoA Metabolism**

| [M + H] m/z | Structure DMCPE | % Area Relative to Standard |
|---|---|---|
| 246 | 3:0 | 206 |
| 274 | pivalyl (internal standard) | 100 |
| 274 | i-5:0 | 26 |
| 288 | 6:0 | 18 |
| 286 | 6:1 | 9 |
| 330 | 9:0 isomer or DC 4:1 | 23 |
| 316 | 8:0 | 16 |

Since no adequate animal model exists for secondary carnitine deficiency, the metabolic basis or underlying cause for the clinical crisis is often not readily apparent. Some years ago, in response to an attending physician's request, we performed carnitine and acylcarnitine analyses on urine, blood, and biopsied samples from a child in crisis. Previous tissue biopsies had shown muscle lipid droplets surrounded by mitochondria, indicative of a carnitine deficiency syndrome. As shown in Table II, the serum levels of free carnitine were approximately 10% of normal values, and the levels of total carnitine were approximately 20% of normal values. Similarly, the free carnitine levels of muscle and liver were 10-15% of control values. Gas chromatographic/mass spectrophotometric analyses of the urinary organic acids were performed by the MSU Mass Spectrophotometry Facility. As shown in the analysis C column of Table II, the 24-hour urinary output of pyruvate and α-ketobutyric acid was greater than 50-fold that of controls, while the 24-hour urinary output of α-ketoisovaleric, α-ketoisocaproic, and α-ketoglutaric acid was greater than 10-fold above control values.

Each of these α-keto acids is a substrate for one or more of the mitochondrial α-keto acid dehydrogenases; their greatly elevated urinary levels strongly indicate an impairment in α-keto acid dehydrogenase function. Although the serum level of total carnitine was approximately 20% of normal, as shown in analysis D of Table II, the total urinary output of carnitine plus acylcarnitines was between 1.3- and 1.9-fold greater than the 24-hour output for controls. These data strongly indicate the patient had muscle and liver carnitine deficiency which compromised the normal functioning of the mitochondrial α-keto acid dehydrogenases, pyruvate dehydrogenase and branched-chain α-keto acid dehydrogenase, as well as α-ketoglutaric acid dehydrogenase of the TCA cycle.

The basis for the elevated excretion of specific α-keto acids metabolized by mitochondria is not known; the acids shown in Table II are the only ones which deviated from normal values. The possibilities are that the reduced carnitine levels resulted in less capacity to adequately buffer the acyl-CoA:CoA ratio with subsequent decreases in free CoA and increases in acyl-CoA, both of which could reduce α-keto acid dehydrogenase activity. Reduced free CoA levels could directly compromise the activity, since it is a substrate for the dehydrogenases and increased acyl-CoA levels could, via short-term metabolic allosteric control, inhibit the dehydrogenases. CoA and acyl-CoA analyses were not done due to the very limited quantities of material. Daily oral carnitine therapy relieved some of the symptoms (personal communication of the physician); consequently, additional diagnostic data were not required. The fact that oral carnitine rapidly relieved the clinical symptoms strongly indicates that carnitine facilitates the action of intramitochondrial $\alpha$-keto acid dehydrogenases. The elevated excretion of substrates for pyruvate dehydrogenase and $\alpha$-ketoglutaric dehydrogenase is surprising and was not anticipated. In contrast, a facilitory role for carnitine in $\alpha$-ketoisovaleric and $\alpha$-ketobutyric acids which are substrates for the branched-chain $\alpha$-keto acid dehydrogenase has previously been proposed (4). These acids are derived from the branched-chain amino acids, leucine and valine, and $\alpha$-ketobutyrate is an intermediate in methionine metabolism.

**TABLE II. Carnitine Status of a Patient with a Renal Carnitine Leak**

| A. | Serum | Total ($\mu$M) | Free L-Carnitine ($\mu$M) |
|---|---|---|---|
| | Normal | 46.1 | 36.7 |
| | Patient (n=3) | 9.6 | 3.5 |
| B. | Tissue | Free L-carnitine (nmol/mg non-collagen protein) | |
| | | Muscle | Liver |
| | Normal | 17.1 | 8.2 |
| | Patient | 2.3 | 1.9 |
| C. | Urine (acids) | Ratio of 24-hour urine: | (Patient) / (Normal) |
| | pyruvic | > 50 X | |
| | $\alpha$-ketobutyric | > 50 X | |
| | $\alpha$-ketoisovaleric | > 10 X | |
| | $\alpha$-ketoisocaproic | > 10 X | |
| | $\alpha$-ketoglutaric | > 10 X | |
| D. | Total Carnitine Output | Ratio of 24-hour urine: | (Patient) / (Normal) |
| | 24-hour urine at weekly intervals | 1.4, 1.3, 1.9 | |

## Effect of Palmitoyl-CoA on CAT Activity

During investigations into the mode of action of a potent inhibitor of CPT, etomoxiryl-CoA, it was shown that this long-chain, hydrophobic acyl-CoA is an inhibitor of CAT (14). This finding raised the possibility that other, including normal long-chain acyl-CoAs might be inhibitory to the short-chain carnitine acetyltransferase. Figure 1 shows the effect of adding 98 $\mu$M palmitoyl-CoA to a reaction mixture containing 98 $\mu$M acetyl-CoA, carnitine, and pigeon breast muscle CAT. Within one minute after addition of the palmitoyl-CoA, greater than 90% inhibition occurs. The response of CAT to varying concentrations of palmitoyl-CoA at a fixed concentration, 98 $\mu$M, of acetyl-CoA is shown in Figure 2. The apparent

$K_i$ for palmitoyl-CoA is 3 $\mu$M, which is the same as the 3 $\mu$M value for etomoxiryl-CoA, using essentially identical incubation conditions and the CATs purified from liver peroxisomes, liver mitochondria, and pigeon breast muscle enzyme (14). Mixed-type inhibition was obtained for the peroxisomal and mitochondrial liver enzyme, and uncompetitive inhibition was obtained for the pigeon breast enzyme. The inhibition type was investigated by determining velocity vs. substrate concentration analyses at palmitoyl-CoA concentrations 1, 2, and 3 times the apparent $K_i$; double reciprocal plots of such data are shown in Figure 3. Palmitoyl-CoA did not behave as an uncompetitive inhibitor; however, additional experiments are needed to determine the exact inhibition type.

The surprisingly potent inhibitory effects of palmitoyl-CoA on CAT raise some interesting questions about the physiological response of CAT to palmitoyl-CoA, particularly in the matrix of mitochondria where long-chain acyl-CoA may be relatively abundant. The data are consistent with fatty acid oxidation in liver mitochondria being ketogenic and with the limited acetylcarnitine production by heart mitochondria from fatty acids when compared to pyruvate as substrate (9,10). The 50% inhibition of CAT by low concentrations of palmitoyl-CoA (in our studies, the $K_i$ was approximately 3 $\mu$M) provides a reasonable explanation for the minimal or limited acetylcarnitine production by mitochondria oxidizing fatty acids under conditions in which acetyl-CoA levels would be expected to be reasonably abundant. These data raise the intriguing possibility that some of the unusual acyl-CoAs, in which carnitine serves as a metabolic detoxicant, might partially inhibit CAT when their levels increase. If such inhibition were sufficiently great to limit the capacity of CAT to modulate or buffer the acyl-CoA:CoA ratio, this could have profound metabolic consequences. One wonders if some of the symptoms, produced by secondary carnitine deficiency that can be alleviated by oral carnitine therapy, are caused by a major imbalance in the acyl-CoA:CoA ratio due to reduced CAT activity as a consequence of unusual acyl-chain inhibition of CAT.

## Conclusions

The data presented herein clearly show a greatly enhanced $\alpha$-keto acid excretion in urine from an individual with secondary carnitine deficiency. These data strongly support an indirect role for carnitine in the mitochondrial metabolism of specific $\alpha$-keto acids, likely via partial inhibition of the mitochondrial $\alpha$-ketoglutarate dehydrogenase, the mitochondrial branched-chain keto acid dehydrogenase, the pyruvate dehydrogenase, as well as the $\alpha$-ketobutyrate dehydrogenase complex. Similarly, the finding that palmitoyl-CoA is a potent inhibitor of CAT strongly indicates that during mitochondrial oxidation of long-chain fatty acids, conversion to acetylcarnitine is restricted and preference is given to the TCA cycle, as well as ketogenesis in liver. These results also provide an explanation for why pyruvate is a superb source of mitochondrial acetylcarnitine efflux, while substrates such as octanoate are poor precursors of acetyl-L-carnitine.

## Acknowledgments

Supported in part by NIH grant DK18427 and by a grant from the American Heart Association of Michigan.

Tissue and urine samples were provided by Dr. Arthur Kornman, La Rabida Children's Hospital, University of Chicago, East 65th Street, Chicago, IL, for diagnostic purposes, as part of the clinical management of one of his patients. Mass spectral analyses of the urinary acids were performed by Dr. John Verbanic of the MSU Mass Spectrometry Facility. Urine from the propionic acidemia patient was provided by Prof. Jean-Marie Saudubray, Clinique de Genetique Medicale, Paris, France. The technical assistance of P. Wagner is gratefully acknowledged.

## Literature Cited

1.   Bremer, J. *Physiol. Rev.* **1983**, *63*, 1420-1468.
2.   Fritz, I. B. *Adv. Lipid Res.* **1963**, *1*, 285-334.
3.   Bieber, L. L.; Emaus, R.; Valkner, K.; Farrell, S. *Fed. Proc.* **1982**, *41*, 2858-2861.
4.   Bieber, L. L. *Ann. Rev. Biochem.* **1988**, *57*, 261-283.
5.   Engel, A. G.; Rebouche, C. J. *J. Inher. Metab. Dis.* **1984**, *7* (Suppl. 1), 38-43.
6.   Gilbert, E. F. *Pathology* **1985**, *17*, 161-169.
7.   Angelini, C.; Trevisan, C.; Isaya, G.; Pegolo, G.; Vergani, L. *Clin. Biochem.* **1987**, *20*, 1-7.
8.   Rebouche, C. J.; Paulson, D. J. *Ann. Rev. Nutr.* **1986**, *6*, 41-66.
9.   Lysiak, W.; Lilly, K.; DiLisa, F.; Toth, P. P.; Bieber, L. L. *J. Biol. Chem.* **1988**, *236*, 1151-1156.
10.   Lysiak, W.; Toth, P. P.; Suelter, C. H.; Bieber, L. L. *J. Biol. Chem.* **1986**, *261*, 13698-13703.
11.   Kerner, J.; Bieber, L. L. *Biochem. Med.* **1982**, *28*, 204-209.
12.   Cederblad, G.; Lindstedt, S. *Clin. Chim. Acta* **1972**, *37*, 235.
13.   Huang, Z.H.; Gage, D.; Bieber, L.L.; Sweeley, C.C. *Anal. Biochem.* **1991**, *199*, 98-105.
14.   Lilly, K.; Chung, C.; Kerner, J.; VanRenterghem, R.; Bieber, L. L. *Biochem. Pharm.* **1992**, *43*, 353-361.

# BIOMARKERS OF ENVIRONMENTAL HEALTH

# Chapter 12

# The Use of Human 101L Cells as a Biomarker, P450 RGS, for Assessing the Potential Risk of Environmental Samples

Jack W. Anderson[1], Kristen Bothner[2], David Edelman[3], Steven Vincent[3], Tien P. Vu[4], and Robert H. Tukey[4]

[1]Columbia Analytical Services, 6060 Corte Del Cedro, Carlsbad, CA 92009
[2]MEC Analytical Systems, Inc., 2433 Impala Drive, Carlsbad, CA 92008
[3]Columbia Analytical Services, 1317 S. 13th Avenue, Kelso, WA 98626
[4]Department of Pharmacology, Cancer Center, University of California at San Diego, La Jolla, CA 92093–0063

A reporter gene system (RGS) assay has been implemented using a human liver cancer cell line, 101L, which carries the human *CYP1A1* gene promoter linked to the firefly luciferase gene. The exposure to these cells with Ah receptor ligands leads to the induction of luciferase activity. From 6 to 18 hours after application of an organic extract of water, tissue, or soil, the reaction is stopped by rinsing, the cells are lysed and the extract is measured for luminescence. Induction of luciferase activity by such compounds as dioxin, dioxin-like PCB congeners (coplanar), and polycyclic aromatic hydrocarbons (PAHs) infers these xenobiotics are present at levels that are potentially toxic, carcinogenic, or mutagenic to organisms. Solvent extracts (using standard extraction methods, EPA 3550) of aquatic sediments, soils, and mussel tissue have been directly applied to these cells. Results produced in the last year show that significant induction from concentrations of inducer compounds, that if present in a typical 40 g sample, would be (in ng/g or ppb): 0.008 for Dioxin; 0.063 for Furan; 6.2-2000 for a range of coplanar PCB congeners; 6.2-7500 for specific PAHs; and 250 for a mixture of PAHs. Comparisons between the EPA Toxic Equivalent Factors (TEFs) and those generated from RGS testing differ considerably. While most PCBs are significantly lower than dioxin (as EPA estimates), some of the PAHs tested exhibit higher TEFs than PCBs. Testing a wide geographic area for the presence of the above chemicals can be performed rapidly, with sensitivity and specificity using the RGS assay.

Dioxins, polychlorinated biphenyls (PCBs), and polycyclic aromatic hydrocarbons (PAHs) are primarily regulated due to their toxicity and persistence in the environment. A human tissue culture cell line, 101L (1), has been implemented as a biological screen (2) to detect these contaminants in environmental samples.

These agents, which are known Ah-receptor (AhR) ligands, bind to the AhR and stimulate a series of signal transduction events that lead to the nuclear uptake of the receptor, culminating in transcriptional activation of AhR responsive genes (3). The *CYP1A1* gene, which encodes a benzo[a]pyrene hydroxylase P450, is a target of the AhR complex, and in the presence of ligand culminates in a significant increase in the rate of gene transcription. In 101L cells, the *CYP1A1* gene promoter and 1800 bases of flanking regulatory DNA that contain AhR enhancer recognition sequences (4) have been linked to the firefly luciferase gene (Reporter Gene System), and this chimeric stably integrated into HepG2 cells. Therefore, any stimulus that activates the AhR and facilitates its migration to the nucleus results in transcriptional activation of the *CYP1A1* gene promoter, which can be quantitated by measuring the relative fold increase in cellular luciferase activity. While the term CYP1A1 reporter gene system is a more accurate description of the assay, recognition of the importance of P450 induction in environmental studies led us to use the term P450 RGS. In this paper and subsequent publications, RGS will be used to aviod conflict with accepted terminology. Since many of the agents that serve as AhR ligands are potentially carcingoenic, the RGS assay can be used as a meaningful measure of relative toxicity and potential carcinogenicity. The advantage of this method over analysis of inducible P4501A1 or quantitation of its mRNA (5-8) is the assays simplicity and the ability to measure multiple samples quickly and accurately. .

Safe (9) has reviewed the biological effects associated with exposure to many chlorinated organics, including dioxins and PCB congeners. The Toxic Equivalent Factors (TEF) proposed by Safe (9) for PCBs, and those for dioxins and furans, used by the Environmental Protection Agency (10) can be used to estimate the overall cancer risk of an environmental sample containing concentrations of these chemicals. First, the products of individual TEF values and their concentrations in the sample are calculated. Then, the sum of these products is used to estimate the overall Toxic Equivalents (TEQ) of a sample. One of the major factors used to establish a TEF for a specific compound is the ability of the chemical to induce benzo[a]pyrene hydroxylase activity. It has been demonstrated that the relative P4501A1 (AHH) induction levels, decrease by about an order of magnitude from dioxin to coplanar PCBs, to mono-ortho PCB congeners. Another factor used to evaluate the toxicity of compounds, such as receptor binding, was also found to decrease by approximately a factor of 100 from dioxin to coplanar PCBs, to mono-ortho coplanars. An order of magnitude difference in the ability of dioxin and PCBs to induce in mouse cells a chimeric reporter gene under the control of AhR enhancer sequences has also been demonstrated (11). Previous studies from our laboratory using 101L cells (RGS) has shown there is a decrease by approximately 100 in CYP1A1 induction potential between dioxin and PCB #126, and about 100 between this congener and the other coplanar PCBs (2).

One of the agencies required to deal with the contamination of sediments from dioxins, PCBs, and PAHs is the U.S. Army Corps of Engineers. McFarland *et al.* (12) have suggested that approaches such as the RGS assay will become an acceptable alternative to the expensive analytical chemical methods of measuring dioxin TEQs. In another report, he and co-workers (13) described the results of intercomparison studies conducted in their laboratory. They have found that the RGS method was more rapid, sensitive, and less expensive than other tests evaluated.

**Objectives.** The objective of this paper is to provide an overview of how a rapid screening technique can be used to estimate the Toxic Equivalent Factors (TEFs) for dioxins, PCBs and PAHs in environmental samples. We believe there is a need to compare these three classes of toxic organic compounds on a equal basis to accurately determine the relationships between the chlorinated and non-chlorniated hydrocarbons. The testing methods, data on standards, and the analysis of field samples will be described.

**Materials and Methods**

**Test System.** A description outlining the construction of the chimerics and the cell line (1) and the use of these cells in assays with standard solutions and sediment extracts (2) has been reported. Cells (101L) used as the *CYP1A1* reporter gene system (RGS) were derived from the human hepatoma cell line HepG2. These cells carry a single copy of a stably integrated plasmid that of the human *CYP1A1* promoter and 5'-flanking sequences fused to the firefly luciferase gene (1). The human CYP1A1 promoter and flanking DNA contain three AhR specific xenobiotic responsive elements that bind the liganded AhR.

Cells were grown in Dulbecco's modified Eagle medium supplemented with 10% fetal calf serum and in 0.4 mg/mL G418. Compounds used as reference inducers (2,3,7,8-dioxin, TCDD and benzo[a]pyrene) were applied in dichloromethane (DCM), and control cells were given DCM alone. Induction from the DCM (at 0.1 % or less) was used as unity in the calculation of fold induction. The RGS fold induction values reported are the means of three replicates. Standard deviations for the three replicates were generally ten percent of the mean or less. Between test variations obtained with a 10 nM (3.2 ng/ mL) concentration of 2,3,7,8-TCDD, used as a reference toxicant, were relatively small (140 fold induction ± 22 S.D.). Samples of soil, sediment or tissue applied to the cells were prepared in dichloromethane (DCM) by the EPA method 3550. When 20 μL or less of DCM was added to 2 mL of medium (0.1%) in the test well, cells remained viable as measured by trypan blue dye exclusion. The reaction time used in this study for induction of cells was 18 hours. Previous studies (1) have shown that reaction times of 6 and 12 hours are also appropriate, but by 24 hours there is a significant decrease in the RGS response on exposure to benzo(a)pyrene and benzo(a)anthracene.

Luciferase assays were as previously described (1,2).  The cells were lysed with a lysis buffer, and cell lysates were centrifuged.  The supernatant (50 µL) was used in luciferase assays started by injection of 100  µL of luciferin.  Light output was measured with a 96-well luminometer (Dynatech ML 2250, Chantilly, VA).  Luciferase activities in the samples were first expressed as relative light units (RLU), and then converted to fold induction vaules, by dividing by the RLU of the solvent control.

**Test Substances.**      Specific PCB congeners used in this study will be discussed by number, but the chemical formulae of the compounds are also listed in Tables I and II.   It should be noted that the Toxic Equivalent Factors (TEFs) for dioxins and specific PCB congeners (Table I) are all related to the toxicity of 2,3,7,8-TCDD, but those of the polyclic aromatic hydrocarbons (PAHs) have only thus far been related to benzo(a)pyrene (14).   A mixture of non-coplanar PCBs was prepared to be certain the test system was specific to coplanar and mono-ortho coplanar compounds.  Table II lists the specific PCBs in this mixture, and the concentrations of each.    Table III lists the concentrations of individual pesticides and three-ring PAHs that have been tested.  Specific PAHs were combined in an experimental mixture (Table IV) to simulate an environmental sample, and this mixture was used to test the response of the RGS.

The levels of specific toxicants added to the cells are expressed as the final weight (ng or µg) in the 2 mL of media, bathing the cells, or as the final concentration (in ng/mL, ppb) in the cell medium.  Since 3.2 ng/mL (10nM) of TCDD is used with the each test, the response to this reference toxicant can be used to convert the average fold induction values of the samples into estimated dioxin (TCDD) equivalents (in ng/g).   We have used this same approach to express the RGS responses in terms of benzo(a)pyrene equivalents (in µg/g).

Normal procedure is to extract forty gram samples of soil or sediment (wet) with DCM, and then reduce the volume to 1 mL.  When 20 µL of these extracts are applied to 2 mL of medium in the wells, the concentration of contaminant in the soil (in ng/g) is a factor of 2.5 greater than the concentration in the exposure solution (in ng/mL).  Therefore, this factor may be used to convert data we present in terms of exposure concentrations to those in the soil that would produce the same response.  We will also present the data in terms of TEF values, producing a specific level of RGS induction.

### Results

**Responses to Standards.**      Initial use of these cells (1) described the induction of CYP1A1-luciferase activity with differing concentrations of dioxin and PAHs, while more recent experiments (2) have documented the use of 5 coplanar and mono-ortho coplanar PCB congeners.  Figure 1 presents these

Table I.  Toxic Equivalent Factors for Dioxins,
PCBs and PAHs

| CHEMICALS | TEFs |
|---|---|
| 2,3,7,8-TCDD | 1 (a) |
| 1,2,3,7,8-PeCDD | 0.5 |
| 1,2,3,4,7,8-HxCDD | 0.1 |
| 1,2,3,6,7,8-HxCDD | 0.1 |
| 1,2,3,7,8,9-HxTCDD | 0.1 |
| 1,2,3,4,6,7,8-HpCDD | 0.01 |
| 1,2,3,4,6,7,8,9-OCDD | 0.001 |
| | |
| 3,3',4,4',5-PCB (#126) (b) | 0.1 (b) |
| 3,3',4,4',5,5'-HCB (#169) | 0.05 |
| 2,3,4,4',5-PCB (#114) | 0.0001 |
| 3,3',4,4'-TCB (#77) | 0.01 |
| 2,3,4,4',5-PCB (#156) | 0.0002 |
| | |
| Benzo(a)pyrene (c) | 1 (c) |
| Benzo(a)anthracene | 0.1 |
| Benzo(b)fluoranthrene | 0.1 |
| Benzo(k)fluoranthrene | 0.01 |
| Chrysene | 0.01 |
| Dibenzo(a,h)anthracene | 1 |
| Indeno(1,2,3-cd)pyrene | 0.1 |

(a) Reference (10)
(b) Reference (9)
(c) TEF values for PAHs are relative to B(a)P,
     and not Dioxin (14).

**Table II.  Non-coplanar PCBs Present in the Test Mixture Applied to the RGS Assay**

| Number | Compound | Concentration (µg/mL) |
|--------|----------|-----------------------|
| 15 | 4,4'-dichlorobiphenyl | 2.37 |
| 28 | 2,4,4'-trichlorobiphenyl | 7.72 |
| 52 | 2,2',5,5'-tetrachlorobiphenyl | 8.99 |
| 49 | 2,2'4,5'- tetrachlorobiphenyl | 5.76 |
| 47 | 2,2',4,4'- tetrachlorobiphenyl | 4.58 |
| 44 | 2,2'3,5'- tetrachlorobiphenyl | 8.38 |
| 37 | 3,4,4'-trichlorobiphenyl | 3.36 |
| 40 | 2,2',3,3'- tetrachlorobiphenyl | 2.12 |
| 74 | 2,4,4',5- tetrachlorobiphenyl | 6.54 |
| 70 | 2,3',4',5- tetrachlorobiphenyl | 12.30 |
| 66 | 2,3',4,4'- tetrachlorobiphenyl | 12.92 |
| 88 | 2,2',3,4,6-pentachlorobiphenyl | 1.76 |
| 60 | 2,3,4,4'- tetrachlorobiphenyl | 10.82 |
| 101 | 2,2',4,5,5'- pentachlorobiphenyl | 8.60 |
| 97 | 2,2',3',4,5- pentachlorobiphenyl | 3.50 |
| 87 | 2,2',3,4,5'- pentachlorobiphenyl | 7.72 |
| 110 | 2,3,3',4',6- pentachlorobiphenyl | 11.24 |
| 82 | 2,2'3,3',4- pentachlorobiphenyl | 3.46 |
| 108 | 2,3,3',4,5'- pentachlorobiphenyl | 3.29 |
| 149 | 2,2',3,4',5'- pentachlorobiphenyl | 4.72 |
| 118 | 2,3',4,4',5- pentachlorobiphenyl | 9.11 |
| 153 | 2,2',4,4',5,5'-hexachlorobiphenyl | 8.30 |
| 105 | 2,3,3',4,4'- pentachlorobiphenyl | 5.31 |
| 141 | 2,2',3,4,5,5'-hexachlorobiphenyl | 2.20 |
| 138 | 2,2',3,4,4',5'- hexachlorobiphenyl | 9.98 |
| 187 | 2,2',3,4',5,5',6-heptachlorobiphenyl | 1.91 |
| 128 | 2,2',3,3',4,4'- hexachlorobiphenyl | 1.97 |
| 180 | 2,2',3,4,4',5,5'-heptachlorobiphenyl | 3.84 |
| | **Total** | **234** |

Table III. Specific Pesticides and Three-Ring Aromatic
Hydrocarbons Tested with the RGS Assay

| Chemical | Concentrations Tested ($\mu$g/mL) |
|---|---|
| 2,6- Dimethylnaphthalene | 0.25, 0.5, 1.0, 2.0, and 4.0 |
| 2,3,5-Trimethylnaphthalene | 0.25, 0.5, 1.0, 2.0, and 4.0 |
| 1-Methylfluorene | 0.25, 0.5, 1.0, 2.0, and 4.0 |
| Phenanthrene | 0.25, 0.5, 1.0, 2.0, and 4.0 |
| 2-Methylphenanthrene | 0.25, 0.5, 1.0, 2.0, and 4.0 |
| 3,6-Dimethylphenanthrene | 0.25, 0.5, 1.0, 2.0, and 4.0 |
| 9-Methylanthracene | 0.25, 0.5, 1.0, 2.0, and 4.0 |
| 2-Methylanthracene | 0.25, 0.5, 1.0, 2.0, and 4.0 |
| 9,10-Dimethylanthracene | 0.25, 0.5, 1.0, 2.0, and 4.0 |
| Chlordane | 2, 10, and 20 |
| Toxaphene | 2, 10, and 20 |
| 2,4'-DDD | 1, 10, and 20 |
| 2,4'-DDE | 1, 10, and 20 |
| 2,4'-DDT | 1, 10, and 20 |

Table IV. PAH Composition of Test Mixture Applied
to the RGS Assay

| Chemicals | Concentration (ug/mL) |
|---|---|
| Naphthalene | 15.0 |
| 1-Methylnaphthalene | 45.7 |
| 2-Methylnaphthalene | 33.5 |
| Acenaphthylene | 4.2 |
| Acenaphthene | 1.7 |
| Fluorene | 3.8 |
| Biphenyl | 14.7 |
| Phenanthrene | 13.6 |
| Anthracene | 5.3 |
| Fluoranthene | 16.1 |
| Pyrene | 29.6 |
| Benzo(a)anthracene | 14.8 |
| Dibenzothiophene | 2.3 |
| Chrysene | 15.9 |
| Benzo(b)fluoranthene | 13.7 |
| Benzo(k)fluoranthene | 13.7 |
| Benzo(a)pyrene | 21.2 |
| Benzo(e)pyrene | 16.4 |
| Indeno(1,2,3-cd)pyrene | 10.5 |
| Dibenzo(a,h)anthracene | 4.7 |
| Benzo(g,h,i)perylene | 13.5 |
| Perylene | 22.4 |
| **Total** | **332.3** |

results on a basis of the level of Toxic Equivalents (TEQs) for each compound detected by RGS.  To produce these values the RGS detection limits for the compound in a soil sample (ng/g) were multiplied by the TEFs for the compound (see Table VIII).  It is important to note that all but two of the PCBs shown in Figure 1 produce significant induction at a dioxin equivalent (TEQ) of 1 or less.

To demonstrate the specificity of the test system, concentrations of the PCB mixture listed in Table II were applied at total concentrations up to 234 ng/mL (ppb).  No induction from this mixture of 28 different non-planar congeners was observed.  There were also no significant RGS responses from even the highest concentrations of 3-ring aromatic hydrocarbons and specific pesticides, listed in Table III.

Because many of the environmental samples we have examined contain significant amounts of contamination from PAHs, it was important to investigate the RGS response to a mixture of PAHs representative of contaminated marine sediments.   Table IV listed the specific compounds present in the mixture and their concentrations.  Figure 2 shows the levels of RGS induction produced when the cells are exposed to concentrations of this mixture between 21 and 665 ng/mL. The dashed line represents a logrithmic regression of the data, with a correlation coeficient (r2) of 0.99.  There was no apparent decrease in induction caused by the possible toxicity of the mixture at the higher concentrations tested.  The maximum induction observed was about 36 fold, and in other testing with marine sediments, we have observed fold induction as high as 900.

Figure 3 summarizes the RGS responses to several individual polycyclic aromatic hydrocarbons with four to six aromatic rings, tested at concentrations up to 5000 ng/mL. Relatively low responses were produced by chrysene, benzofluorene, and benzo(a)pyrene. Benzo(k)fluoranthrene, indeno(1,2,3-cd)pyrene, and dibenz(a,h)anthracene produced the strongest induction, and benz(a)anthracene was intermediate in response.   These results will be considered further in the discussion section, as the relative responses of these PAHs are compared to those of dioxin and coplanar PCBs.

**Responses to Extracts of Environmental Samples.**  Even when sample extracts apparently do not contain enough contaminant to produce a strong induction of the RGS, there is a background or baseline response. Natural and anthropogenic compounds present in soils, sediments and tissues at low levels produce a degree of induction, suggesting a reference sample (from a presumed clean location) should be used to determine the background RGS response for a given area or tissue type.  We have found that DCM volumes from extracts of soils, sediments, and mussel tissue between 2 and 20 μL give a background response of 2 to 5 fold induction.

In as many cases as possible we attempt to correlate the RGS results with those produced from the chemical analysis of the same sample (split

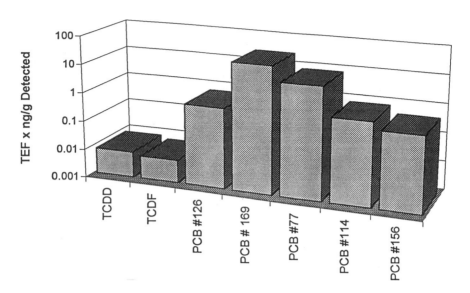

**Figure 1.** *Toxic Equivalents (TEQs) Detected in a Soil Sample by the RGS Assay.* Bars represent the product of TEF values for 2,3,7,8-TCDD, 2,3,7,8-TCDF, and specific PCB congeners, and the level of detection (ng/g) of the RGS. TEFs are listed in Table I and detection limits are in Table VIII.

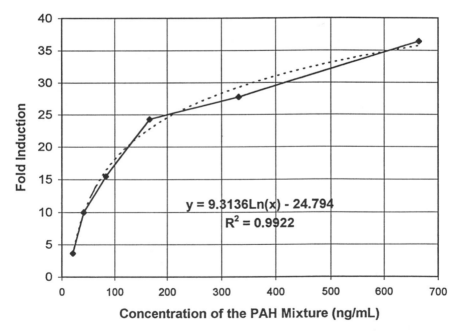

**Figure 2.** *Responses of the RGS to varying concentrations (ng/mL) of a mixture of PAHs.* Solid line connects data points, and the dashed line is the computed log regression of the data, with the equation and correlation coefficient ($r^2$) shown. The composition of the mixture is listed in Table IV.

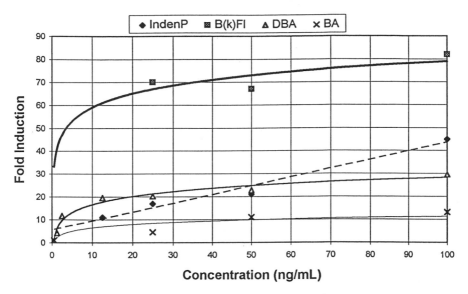

**Figure 3.** *Responses of the RGS to varying concentrations of specific PAHs.*
**A.** RGS Responses to Low Levels of PAHs. IND-Indeno(1,2,3-cd)pyrene;
DBA-Dibenzo(a,h)anthracene; BA-Benzo(a)anthracene; BkF-
Benzo(k)fluoranthrene.

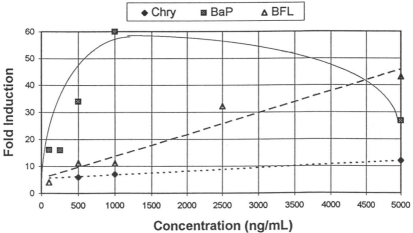

**B.** RGS Responses to High Levels of PAHs. BaP-Benzo(a)pyrene; BFL-
Benzofluorene; CHR-Chrysene. Curves shown are computer generated linear
or log regressions of the dose-response data, except for BaP, which was hand
drawn.

Table V.  Results of Chemical Analyses and TEQ Calculations for Soil Samples Tested by RGS

| CHEMICALS | TEFs | 9SW-7.0 (pg/g) | TEFs | 8SW-6.5 (pg/g) | TEFs | S1/S2 (pg/g) | TEFs |
|---|---|---|---|---|---|---|---|
| TCDDs (total) | 1 | 16 | 0.016 | 4.8 | 0.0048 | | |
| 2,3,7,8-TCDD | 1 | 2.6 | 0.003 | 1.8 | 0.0018 | 0.11 | 0.11 |
| 1,2,3,7,8-PeCDD | 0.5 | | 0 | | 0 | 5.3 | 2.65 |
| HxCDDs (total) | 0.1 | 43 | 0.004 | 37 | 0.0037 | | 0 |
| 1,2,3,4,7,8-HxCDD | 0.1 | | 0 | | 0 | 14 | 1.4 |
| 1,2,3,6,7,8-HxCDD | 0.1 | | 0 | | 0 | 58 | 5.8 |
| 1,2,3,7,8,9-HxTCDD | 0.1 | | 0 | | 0 | 39 | 3.9 |
| HpCDDs (total) | 0.01 | 140 | 0.001 | 120 | 0.0012 | | 0 |
| 1,2,3,4,6,7,8-HpCDD | 0.01 | 59 | 0.0006 | 52 | 0.0005 | 1300 | 13 |
| 1,2,3,4,6,7,8,9-OCDD | 0.001 | 440 | 0.0004 | 390 | 0.0004 | 15000 | 15 |
| DIOXINS | TOTAL | 700.6 | 0.0253 | 605.6 | 0.0124 | 16416.4 | 41.86 |
| | | | | | | | |
| TCDFs (total) | 0.1 | 4.2 | 0.0004 | | 0 | | |
| 2,3,7,8-TCDF | 0.1 | | 0 | | 0 | 1.6 | 0.16 |
| PeCDFs (total) | | | 0 | | 0 | | 0 |
| 1,2,3,7,8-PeCDF | 0.05 | | 0 | | 0 | 0.22 | 0.011 |
| 2,3,4,7,8-PeCDF | 0.5 | | 0 | | 0 | 10 | 5 |
| HxCDFs (total) | 0.1 | 26 | 0.0026 | 28 | 0.0028 | | 0 |
| 1,2,3,4,7,8-HxCDF | 0.1 | | 0 | | 0 | 41 | 4.1 |
| 1,2,3,6,7,8-HxCDF | 0.1 | | 0 | | 0 | 16 | 1.6 |
| 2,3,4,6,7,8-HxCDF | 0.1 | | 0 | | 0 | 21 | 2.1 |
| 1,2,3,7,8,9-HxCDF | 0.1 | | 0 | | 0 | 6.6 | 0.66 |
| HpCDFs (total) | 0.01 | 77 | 0.0008 | 73 | 0.0007 | | 0 |
| 1,2,3,4,6,7,8-HpCDF | 0.01 | | 0.0000 | 20 | 0.0002 | 400 | 4 |
| 1,2,3,4,7,8,9-HpCDF | 0.01 | | 0.0000 | | 0.0000 | 24 | 0.24 |
| 1,2,3,4,6,7,8,9-OCDF | 0.001 | 48 | 0.0000 | 42 | 0.0000 | 1500 | 1.5 |
| FURANS | TOTAL | 155.2 | 0.0038 | 163 | 0.0038 | 2020.42 | 19.371 |
| | | | TEQ | | TEQ | | TEQ |
| TOTALS | | 0.8558 ng/g | 0.02917 | 0.7686 ng/g | 0.0162 | 18.4368 ng/g | 61.231 |

sample) or one collected nearby. Table V shows the analytical results for three samples that were first screened by RGS assays.    The first two samples contained quite low levels of individal dioxins and furans, with totals of about 0.8 ng/g.    When these individual concentrations have been multiplied by the individual TEFs, and the sum produced, the total TEQ is only about 0.03.    The RGS assay indicated these samples contained significant amounts of toxic organics equivalent to 0.2 ng/g of dioxin.  This value compares favorably with the sum of measured dioxins and furans, but not after these have been multiplied by their individual TEFs, and the sum of these expressed as dioxin Toxic Equivalents (TEQ).    There are many dioxin and furan congeners, which have not yet been asigned a TEF, and thus the contribution of these compounds are not included in the final TEQs.

The third sample (S1/S2) was taken in between samples S1 and S2, for which in RGS produced estimated dioxin equivalents of 51 and 52 ppb. The RGS induction was conservatively high compared to the chemical analyses (18 ng/g dioxins and furans) of the S1/S2 sample.    Table V shows that the chemical analyses and calculation of individual TEFs produced a total TEQ of 61 for S1/S2.  These comparative data are shown again in Table VI, combined with additional data available from projects that have followed our RGS screening test with chemical analyses of selected samples.  Samples marked SAC1 through SAC10 were first screened by RGS, and later 6 samples collected nearby these 6 earlier samples were analyzed for dioxins and furans. There is reasonably good agreement between the two assays for five of the stations, but sample SAC4 was predicted by RGS to have significantly higher dioxin equivalents than chemical analyses showed.  Extracts from both samples SAC4 and SAC7 had a very dark appearance, which often indicates substantial contamination from PAHs.

Several projects have been conducted with the RGS assay, where the suspected contaminants were PAHs.  Table VII presents a reduced data set from one such investigation. Since the first set of tests with RGS, using 10 µL of extract produced very high fold induction, we later used 2  µL of the extracts to confirm the findings.    Also shown in Table VII are the analytical data, produced much later, for total PAHs and PCBs.  Since congener specific analyses of PCBs were not conducted, we can not assess the possible contribution of coplanar PCB induction of the RGS.  The data have also been converted to estimated benzo(a)pyrene (BaP) equivalents.  The estimated µg/g levels of benzo(a)pyrene equivalents correlates well ($r^2 = 0.85$) with the total PAH values measured in these sediments.  Since there are several PAHs that induce RGS more strongly than B(a)P, the RGS estimates of B(a)P equivalents can be higher or lower than chemical analytical data depending upon the levels of specific compounds present in the extract.

The final aspect of our testing results relates to the determination of RGS fold induction from extraction and analyses of animal tissues.  Spiking and extraction studies have been successfully conducted with fish, shrimp and mussel tissues. Field deployment of the mussel (*Mytilus edulis*) has been going on for about 20 years (mussel-watch), as these organisms are known to be excellent at accumulating various types of marine pollutants. Some of the data from a 2-month deployment of mussels at various sites in San Diego Bay are shown in Figure 4. From chemical analyses it was clear that PAHs were the primary contaminants accumulated by these animals. Figure 4 demonstrates a very good fit of the RGS responses ($r^2 = 0.87$) to the total PAH concentrations found in the mussels.

## Discussion

**Toxic Equivalency.** Table VIII provides an overview of the EPA TEF values for dioxin (10), and those of Safe (9) for the 5 PCB congeners tested previously (2). In addition, the table includes the recently tested PAHs, which are listed by the EPA in there most recent assessment of potential risk from PAHs (14). Nisbet and LaGoy (15) provided an additional evaluation of the studies comparing the carcinogenicity of individual PAHs, and proposed TEFs somewhat different than those listed by EPA (Table VIII). It is important to note that to our knowledge there has been no attempt to compare the relative risk of dioxins and coplanar PCBs to specific PAHs, using the same biological assay. Since we have tested each of these compounds with the RGS assay, it is possible to list the concentrations of each compound found to produce a significant (10 fold) induction. The range ($10^6$) of concentrations is from a low of 0.003 ng/mL for 2,3,7,8-dioxin to a high of 3,000 ng/mL for chrysene. As measured by the RGS test, the coplanar PCBs would be ranked orders of magnitude more distant from TCDD, than the present TEFs indicate. It is most interesting to find that PCB # 126, which is the strongest inducing PCB, is approximately equal to dibenzo(a,h)anthracene in *CYP1A1* induction potency. Two other PAHs (benzo(k)fluoranthene and indeno[1,2,3-cd]pyrene) induce transcription at lower concentrations than the other four coplanar PCBs (77, 114, 156, and 169).

The RGS responses are first ranked for TEFs in comparison with dioxin, as is the normal procedure. These estimated TEF values merely reflect the strength of induction discussed above, but also place the PAHs in this ranking well above four of the five coplanar PCBs tested. The last two columns compare our experimentally produced TEFs, using the RGS responses to PAHs, with the TEFs for these compounds previously published by the EPA (14) and proposed by Nisbet and LaGoy (15). Dibenzo(a,h)anthracene induces *CYP1A1* at concentrations considerably lower than those of benzo(a)pyrene, in disagreement with the EPA ranking, but closer to other estimates (15). In fact, benzo(k)fluoranthene, indeno(1,2,3-cd)pyrene, and benzo(a)anthracene produced a 10 fold induction of *CYP1A1* at lower concentrations than B(a)P.

Table VI. TEQs from RGS and Chemical Analyses of Soil and Water Samples

| Sample | RGS Estimates of TCDD Equiv. | | | | |
|---|---|---|---|---|---|
| | Fold Induction | Extract (ng/mL) | TEQ by RGS Estimate (ng/g; ng/L) | TEQ by Chemical Analysis (ng/g; ng/L) | Total Dioxins & Furans Measured (ng/g; ng/L) |
| 9SW-7.0 | 3 | 6.85 | 0.17 | 0.00003 | 0.86 |
| 8SW-6.5 | 3 | 6.85 | 0.17 | 0.00002 | 0.77 |
| S1/S2 | 43 | 196.60 | 51.00 | 0.06 | 18.44 |
| SAC1 | 25 | 2370.50 | 119.50 | | NC |
| SAC2 | 29 | 2701.10 | 135.10 | | NC |
| SAC3 | 78 | 7309.20 | 365.50 | 0.54 | 250.50 |
| SAC4 | 213 | 19889.10 | 995.00 | 8.40 | 4885.00 |
| SAC5 | 31 | 2893.20 | 144.70 | | NC |
| SAC6 | 117 | 13903.10 | 545.20 | 0.44 | 18.97 |
| SAC7 | 49 | 4540.40 | 227.00 | 0.33 | 145.60 |
| SAC8 | 17 | 1576.60 | 78.80 | | NC |
| SAC9 | 8 | 740.20 | 37.00 | | NC |
| NW1 | 116 | | 265.40 | 0.20 | 11.37 |
| NW2-WATER | 6 | | 13.22 | 0.02 | 2.61 |
| NW3-WATER | 3 | | 5.83 | 0.03 | 3.58 |

NC: No chemical analyses available

Table VII. RGS Responses to Marine Sediment Samples (40g)
Expressed as Benzo(a)pyrene Equivalents

| Station | PAHs (ug/g) | PCBs (ug/g) | Ave. Fold Induc. (10 uL) | (2 uL) | Equivalents BaP(ug/g) |
|---|---|---|---|---|---|
| 11 | 5.18 | 0.463 | 319 | 65.0 | 1.95 |
| 1 | 5.68 | 0.700 | 494 | 80.8 | 2.42 |
| 9 | 6.73 | 0.489 | 783 | 96.6 | 2.90 |
| 7 | 8.26 | 0.555 | 806 | 102.2 | 3.07 |

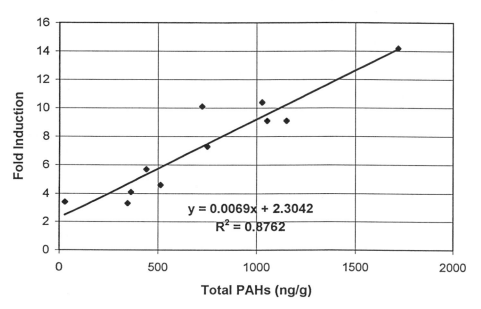

Figure 4. *Correlation of RGS responses to the application of mussel tissue extracts and the concentrations of total PAH in the tissue.* After two months at various locations, composite mussel tissue samples were extracted, and 20 μL of the dichloromethane extracts were applied to the RGS. Chemical analyses were provided by J. Means, Louisiana State Univer., Baton Rouge, LA.

Table VIII. Comparison of Concentrations of TCDD, TCDF, PCBs, and PAHs Required to Produce a 10 Fold Induction in P450 RGS and Relationship to EPA values for TEFs

| CHEMICAL | CONCENTRATION (ng/mL) | (ng/g soil) | RGS Relative to Dioxin | RGS Estimated TEFs (to Dioxin) | EPA TEFs | (Ref 15) Proposed TEFs | RGS Estimated TEFs | |
|---|---|---|---|---|---|---|---|---|
| | | | | | | | | |
| 2,3,7,8-TCDD | 0.003 | 0.008 | 1 | 1 | 1 | | | |
| 2,3,7,8-TCDF | 2.5 | 6.3 | 781 | 0.001280 | 0.10 | | | |
| PCB #126 | 2.5 | 6.3 | 781 | 0.001280 | 0.10 | | | |
| PCB #169 | 200.0 | 500.0 | 62,500 | 0.000016 | 0.05 | | | |
| PCB #77 | 300.0 | 750.0 | 93,750 | 0.000011 | 0.01 | | | |
| PCB #114 | 300.0 | 750.0 | 93,750 | 0.000011 | 0.001 | | | |
| PCB #156 | 800.0 | 2000.0 | 250,000 | 0.000004 | 0.0002 | | | |
| | | | | | (to BaP) | (to BaP) | (to BaP) | (to DBA) |
| Dibenzo(a,h)anthracene | 2.5 | 6.3 | 781 | 0.001280 | 1.00 | 5.00 | 100 | 1 |
| Benzo(k)fluoranthene | 10.0 | 25.0 | 3,125 | 0.000320 | 0.01 | 0.10 | 25 | 0.25 |
| Indeno(1,2,3-cd)pyrene | 12.5 | 31.3 | 3,906 | 0.000256 | 0.10 | 0.10 | 20 | 0.20 |
| Benzo(a)anthracene | 50.0 | 125.0 | 15,625 | 0.000064 | 0.10 | 0.10 | 5 | 0.05 |
| Benzo(a)pyrene | 250.0 | 625.0 | 78,125 | 0.000013 | 1.00 | 1.00 | 1.00 | 0.010 |
| Benzofluorene | 500.0 | 1250.0 | 156,250 | 0.000006 | NG | NG | 0.50 | 0.005 |
| Chrysene | 3000.0 | 7500.0 | 937,500 | 0.000001 | 0.01 | 0.01 | 0.08 | 0.00083 |

TCDF-2,3,7,8-Tetrachlorodibenzofuran    NG - Not Given

DBA - Dibenzo(a,h)anthracene

**Environmental Samples.**    Using the data generated in the various studies with RGS, it appears that the assay (with 20 µL) of an organic extract from a 40 gram environmental sample of sediment, soil or tissue, would detect as little as 0.008 ng/g of 2,3,7,8-TCDD. The most toxic PCB coplanar (#126) would be detected at about 6.2 ng/g. At about two orders of magnitude higher concentration (200 to 800 ng/g) the other specific PCB congeners would be detected in environmental samples. At concentrations of about 31 ng/g or less, three of the strongest inducing PAHs would produce a RGS induction of about 5 times background. The range of PAH induction indicates the RGS could detect some at 6 ng/g, but others would not be detected until the level reached 7.5 µg/g (chrysene).

The mixture of PAHs used as representative of environmental contamination could be detected in sediment samples at about 100 ng/g. Water samples are generally of 1 liter volumes, which means the levels that could be detected would be approximately 0.04 times (40g soil/ 1000g water) the detection limits discussed. This enhanced detection would likely be necessay, since most of the inducing compounds are quite low in water solubility.

We presently do not know which of the approximately 40 specific PAHs often quantified in environmental samples are potential CYP1A1 inducers. Those tested thus far that have exhibited induction (4- to 6-rings) are not necessarily the most prominent of PAHs at contaminated sites. Since petroleum contains a higher proportion of alkylated compounds, than the parent (ring structure only) structure, we would expect to find higher levels of the former in environmental samples. It likely that the complex mixture of compounds found in environmental samples contain some inducers, some that have no affect on the system, and others that inhibit to a degree the cellular response to inducers. Studies using various combinations of specific PAH should be conducted to evaluate this possibility. In the many sediment mixtures of PAH we have analyzed by RGS, there would seem to be little inhibition demonstrated, as the induction correlates well with the concentrations of total meansured PAHs. Testing with combinations of PAHs, coplanar PCBs and dioxin has shown the responses vary between about 30 and 100% of the response that would be expected if the effects were additive (J.W. Anderson, unpublished data). The mean response for these tests was 58% of expected additive effects.

An apparent weakness of the RGS method is its inability to identify specifically which components in extracts were responsible for the observed strong CYP1A1 induction. However, when working with environmental samples, which seldom contain only one type of contaminant, it is very cost-effective to use a screening tool such as RGS to determine which samples are of major concern and require further investigation. In these cases, the detection of a wide range of possible xenobiotics can be considered a strength of the approach. The purpose of these screening tests is to determine if one or some combination of compounds in the sediments has the potential for causing DNA-receptor mediated binding, which would have deleterious effects on organisms exposed to these substrates. This signal must then be followed by detailed chemical characterization to identify the responsible substances.

## Conclusions

The present approachs which use Toxic Equivalents (TEQs) to evaluate the potential risk from dioxins and coplanar PCBs use rather expensive chemical analyses to weight the different congeners on a basis of potential risk, and finally derive an overall risk assessment for the sample (10). The RGS has several advantages in that it is rapid, inexpensive, and it integrates the effects from multiple xenobiotics present in the extract. This study has demonstrated that PAH contamination of environmental samples may represent the same level of risk as PCB contamination. PCBs have been shown to biomagnify in the food web, unlike PAHs, but the overall concern should relate to impacts on biological systems, including humans. Considering the ubiquitous nature of PAH contamination and the comparative induction responses measured in this investigation, it is possible these compounds should be given attention at least equal to that of PCBs.

Chemical analyses have the advantage of being capable of measuring very low concentrations of specific dioxins, PCBs and PAHs, but because of time and cost limitations, few samples are completely characterized. For screening large sites, both spatially and vertically, and for most evaluations of biological significance, the detection limits offered by the RGS may be satisfactory. We believe it is more likely that a hot spot will be overlooked (in an effort to decrease costs) when composite samples or widely separated samples are characterized chemically, than when a screening test is run on numerous samples to narrow the selection for detailed chemical characterization

*Acknowledgements.* This research was conducted in part through support by the United States Health Service grant CA37139.

## Literature Cited

(1) Postlind, H., T. P. Vu, R. H. Tukey, and L. C. Quattrochi. 1993. Response of human CYP1-luciferase plasmids to 2,3,7,8-tetrachlorodibenzo-p-dioxin and polycyclic aromatic hydrocarbons. Toxicol. Appl. Pharmacol. 118:255-262.

(2) Anderson, J. W., S. S. Rossi, R. H. Tukey, T. Vu, and L. C. Quattrochi. 1995. A biomarker, P450 RGS, for assessing the induction potential of environmental samples. Environ. Toxicol. Chem. 14:1159-1169.

(3) Hankinson, O. 1995. The aryl hydrocarbon receptor complex. Annu. Rev. Pharmacol. Toxicol. 35:307-340.

(4) Denison, M. S., J. M. Fisher, and J. P. J. Whitlock. 1988. The DNA recognition site for the dioxin-Ah receptor complex. Nucleotide sequence and functional analysis. J. Biol. Chem. 263:17221-17224.

(5) Hahn, M. E., A. Poland, E. Glover, and J. J. Stegeman. 1987. The Ah receptor in marine animals:Phylogenetic distribution and relationship to cytochrome P4501A inducibility. Marine Environ. Res. 34:87

(6) Stegeman, J., M. Brouwer, R. T. Di Giulio, L. Forlin, B. A. Fowler, B. M. Sanders, and P. A. Van Veld. 1992. Molecular responses to environmental contamination: Enzyme and protein systems as indicators of chemical exposure and effect. In Biomarkers: Biochemical, Physiological and Histological Markers of Anthropogenic Stress. R. J. Huggett, R. A. Kimerle, P. M. J. Mehrle, and H. L. Bergman, editors. Lewis Publishers, Chelsea, MI.

(7) Collier, T. K., S. D. Connor, B. L. Eberhart, B. F. Anulacion, A.Goksyr, and U. Varanasi. 1992. Using cytochrome P450 to monitor the aquatic environment:initial results from regional and national surveys. Marine Environ. Res. 34:195-199.

(8) Goksyr, A. and A. Husy. 1992. The cytochrome P450 1A1 response in fish: Application of immunodetection in environmental monitoring and toxicological testing. Marine Environ. Res. 34:147-150.

(9) Safe, S. 1990. Polychlorinated biphenyls (PCBs), dibenzo-p-dioxins (PCDDs), dibenzofurans (PCDFs), and related compounds: Environmental and mechanistic considerations which support the development of toxic equivalency factors (TEFs). CRC. Crit. Rev. Toxicol. 21:51-88.

(10) Barnes, D.G. 1991. Toxicity Equivalents and EPA's risk assessment of 2,3,7,8-TCDD. Science Total Environment 104:73.

(11) El-Fouly, M. H., C. Richter, J. P. Giesy, and M. S. Denison. 1995. Production of a novel recombinant cell line for use as a bioassay system for detection of 2,3,7,8-tetrachlorodibenzo-p-dioxin-like chemicals. Environ. Toxicol. Chem. 13:1581-1588.

(12) McFarland, V. A., J. U. Clarke, and P. W. Ferguson. 1993. Dioxin in Sediments: Application of toxic equivalents based on international toxic equivalency factors to regulation of dredged material. Environmental Effects of Dredging Technical Notes. U.S. Army Enginer Waterways Experiment Station, Vicksburg, MS. EEDP-04-18:

(13) McFarland, V. A., J. U. Clarke, and P. W. Ferguson 1993. Initial comparisons of six assays for the assessment of sediment genotoxicity. Environmental Effects of Dredging TEchical Notes. U.S. Army Engineer Waterways Experimental Station, Vicksburg, MS. EEDP-04-20:

(14) Provisional guidance for quantitative risk assessment of polycyclic aromatic hydrocarbons. 1993. Office of Research and Development, Washington, DC. United States Environmental Protection Agency. EPA/600/R-93/089.

(15) Nisbet, I.C.T. and P.K. LaGoy. 1992. Toxic equivalency factors (TEFs) for polycyclic aromatic hydrocarbons (PAHs). Regulatory Toxicol. Pharmacol. 16:290-300.

# Chapter 13

# Field Application of Fluorescence-Based Catalytic Assays for Measuring Cytochrome P450 Induction in Birds

**J. A. Davis[1], D. M. Fry, and B. W. Wilson**

**Center for Ecological Health Research, 139 Hoagland Hall, University of California, Davis, CA 95616**

A variety of fluorometric catalytic assays have been developed for use in mammals to specifically measure cytochrome P450 (P450) induction due to two distinct classes of pollutants: dioxin-like compounds and certain organochlorine pesticides and PCBs. These substrates have also been employed in studies of P450 induction in birds. In birds, however, the specificity of many of the assays is altered and the information they yield is difficult to relate with exposure to specific pollutants. This report discusses the utility of resorufin- and coumarin-based assays in birds based on published data and results from our own laboratory and field studies. Only one of these assays, the ethoxyresorufin-o-deethylase (EROD) assay, has been demonstrated to have the sensitivity and specificity required in a biomarker for contaminant exposure in wild bird populations.

Induction of cytochrome P450 (P450) enzymes is one of the most widely measured biomarkers of pollutant exposure in wild animals (1,2). P450 enzymes catalyze the oxidation of both endogenous molecules and environmental toxicants. Exposures to molecules that are substrates for P450 induce tissues to manufacture more of these enzymes. The increased concentrations of P450 in tissues can be detected and used as a marker of exposure to the inducing chemicals. In mammals a variety of fluorometric catalytic assays can be used to specifically measure induction due to two distinct classes of pollutants: dioxin-like compounds and certain organochlorine pesticides and PCBs. To a limited extent, these substrates have also been employed in laboratory and field studies of P450 induction in birds. In birds, however, the specificity of many of the assays is altered and the information they yield is difficult to relate with exposure to specific pollutants. This review

[1]Current address: San Francisco Estuary Institute, 1325 South 46th Street, No. 180, Richmond, CA 94804

0097–6156/96/0643–0169$15.00/0

discusses the utility of fluorometric assays in studies of P450 induction in birds based on published data and results from our own studies. Desirable properties in biomarkers applied in the field are also discussed based on our detailed studies of EROD induction in wild bird populations.

## P450 Enzymes

Two points regarding the P450 enzyme family must be considered in interpretation of the catalytic activities of these enzymes in different species. First, catalytic assays may measure the activity of more than one P450 isozyme. The label P450 applies to a diverse collection of over 150 different isozymes, with some representatives found in every biological kingdom (3). Individual organisms can also have many different P450 enzymes in their tissues. The rat expresses more than 40 different P450 genes, and most of these are expressed in a single organ, the liver (4). Further complicating matters, P450 enzymes include constitutive and inducible forms, and both of these types can contribute to observed catalytic activity. In interpreting patterns of catalytic activity among different species it is important to consider that the patterns may be a product of the activities of a multiplicity of inducible and constitutive isozymes.

Another important point relating to species differences in catalytic activity is that minor changes in amino acid sequences in proteins can make qualitative differences in catalytic activity. For example, a single amino acid change in a mouse P450 peptide of 494 residues was observed to change the peptide's catalytic activity from coumarin 7-hydroxylase to testosterone 15-α-hydroxylase (5). Considering this observation, one might expect potentially substantial variation in catalytic activity among species, and even among populations and individuals.

In the present context the P450 isozymes of greatest interest are those that are induced by pollutants because they are potential biomarkers. In mammals, the best studied animal Class, there are two major P450 families that are induced by pollutants: the CYP1A family, which is induced by dioxins (e.g., 2,3,7,8-tetrachlorodibenzo-p-dioxin [TCDD]), dibenzofurans, polychlorinated biphenyls (PCBs), polycyclic aromatic hydrocarbons, and other compounds that bind to the Ah receptor; and the CYP2B family, which is induced by phenobarbital (PB), certain organochlorine pesticides (e.g., DDT and α-hexachlorocyclohexane), PCBs, and other compounds. Analogues to these two families are present in birds. In this paper the CYP1A analogues are referred to as TCDD-induced P450 and the CYP2B analogues as PB-induced P450.

**Avian TCDD-induced P450.** Two P450 isozymes are induced in chickens exposed to TCDD or other compounds with affinity for the Ah receptor (6,7). The mechanism of enzyme induction mediated by the Ah receptor has received a great deal of attention (8-11), and appears to be consistent among birds and mammals (12). Ah receptor agonists bind to the receptor and initiate a sequence of events that leads to increased transcription of mRNA and protein synthesis. One of the two P450 isozymes induced by TCDD in chickens (6) has a high capacity for oxidation as measured with the ethoxyresorufin-o-deethylase (EROD) and aryl hydrocarbon hydroxylase (AHH) assays and is currently identified as TCDD$_{AHH}$. The other isozyme is efficient at epoxygenation of arachidonic acid and is identified as TCDD$_{AA}$. Conventional "CYP" designations will be given to these enzymes when they have been cloned and sequenced (*6*).

**Avian PB-induced P450.** Three proteins are induced by PB in chickens (*7,13*). Genes for two of these proteins have been sequenced and designated CYP2H1 and CYP2H2 (*13*). Unlike TCDD induction of P450, the mechanism of enzyme induction by PB remains unclear but may involve a labile repressor protein that modulates transcription of the CYP2H genes (*13*). The PB-induced isozymes metabolize some substrates that are not metabolized by the TCDD-inducible forms (*e.g.*, aminopyrine and aldrin) and some that are (arachidonic acid and ethoxycoumarin) (*7,14*). Catalytic activity specific to the PB-induced isozymes (aminopyrine demethylase) is present in all three isozymes (*7*). Since organochlorine pesticides also cause PB-type induction, these PB-inducible P450 isozymes are potential biomarkers for these contaminants.

**Fluorescence-based Catalytic Assays for Measuring P450 Induction**

Induction of P450 isozymes can be quantified by measurement of increases in catalytic activity, immunochemical detection of P450 protein, and detection of mRNA using cDNA probes. These approaches complement each other. Catalytic assays are valuable because they provide an index of the actual metabolic capacity of functional P450 protein.

Our studies of P450 induction have employed fluorescence-based catalytic assays for several reasons. Fluorometric assays such as those based on alkylated homologues of resorufin and coumarin involve a relatively minimal amount of sample manipulation. Fluorescence of the metabolized substrate in these assays can be read directly in the vessel in which the reaction takes place. This is particularly useful in assays of cultured cells where the assay can be performed in the wells of multi-well culture plates. Furthermore, multiple fluorescent endpoints can be measured in the same well (*15,16, 51*). The advent of microplate fluorometers has also greatly reduced the sample volumes needed for fluorometric assays and vastly improved the efficiency with which these assays can be performed (*15*). Microplate fluorometers are sensitive enough to detect low rates of uninduced P450 activity in reaction volumes on the order of 100 ul.

A variety of fluorogenic substrates have been developed that can be used to specifically measure either TCDD- or PB-induced P450 isozymes in mammals. This specificity is of great value in field applications because it makes the induced enzyme a marker for a specific class of contaminant rather than just a marker for general contamination. To a limited extent, these substrates have also been employed in laboratory and field studies of cytochrome P450 induction in birds. In birds, however, the specificity of many of the assays is altered and the information they yield is difficult to relate with exposure to specific pollutants.

**Alkoxyresorufin Substrates**. The EROD assay yields consistent results in birds and mammals. In rats TCDD-type inducers cause approximately a 100-fold increase in hepatic EROD activity, while PB causes only a 3- to 5-fold increase (*17, 18*). Studies of hepatic EROD induction in many species of birds (chicken [*Gallus gallus*], pigeon [*Columba livia*], double-crested cormorant [*Phalacrocorax auritus*], great blue heron [*Ardea herodias*], black-crowned night heron [*Nycticorax nycticorax*], and eider duck [*Somateria mollissima*]) exposed to TCDD-type inducers indicate that the sensitivity of this response is similar to that seen in mammals (Table I). Exposure to TCDD or TCDD-type inducers in these species caused increases of 20- to 200-fold in hepatic EROD activity. There are exceptions to this rule, however. In some, but not all, studies Japanese quail (*Coturnix coturnix japonica*) have been relatively unresponsive to TCDD-type inducers (*27, 29, 30*). American kestrels (*Falco sparvarius*) (*33*) and domestic mallards (*Anas platyrhynchos*) (*34*) were also relatively unresponsive. The specificity of EROD as a measure of TCDD-induced P450 in birds is similar to that in mammals, with PB exposure causing only slight increases in activity (Table I). Detailed studies in chickens have identified one TCDD-inducible isozyme, TCDD$_{AHH}$, that is responsible for the increased EROD activity observed in induced individuals (*6, 7*).

Unlike EROD, data for other resorufin-based assays in birds indicate that they are either not specific for either class of P450 enzyme or conflict with the data from mammals. For example, Lubet *et al.* (*18*) found that pentoxyresorufin-o-deethylase (PROD) in rats is a sensitive and specific marker for PB-type induction. PB in this study increased PROD activity 41-fold over controls while the TCDD-type inducer β-naphthoflavone (BNF) caused only a negligible increase. In contrast, studies of PROD induction in 4 species of birds exposed to PB all indicate that PROD activity is induced by TCDD-type inducers, but not by PB (Table II). Lubet *et al.* (*18*) also found that in rats benzyloxyresorufin-o-deethylase (BROD) was sensitive (47-fold increase) and specific for PB-type induction. In birds, however, BROD can be induced by both TCDD- and PB-type inducers, with a more sensitive response to TCDD-type compounds (Table III). Therefore, of these three most commonly used alkoxyresorufin assays, only EROD retains the specificity observed in mammals. Although PROD and BROD have been measured along with EROD in wild bird populations, available evidence indicates that the lack of specificity of PROD and BROD makes it impossible to identify the pollutants causing whatever variation is observed in these activities.

**Alkoxycoumarin Substrates**. Fluorometric assays for P450 induction using alkoxycoumarin substrates have also been developed in rats. Assays of ethoxycourmarin-o-deethylase (ECOD) activity (*35*) have been commonly employed. In such studies ECOD activity is a general index of P450 induction, since ECOD is induced in rats by both TCDD-type and PB-type inducers (*35, 36*). Hepatic methoxycoumarin-o-deethylase (MCOD) activity has also been studied in rats, but with some contradictory results. One study (*37*) found that MCOD was specifically, though not markedly, induced by PB about 4- to 8-fold over controls. In contrast, another study found MCOD activity to be inducible by both BNF (a TCDD-type inducer) and PB (*36*).

**Table I. Summary of literature on experimental induction of hepatic EROD activity in birds**

| Species[a] | Age | Inducer | Dose | Preparation[b] | Induced activity[c] | Reference |
|---|---|---|---|---|---|---|
| Chicken | 17 d | BNF | 3 mg/egg | M | 34 | 20 |
| Chicken | 19 d | PCB 77 | 500 nmol/egg | S9 | 48 | 21 |
| | | PCB 169 | 500 nmol/egg | | 31 | |
| | | TCDD | 6 nmol/egg | | 58 | |
| Chicken | 10 d | PCB 126 | 4 ug/kg egg | H | 30 | 22 |
| | 8 d | PCB 77 | 4.3 ug/kg egg | | 74 | |
| Chicken | 18 d | BNF | 6.7 mg/egg | M | 100 | 7 |
| | | PB | 20 mg/egg | | 1.1 | |
| Chicken | 19 d | TCDD | 10 nmol/egg | S9 | 200 | 23 |
| Chicken | 16 d | TCDD | 3 ug/kg egg | M | 28 | 24 |
| Chicken | 18 d | BNF | 2 mg/egg | M | 169 | 25 |
| | | PB | 10 mg/egg | | 1.5 | |
| Chicken | 3 wk | TCDD | 100 ug/kg i.p. | M | 40 | 26 |
| Quail (F) | 8-12 wk | BNF | 7.3 mg/bird, i.p. | S9 | 20 | 19 |
| | | PB | 0.1% in drinking water | | 1 | |
| Quail (F) | Adult | 3MC | 20 mg/kg d, i.p. | M | 2.5 | 27 |
| | | BNF | 30 mg/kg d, i.p. | | 2.8 | |
| | | PB | 160 mg/kg d, i.p. | | 1 | |
| Quail (F) | Adult | BNF | 150 mg/kg/d, i.p. | M | 13 | 28 |
| | | PB | 70 mg/kg d, i.p. | | 1 | |
| Quail (M) | Adult | 3MC | 15 mg/kg d, i.p. | M | 0.5 | 29 |
| Quail (F) | Adult | PCB 126 | 50 ug/kg d, oral | M | 4.5 | 30 |
| | | PCB 105 | 3 mg/kg d, oral | | 36 | |
| Pigeon | 14 d | TCDD | 100 ug/kg egg | M | 30 | 24 |
| Heron | 22 d | TCDD | 100 ug/kg egg | M | 35 | 24 |
| Cormorant | 22 d | TCDD | 100 ug/kg egg | M | 22 | 24 |
| Night heron | 21 d | 3MC | 200 ug/egg | M | 14 | 31 |
| | | PB | 2 mg/egg | | 2 | |
| Eider duck | 4-5 wk | PCB 77 | 50 mg/kg i.p. | M | 57 | 32 |
| Kestrel | 1 yr | PCB 126 | 0.05 mg/kg d, oral | M | 1.9 | 33 |
| | | PCB 105 | 3 mg/kg d, oral | | 0.7 | |
| Mallard | Adult (6-8 wk) | BNF | 25 mg/kg d, oral | M | 5 | 34 |
| | | PB | 80 mg/kg, oral | | 5 | |

[a]See text for full species names. F=female, M=male.

[b]M=microsomes, S9=9000g supernatant, H=liver homogenate

[c]Induced activity expressed as fold-increase over control activity

**Table II.    Summary of literature on experimental induction of hepatic PROD activity in birds**

| Species[a] | Age | Inducer | Dose | Preparation[b] | Induced activity[c] | Reference |
|---|---|---|---|---|---|---|
| Chicken | 18 d | PB | 10 mg/egg | M | 1.3 | 25 |
|  |  | BNF | 2 mg/egg |  | 13 |  |
| Quail (F) | 8-12 wk | PB | 0.1% in drinking water | S9 | 1 | 19 |
|  |  | BNF | 7.3 mg/bird, i.p. |  | 4 |  |
| Night heron | 21 d | PB | 2 mg/egg | M | 1 | 31 |
|  |  | 3MC | 200 ug/egg |  | 12 |  |
| Mallard | Adult (6-8 wk) | PB | 80 mg/kg, oral | M | 0.8 | 34 |
|  |  | BNF | 25 mg/kg, oral |  | 6.3 |  |

[a]See text for full species names. F=female.

[b]M=microsomes, S9=9000g supernatant

[c]Induced activity expressed as fold-increase over control activity

Assays using alkoxycoumarin substrates have been applied much less frequently in studies of birds than have the alkoxyresorufins. Results from these studies are consistent with those from rats. Nakai *et al.* (7) purified the P450 isozymes induced in chick embryo liver by both BNF and PB. ECOD activity was found in both of the TCDD-inducible and in one of the PB-inducible P450 isozymes. In domestic mallards ECOD is also induced by both BNF and PB (34). One study of wild birds (black-crowned night herons) did not find significantly elevated ECOD after exposure to either 3-methylcholanthrene (a TCDD-type inducer) or PB (31).

The very few data available on induction of MCOD activity in birds are contradictory. Data from Riviere (34) indicate that this substrate is specific for PB-induced P450 in mallard ducks, with a slight but significant 1.4-fold increase in PB-treated animals and no increase in BNF-treated animals. Following up on the data of Riviere, we examined MCOD induction in chickens and found that both BNF and PB caused increases in activity of 1.5- and 1.6-fold, respectively. Even if MCOD is a measure of PB-induced activity in birds, the small increase observed in the induced state would make it difficult to apply this response as a biomarker in field studies.

**Summary of Fluorescence-based Catalytic Assays for Measuring P450 Induction**. One alkoxyresorufin assay, EROD, has the sensitivity and specificity needed in a biomarker for field studies of P450 induction in birds due to environmental contamination. This assay is a very informative marker of environmental exposure of birds to TCDD-like compounds, and has been used in many studies (31, 32, 39-44). Unfortunately, other resorufin- or coumarin-based P450 assays lack either the sensitivity or the specificity required to make them

**Table III. Summary of literature on experimental induction of hepatic BROD activity in birds**

| Species[a] | Age | Inducer | Dose | Preparation[b] | Induced activity[c] | Reference |
|---|---|---|---|---|---|---|
| Quail (F) | 8-12 wk | PB | 0.1% in drinking water | S9 | 3.6 | 19 |
| | | BNF | 7.3 mg/bird, i.p. | | 13 | |
| Night heron | 21 d | PB | 2 mg/egg | M | 3 | 31 |
| | | 3MC | 200 ug/egg | | 14 | |

[a]See text for full species names. F=female.

[b]M=microsomes, S9=9000g supernatant

[c]Induced activity expressed as fold-increase over control activity

useful in field studies of bird populations. At present there is no fluorogenic substrate for measurement of induction of PB-induced P450, a potential enzyme marker for organochlorine insecticides. It should also be noted that assays based on other types of measurements have not been established as markers of PB-induced P450, and that many of these non-fluorometric assays suffer from the same lack of sensitivity and specificity seen in the fluorescence-based assays.

**Case Study: EROD in Wild Avian Embryos**

In an ongoing study we have measured EROD activity in liver of double-crested cormorant (*Phalacrocorax auritus*) embryos (Davis *et al.*, in preparaton). One objective of this project is to examine hepatic EROD as an indicator of accumulation of TCDD-like compounds at the top of the aquatic food web of San Francisco Bay. Several studies have demonstrated that EROD activity in hepatic microsomes of wild bird species inhabiting polluted environments is highly correlated with accumulated concentrations of TCDD-like compounds (*31, 40, 42, 43, 50*). Sanderson *et al.* (*43*) specifically established this relationship for double-crested cormorant embryos, finding a significant regression of hepatic EROD activity on log-transformed TCDD toxic equivalents ($R^2$=0.69, p<0.00005). Because of this strong relationship between hepatic EROD activity and body burden of TCDD-equivalents, measurement of hepatic EROD represents a cost-effective alternative to expensive chemical analysis of dioxins, dibenzofurans, and coplanar PCBs.

In a two year period (1993-1994) we found the lowest median activity in 1993 samples from Hunters Island, OR (54 pmol/min/mg) (Table IV). In 1993 and 1994 median activity in samples from the 2 locations in San Francisco Bay was elevated 4- to 8-fold over Hunters Island and these differences were statistically significant. For both locations in San Francisco Bay median activity in 1994 was higher than in 1993, but due to the large variance in the data these increases were not statistically significant. The 4- to 8-fold elevation of median EROD activity observed in San Francisco Bay colonies is comparable to the degree of elevation

**Table IV. Summary of hepatic microsomal EROD in double-crested cormorant embryos**

| Location | Year | Count | Median[a] | Minimum | Maximum |
|----------|------|-------|-----------|---------|---------|
| Hunter Island, OR | 1993 | 10 | 54 (a) | 13 | 103 |
| Richmond Bridge, San | 1993 | 11 | 206 (b) | 87 | 801 |
| Francisco Bay, CA | 1994 | 10 | 370 (b) | 90 | 467 |
| San Mateo Bridge, San | 1993 | 8 | 202 (b) | 95 | 427 |
| Francisco Bay, CA | 1994 | 13 | 393 (b) | 48 | 1095 |

[a]Medians not sharing letter were significantly different in Kruskal-Wallis ANOVA and multiple comparison procedure, overall $p<.05$ for all comparisons.

found in other studies of birds inhabiting environments contaminated with high concentrations of TCDD-like compounds (*31, 39-43*).

The elevated microsomal EROD activities observed in colonies in San Francisco Bay is consistent with many other datasets indicating contamination of this ecosystem with polychlorinated biphenyls (*45-48*). We are currently analyzing PCB residues in yolk sacs from these embryos.

In experimental injections of double-crested cormorant eggs with TCDD, Sanderson and Bellward (*24*) established a relationship between TCDD dose and hepatic microsomal EROD activity in hatchlings. At a dose of 3 ug TCDD/kg egg, cormorant embryos showed a 5-fold increase in activity over controls. Sanderson and Bellward (*24*) also found that 3 ug TCDD/kg egg was the lowest dose at which mortality and subcutaneous edema, characteristic consequences of exposure to TCDD-like compounds (*49*), began to appear in cormorant embryos. These findings suggest that wild double-crested cormorant embryos in San Francisco Bay, which have a 4- to 8-fold greater activity than embryos from a reference location, are exposed to a concentration of TCDD-equivalents at the threshold for toxic effects in this species. Comparison of a biomarker response observed in the field with responses observed in controlled exposures, as made possible in this case by the work of Sanderson and Bellward (*24*), can provide an appreciation of the toxicological significance of the responses measured in wild populations.

This study demonstrates the utility of avian hepatic EROD measurement in studies of contamination due to TCDD-like compounds. Hepatic EROD possesses two important features that make it a useful biomarker in field applications. First, the sensitivity of the EROD response allows for statistical differentiation of populations with moderate differences in levels of exposure. Second, the specificity of EROD for TCDD-induced P450 allows conclusions to be drawn on the identity of the inducing substances.

**Acknowledgments**

Supported by the USEPA Center for Ecological Health Research (R814709), the NIEHS Center for Environmental Health Sciences (P30-ES-05707), and the US Air Force (AFOSR-91-0226).

## Literature Cited

1 Stegeman, J.J.; Brouwer, M.; DiGiulio, R.T.; Forlin, L; Fowler, B.A.; Sanders, B.M.; Van Veld, P.A. In *Biomarkers: Biochemical, Physiological, and Histological Markers of Anthropogenic Stress*; Editors, Huggett, R.J.; Kimerle, R.A.; Mehrle, P.M. Jr.; Bergman, H.L.; Lewis Publishers: Boca Raton, FL, 1992; pp 235-335.

2 Peakall, D.B. Animal Biomarkers as Pollution Indicators; Chapman & Hall: New York, N.Y., 1992.

3 Nelson, D.R., Kamataki, T., Waxman, D.J., Guengerich, F.P., Estabrook, R.W., Feyereisen, R., Gonzalez, F.J., Coon, M.J., Gunsalus, I.C., Gotoh, O., Okuda, K., and Nebert, D.W. *DNA Cell Biol.* **1993**, *12*, 1-51.

4 Guengerich, F.P. *Amer. Scientist* **1993**, *81*, 440-447.

5 Lindberg, R.L.P.; Negishi, M. *Nature* **1989**, *339*, 632-634.

6 Rifkind, A.B.; Kanetoshi, A.; Orlinick, J.; Capdevila, J.H.; Lee, C. *J. Biol. Chem.* **1994**, *269*, 3387-3396.

7 Nakai, K.; Ward, A.M.; Gannon, M.; Rifkind, A.B. *J. Biol. Chem.* **1992**, *267*, 19503-19512.

8 Denison, M.S. *Chemosphere* **1991**, *23*, 1825-1830.

9 Landers, J.P; Bunce, N.J. *Biochem. J.* **1991**, *276*, 273-287.

10 Okey, A.B. *Pharmac. Ther.* **1990**, *45*, 241-298.

11 Whitlock, J.P. Jr. *Annu. Rev. Pharmacol. Toxicol.* **1990**, *30*, 251-277.

12 Denison, M.S.; Okey, A.B.; Hamilton, J.W.; Bloom, S.E.; Wilkinson, C.F. *J. Biochem. Toxicol.* **1986**, *1*, 39-49.

13 Dogra, S.C.; Hahn, C.N.; May, B.K. Arch. *Biochem. Biophys.* **1993**, *300*, 531-534.

14 Lorr, N.A.; Bloom, S.E.; Park, S.S.; Gelboin, H.V.; Miller, H.; Friedman, F.K. *Mol. Pharmacol.* **1989**, *35*, 610-616.

15 Kennedy, S.W.; Lorenzen, A.; James, C.A.; Collins, B.T. *Anal. Biochem.* **1993**, *211*, 102-112.

16 Kennedy, S.W.; Jones, S.P. *Anal. Biochem.* **1994**, *222*, 217-223.

17 Burke, M.D.; Thompson, S.; Elcombe, C.R.; Halpert, J.; Haaparanta, T.; Mayer, R.T. *Biochem. Pharmacol.* **1985**, *34*, 3337-3345.

18 Lubet, R.A.; Guengerich, F.P.; Nims, R.W. *Arch. Environ. Contam. Toxicol.* **1990**, *19*, 157-163.

19 Lubet, R.A.; Syi, J.L.; Nelson, J.O.; Nims, R.W. *Chem.-Biol. Interactions.* **1990**, *75*, 325-339.

20 De Matteis, F.D.; Trenti, T.; Gibbs, A.H.; Greig, J.B. *Mol. Pharmacol.* **1989**, *35*, 831-838.

21 Rifkind, A.B.; Sassa, S.; Reyes, J.; Muschick, H. *Toxicol. Appl. Pharmacol.* **1985**, *78*, 268-279.

22 Brunstrom, B. *Chem.-Biol. Interactions* **1991**, *81*, 69-77.

23 Quilley, C.P.; Rifkind, A.B. *Biochem. Biophys. Res. Comm.* **1986**, *136*, 582-589.

24 Sanderson, J.T.; Bellward, G.D. *Toxicol. Appl. Pharmacol.* **1994**, *132*, 131-145.

25 Davis, J.A.; Fry, D.M.; Wilson, B.W. Unpublished data.

26 Sawyer, T.; Jones, D.; Rosanoff, K.; Mason, G.; Piskorska-Pliszczynska, Safe, S. *Toxicol.* **1986**, *39*, 197-206.

27 Riviere, J.L. *Les monoxygenase hepatiques et duodenales de la caille japonaise: Mise en evidence, induction et inhibition*; Thesis; University de Paris Sud: Centre Diorsay, France, 1984.

28 Carpenter, H.M.; Williams, D.E., Buhler, D.R. *J. Toxicol. Environ. Health* **1985**, *15*, 93-108.

29 Neal, G.E.; Judah, D.J.; Green, J.A. *Toxicol. Appl. Pharmacol.* **1986**, *82*, 454-460.

30 Elliott, J.E.; Kennedy, S.W.; Peakall, D.B.; Won H. *Comp. Biochem. Physiol.* **1990**, *96C*, 205-210.

31 Rattner, B.A.; Melancon, M.J.; Custer, T.W.; Hothem, R.L.; King, K.A.; LeCaptain, L.J.; Spann, J.W.; Woodin, B.R.; Stegeman, J.J. *Environ. Toxicol. Chem.* **1993**, *12*, 1719-1732.

32 Murk, A.J.; Van den Berg, J.H.J.; Fellinger, M.; Rozemeijer, M.J.C. *Environ. Pollut.* **1994**, *86*, 21-30.

33 Elliott, J.E.; Kennedy, S.W.; Jeffrey, D.; Shutt, L. *Comp. Biochem. Physiol.* **1991**, *99C*, 141-145.

34 Riviere, J.L. *Ecotoxicol.* **1992**, *1*, 117-135.

35 Greenlee, W.F.; Poland, A. *J. Pharmacol. Exper. Ther.* **1978**, *205*, 596-605.

36 Fry, J.R.; Garle, M.J.; Lal, K. *Xenobiotica* **1992**, *22*, 211-215.

37 Reen, R.K.; Ramakanth, S.; Wiebel, F.J.; Jain, M.P.; Singh, J. *Anal. Biochem.* **1991**, *194*, 243-249.

38 Boersma, D.C.; Ellenton, J.A.; Yagminas, A. *Environ. Toxicol. Chem.* **1986**, *5*, 309-318.

39 Fossi, C; Leonzio, C; Focardi, S. *Bull. Environ. Contam. Toxicol.* **1986**, *37*, 538-543.

40 Bellward, G.D.; Norstrom, R.J.; Whitehead, P.E.; Elliott, J.E.; Bandiera, S.M.; Dworschak, C.; Chang, T.; Forbes, S.; Hart, L.E.; Cheng, K.M. *J. Toxicol. Environ. Health* **1990**, *30*, 33-52.

41 van den Berg, M.; Craane, B.L.H.J.; Sinnige, T.; van Mourik, S.; Dirksen, S.; Boudewijn, T.; van der Gaag, M.; Lutke-Schipholt, I.J.; Spenkelink, B.; Brouwer, A. *Environ. Toxicol. Chem.* **1994**, *13*, 803-816.

42 Bosveld, A.T.C.; Gradener, J.; Murk, A.J.; Brouwer, A.; van Kampen, M.; Evers, E.H.G.; Van den Berg, M. *Environ. Toxicol. Chem.* **1995**, *14*, 99-115.

43 Sanderson, J.T.; Norstrom, R.J.; Elliott, J.E.; Hart, L.E.; Cheng, K.M.; Bellward, G.D. *J. Toxicol. Environ. Health* **1994**, *41*, 247-265.

44 Sanderson, J.T.; Elliott, J.E.; Norstrom, R.J.; Whitehead, P.E.; Hart, L.E.; Cheng, K.M.; Bellward, G.D. *J. Toxicol. Environ. Health* **1994**, *41*, 435-450.

45 San Francisco Bay Regional Water Quality Control Board. *Contaminant Levels in Fish Tissue from San Francisco Bay, Final Draft Report*; San Francisco Bay Regional Water Quality Control Board: Oakland, CA, 1994.

46 San Francisco Estuary Institute. *1993 Annual Report: San Francisco Estuary Regional Monitoring Program for Trace Substances*; San Francisco Estuary Institute: Richmond, CA, 1994.

47 Davis, J.A.; Gunther, A.J.; Richardson, B.J.; O'Connor, J.M.; Spies, R.B.; Wyatt, E.; Larson, E.; Meiorin, E.C. *Status and Trends Report on Pollutants in the San Francisco Estuary*; San Francisco Estuary Project: Oakland, CA, 1991.

48 Phillips, D.J.H.; Spies, R.B. *Mar. Pollut. Bull.* **1988**, *19*, 445-453.

49 Gilbertson, M.; Kubiak, T.; Ludwig, J.; Fox, G. *J. Toxicol. Environ. Health* **1991**, *33*, 455-520.

50 Rattner, B.A.; Hatfield, J.S.; Melancon, M.J.; Custer, T.W.; Tillitt, D.W. *Environ. Toxicol. Chem.* **1994**, *13*, 1805-1812.

51 Kennedy, S.W.; Jones, S.P.; Bastien, L.J. *Anal. Biochem.* **1995**, *226*, 362-370.

Chapter 14

# Initial Results for the Inductively Coupled Plasma–Mass Spectrometric Determination of Trace Elements in Organs of Striped Bass from Lake Mead, U.S.A.

Vernon Hodge[1], Klaus Stetzenbach[2], and Kevin Johannesson[2]

[1]Department of Chemistry and [2]Harry Reid Center for Environmental Studies, University of Nevada, Las Vegas, NV 89121–4003

Twenty-two elements (Li, Be, V, Cr, Mn, Co, Ni, Ga, As, Se, Rb, Sr, Ag, Cd, In, Te, Cs, Ba, Tl, Pb, Bi, U) were determined in tissues from nine organs of striped bass (*Morone saxatilis*) from Lake Mead by ICP-MS. Similarities and differences are presented in the concentrations of these elements and in chemical families of these elements. Results for rare earth elements (REEs) are presented for the intestinal contents of the fish, for filtered lake water, and for particulate matter filtered from the water. The shale-normalized patterns for the REEs in the intestinal contents and in the water-borne particulates show a similar depletion in the heavy REEs. This suggests that particulate matter may be a significant source of the REEs in the fish's diet.

The highest concentrations of natural radioactivity ever reported for any biological tissue were found in the digestive organs of oceanic fish (*1*). Almost one million times more $^{210}$Po was found in the caecum of an albacore than was found in seawater. However, high concentrations of $^{210}$Po are not limited to oceanic fish, but are also found in the digestive organs of striped bass and catfish from Lake Mead (*2*). About 500,000 times more $^{210}$Po was found in the intestine of a striped bass than in the unfiltered lake water. The question posed by these remarkable accumulations was: Are elements in the same chemical family, such as selenium and tellurium, also highly concentrated (bioaccumulated) in the digestive tract? The availability of an inductively coupled plasma-mass spectrometer (ICP-MS) provided the analytical tool to determine not only the concentrations of selenium and tellurium, but a total of 36 elements in the acid digests of several different tissues from striped bass caught in Lake Mead. The rare earth elements were also measured in filtered lake water (0.45 $\mu$m) and in the associated particulate material.

0097–6156/96/0643–0180$15.00/0

## Experimental

The fish were caught during the summer of 1994 in Lake Mead, placed in plastic bags, and frozen until needed for analysis (less than one week). The fish were dissected, and all or portions of the desired organs from three healthy fish were weighed into acid-washed Teflon beakers and dried on a hotplate. Twenty-five mL of ultrapure nitric acid (Seastar Chemicals, Seattle, WA) was added to each of the samples. They were covered with a Teflon watch glass and allowed to react at room temperature. Subsequently, the samples were digested for two days at reflux until they remained light yellow or colorless. Additional nitric acid was added periodically to all samples during the digestion process to replace that which evaporated. Finally, the samples were evaporated to about 10 mL and diluted to 100 mL with 1% (v/v) nitric acid for analysis. Two reagent blanks were carried along with every 10 samples. Ultrapure water was prepared from tap water which was passed through a reverse osmosis membrane, mixed-bed ion exchange resins, and distilled in an all glass still. The purified water was used to make all solutions (*3*).

Particulate matter was collected from lake water by pumping 35 L of water through a polysulfone filter cartridge, 0.45 $\mu$m (Gelman Sciences, Ann Arbor, MI), or by vacuum filtering a 50 L sample through acid-washed paper filters (Whatman, No. 41, Whatmann International Ltd., Maidstone, England). Four liters of the filtered water was acidified with ultrapure nitric acid to 1% (v/v) and saved for analysis. The particulates were removed from the filters by refluxing the filters with aqua regia for 24 hours. The aqua regia was evaporated to dryness in 1000-mL Teflon™ beakers. After several treatments with concentrated nitric acid followed by hydrochloric acid (Seastar Chemicals, Seattle, WA), the residues were dissolved in 1% (v/v) hydrochloric acid, filtered, and the REEs preconcentrated for ICP-MS analysis by cation exchange in the same manner as the filtered water samples (*3*). The final residues from the columns were dissolved in 1 mL of hot concentrated nitric acid and diluted to 20 mL with ultrapure water for ICP-MS analysis. The analysis included blanks for the filters and the ion exchange procedure.

All samples were assayed using a Perkin-Elmer Corporation (Norwalk, CT) Elan-5000 ICP-MS (*3*). Prior to analysis, 20 mL of the sample was placed into acid-washed polyethylene bottles and spiked with enough terbium to produce a 10 ppb signal. The terbium served as an internal standard for the analyses of all elements excepting the REEs. Platinum was used as the internal standard for the determination of the REEs. Analysis of National Institute of Standards and Technology (NIST) "Trace Elements in Water" (SRM 1643c) and other solutions prepared from NIST-traceable multielement solutions (High-Purity Standards, Charleston, SC) served as a check on the accuracy of the calibration. The NIST solution contains 15 of the 22 elements measured in the striped bass organs covering the mass range from 7 (Li) to 208 (Pb). Details of the analysis procedure are reported elsewhere (*3*). All results have been corrected for the appropriate blank analysis.

## Results and Conclusions

The average concentrations of 22 elements in the organs of the striped bass are listed in Table I. The REEs in the intestinal contents, filtered water, and particulates are shown in Table II. In general, the REEs were at or below the ICP-MS detection limits for the 100-mL analysis solutions of most of the tissue samples. Preconcentration by cation exchange was used to obtain the values reported for the water and particulate samples.

The highest concentrations of lithium are in samples of the fin + skin, the gill, and the digestive tract (27 ppb to 234 ppb). The same general trend is seen for vanadium (5.4 ppb to 135 ppb), manganese (527 ppb to 3300 ppb), nickel (31 ppb to 1900 ppb), strontium (1730 ppb to 135,000 ppb), barium (129 ppb to 7900 ppb), and uranium (2.8 ppb to 42.1 ppb). The high concentrations of strontium and barium, and possibly lithium, suggest that these elements are associated with the bony material in the fish and its food. Beryllium, silver, indium, and tellurium are found in measurable concentrations only in the sample of the intestine + contents: 5 ppb beryllium (five times the standard deviation of the measurement), 0.01 ppb indium (six times the standard deviation), and 13 ppb tellurium (thirteen times the standard deviation). Preliminary measurements show that there is about 0.005 ppb tellurium in the Lake Mead water analyzed, giving a bioconcentration factor of about 3000 for the intestine, more than 100 times less than that for $^{210}$Po, suggesting that these two elements are not similarly accumulated in the intestines of striped bass.

Gallium concentrations range from 4 ppb to 290 ppb, with the highest concentration found in the sample of fin + skin. Arsenic concentrations range from 283 ppb to 1340 ppb, with the highest concentrations in the eye, liver, and digestive tract. Selenium concentrations peak in the liver tissue (12,200 ppb), with high concentrations also in the gill, muscle, kidney, and intestine without contents, all having about 5000 ppb. Preliminary measurements suggest that there was about 5 ppb of selenium in the Lake Mead water analyzed, which gives a bioconcentration factor for selenium in the intestine of about 500, about 1000 times less than that of $^{210}$Po. Cadmium appears in the liver (64 ppb) and digestive tract (from 6.5 ppb, in the empty stomach, to 48 ppb in the intestinal contents). Measurable lead concentrations were found in the eye (55 ppb), fin + skin (120 ppb), intestine + contents (24.8 ppb), empty intestine (52 ppb), and the empty stomach (14 ppb). While the fish were alive, the eye and fin + skin samples were directly exposed to the lake water, and the digestive tract was also in intimate contact with the water via the fish's food. However, the gill does not show measurable lead concentrations. It would be expected that this organ would have measurable lead, if the lead came directly from the water. Moreover, since the digestion blanks did not contain measurable concentrations of lead, contamination of the samples during the lengthy digestion procedure does not appear likely. Bismuth is found in samples of the heart (20.5 ppb), muscle (1.5 ppb), liver (0.7 ppb), and intestinal tract (1.2 ppb to 2.7 ppb).

The concentrations of strontium and barium, two alkaline earth elements, peak in the sample of fin + skin (135,000 ppb and 7900 ppb, respectively) and reach their lowest values in the muscle and liver tissue samples (about 500 ppb and about 40 ppb respectively).  The ratio of strontium/barium is not constant and ranges from 3.7 in the stomach contents to 41 in the heart.  Excluding those organs, the eye and the intestine + contents, the average strontium/barium ratio is 14 ± 3.

Lithium, rubidium, and cesium are alkali metals.  Except for the sample of intestine + contents, the stomach without contents, and the stomach contents, the rubidium/cesium ratio is fairly uniform, 66 ± 9, or, excluding all of the digestive tract samples, 70 ± 5.   However, the rubidium/lithium and cesium/lithium ratios are quite variable, ranging from 1.1 to 770, and 0.10 to 3.6, respectively.  The rubidium/lithium ratio is the highest in samples of the muscle (770), kidney (270), heart (190), and liver (120), while the cesium/lithium ratio is highest in the kidney (3.6), heart (2.7) and liver (1.6), and near its lowest in the muscle (0.13).  All of the lowest values for both ratios are in the digestive tract, the intestine, and the stomach tissue.

The fourteen rare earth elements, including lanthanum (promethium was excluded because it does not occur in nature), were measured in all of the samples, but found only in high concentrations in the intestinal contents. Yttrium and thorium are also reported (Table II).  The REEs have been used to fingerprint ocean waters (4) and groundwaters (5) by the patterns that result when the absolute concentrations are normalized to the REE concentrations found in a standard material, such as a shale (4).  When the concentrations of the elements in the intestinal contents are compared to the intestine from which they were squeezed, it is apparent that most of the elements are more highly concentrated in this exudate than in the intestine tissue itself.  Therefore, it is apparent that much of the biomagnification has occurred before reaching the fish, and may include the particulate (detritus and minute organisms) in the water.

The concentrations of the REEs in the intestinal contents are about 10,000 times greater than the concentrations in the filtered water and only about 150 times the concentrations in the particulates.   Thus, the build-up factor, bioconcentration or bioaccumulation factor, will be very different for the REEs or possibly any element, depending on which is considered the source of the elements of interest.  Figure 1 shows the shale-normalized ratios for the REEs in the intestinal content, the filtered water, and the captured particulates.  The curves for the intestinal contents and the particulate matter appear to have similar shapes and are different from the pattern for the REEs in the filtered water, or the "dissolved" REEs.   The normalized values for the intestinal contents and the particulates show a trend toward depletion in the heavy REEs (HREEs).  In contrast, the filtered water shows a "w shaped" pattern, with a possible increase toward the HREEs.  These initial results suggest that the bioaccumulation factors for elements in the sample of intestinal contents will probably be more informative if calculated relative to the concentrations of the elements in the particulate matter.  It may be possible, with preconcentration, to establish whether or not the "dissolved" fraction of the REEs is an important

Table I.  Concentrations of Twenty-two Trace Elements in Selected Tissues

Trace Element Concentrations in ppb* ($\mu$g/kg) for Wet Tissue

| Element | Eye | | | Fin & Skin | | | Heart | | |
|---------|------|---|------|---------|---|------|------|---|------|
| Li | 15 | ± | 1 | 91 | ± | 4 | 9 | ± | 1 |
| Be | 2 | ± | 1 | 2 | ± | 1 | < 0.5 | | |
| V | 6.8 | ± | 0.3 | 80 | ± | 1 | 12 | ± | 1 |
| Cr | 310 | ± | 20 | 370 | ± | 40 | 140 | ± | 60 |
| Mn | 233 | ± | 3 | 3300 | ± | 40 | 306 | ± | 7 |
| Ni | 50 | ± | 10 | 1900 | ± | 100 | 50 | ± | 30 |
| Co | 5.8 | ± | 0.3 | 54 | ± | 5 | 20 | ± | 1 |
| Ga | 19 | ± | 1 | 290 | ± | 30 | 19 | ± | 1 |
| As | 1160 | ± | 40 | 610 | ± | 20 | 394 | ± | 7 |
| Se | 3700 | ± | 100 | 3540 | ± | 70 | 3720 | ± | 20 |
| Rb | 663 | ± | 4 | 1380 | ± | 6 | 1690 | ± | 20 |
| Sr | 4250 | ± | 30 | 135,000 | ± | 6000 | 1760 | ± | 20 |
| Ag | <0.4 | | | < 0.3 | | | < 1 | | |
| Cd | <1 | | | < 0.7 | | | < 5 | | |
| In | <0.01 | | | < 0.01 | | | < 0.02 | | |
| Cs | 10.1 | ± | 0.3 | 19.0 | ± | 0.2 | 24.7 | ± | 0.2 |
| Ba | 650 | ± | 10 | 7900 | ± | 80 | 43 | ± | 3 |
| Tl | 1.8 | ± | 0.2 | 3.3 | ± | 0.1 | 5.0 | ± | 0.2 |
| Pb | 55 | ± | 1 | 120 | ± | 2 | < 2 | | |
| Bi | 0.2 | ± | 0.1 | 0.10 | ± | 0.05 | 20.5 | ± | 0.7 |
| U | 0.6 | ± | 0.1 | 9.4 | ± | 0.7 | 0.5 | ± | 0.2 |
| Te | < 0 .1 | | | < 0.1 | | | < 0.1 | | |

* ± One standard deviation

of Striped Bass from Lake Mead, USA, as Determined by ICP-MS.

## Trace Element Concentrations in ppb* (μg/kg) for Wet Tissue

| Gill | Muscle | Kidney | Liver |
|---|---|---|---|
| 76 ± 3 | 3.4 ± 0.4 | 12 ± 1 | 14 ± 1 |
| 5 ± 2 | <0.5 | <0.6 | 3 ± 1 |
| 56 ± 1 | 0.8 ± 0.1 | 4.8 ± 0.2 | 35 ± 1 |
| 600 ± 40 | 150 ± 10 | 130 ± 20 | 450 ± 30 |
| 2370 ± 10 | 71 ± 2 | 228 ± 3 | 1250 ± 20 |
| 610 ± 20 | <3 | 78 ± 9 | <5 |
| 20 ± 1 | 2.1 ± 0.1 | 47.5 ± 0.4 | 151 ± 2 |
| 41 ± 3 | 29 ± 1 | 54 ± 2 | 42 ± 1 |
| 650 ± 20 | 283 ± 3 | 295 ± 9 | 1340 ± 20 |
| 5100 ± 100 | 4900 ± 200 | 5200 ± 300 | 12,200 ± 800 |
| 1010 ± 8 | 2610 ± 20 | 3250 ± 40 | 1640 ± 20 |
| 55,300 ± 800 | 505 ± 5 | 4460 ± 60 | 562 ± 4 |
| <1 | <0.4 | <0.6 | <0.4 |
| <1 | <0.8 | < 2 | 64 ± 2 |
| < 0.02 | <0.05 | <0.02 | <0.02 |
| 13.7 ± 0.3 | 43.6 ± 0.1 | 43.7 ± 0.6 | 22.3 ± 0.02 |
| 3350 ± 40 | 39 ± 1 | 254 ± 6 | 35 ± 1 |
| 3.4 ± 0.2 | 3.3 ± 0.2 | 12.2 ± 0.2 | 9.6 ± 0.1 |
| <2 | <1 | <1 | 2 ± 1 |
| 0.3 ± 0.1 | 1.5 ± 0.1 | <0.8 | 0.8 ± 0.1 |
| 9.4 ± 0.8 | <0.9 | 1.1 ± 0.2 | 0.7 ± 0.1 |
| <0.1 | <0.1 | <0.1 | 3 ± 1 |

Table I. (Continued) Concentrations of Twenty-two Trace Elements in Selected Tissues

Trace Element Concentrations in ppb* ($\mu$g/kg) for Wet Tissue

| Element | Intestine + Contents | | | Intestine - Contents** | | | Intestinal Contents | | |
|---------|------|---|-----|------|---|-----|------|---|-----|
| Li | 234 | ± | 5 | 27 | ± | 1 | 113 | ± | 5 |
| Be | 5 | ± | 1 | 4 | ± | 2 | 2 | ± | 1 |
| V | 135 | ± | 1 | 15.0 | ± | 0.7 | 113 | ± | 1 |
| Cr | 250 | ± | 10 | 67 | ± | 4 | 100 | ± | 30 |
| Mn | 2300 | ± | 20 | 629 | ± | 6 | 2360 | ± | 30 |
| Ni | 142 | ± | 4 | 45 | ± | 8 | 200 | ± | 10 |
| Co | 48.9 | ± | 0.5 | 27.8 | ± | 0.6 | 49.6 | ± | 0.7 |
| Ga | 33 | ± | 1 | 32 | ± | 1 | 46 | ± | 1 |
| As | 470 | ± | 5 | 1330 | ± | 30 | 450 | ± | 10 |
| Se | 2100 | ± | 100 | 6700 | ± | 200 | 1160 | ± | 20 |
| Rb | 262 | ± | 1 | 1920 | ± | 5 | 2340 | ± | 20 |
| Sr | 18,100 | ± | 100 | 1730 | ± | 10 | 9900 | ± | 100 |
| Ag | <0.2 | | | < 0.6 | | | < 1 | | |
| Cd | 42 | ± | 2 | 18 | ± | 2 | 48 | ± | 2 |
| In | 0.06 | ± | 0.01 | < 0.02 | | | < 0.07 | | |
| Cs | 26.5 | ± | 0.3 | 29.4 | ± | 0.6 | 49.3 | ± | 0.7 |
| Ba | 4060 | ± | 20 | 129 | ± | 3 | 900 | ± | 10 |
| Tl | 5.3 | ± | 0.3 | 4.6 | ± | 0.2 | 9.8 | ± | 0.2 |
| Pb | 24.8 | ± | 0.8 | 52 | ± | 3 | < 2 | | |
| Bi | 1.2 | ± | 0.1 | 2.7 | ± | 0.1 | 1.6 | ± | 0.1 |
| U | 42.1 | ± | 0.3 | 2.8 | ± | 0.8 | 12.7 | ± | 0.5 |
| Te | 13 | ± | 1 | 4 | ± | 1 | 3 | ± | 1 |

* ± One standard deviation

of Striped Bass from Lake Mead, USA, as Determined by ICP-MS.

Trace Element Concentrations in ppb* ($\mu$g/kg) for Wet Tissue

| Stomach + Contents | | | Stomach - Contents** | | | Stomach Contents | | |
|---|---|---|---|---|---|---|---|---|
| 38 | ± | 1 | 60 | ± | 2 | 122 | ± | 2 |
| 3 | ± | 1 | 2 | ± | 1 | 1 | ± | 1 |
| 8.7 | ± | 0.3 | 11.6 | ± | 0.2 | 5.4 | ± | 0.8 |
| 430 | ± | 20 | 490 | ± | 30 | 130 | ± | 30 |
| 1030 | ± | 10 | 527 | ± | 3 | 1980 | ± | 10 |
| 31 | ± | 3 | 39 | ± | 4 | 210 | ± | 10 |
| 10.0 | ± | 0.3 | 24.1 | ± | 0.4 | 39.6 | ± | 0.8 |
| 4 | ± | 1 | 23 | ± | 1 | 20 | ± | 1 |
| 1040 | ± | 20 | 450 | ± | 10 | 288 | ± | 5 |
| 5200 | ± | 200 | 2600 | ± | 100 | 1470 | ± | 30 |
| 1510 | ± | 20 | 170 | ± | 1 | 208 | ± | 2 |
| 2210 | ± | 20 | 2900 | ± | 6 | 18,800 | ± | 100 |
| | <0.3 | | | <0.3 | | | <1 | |
| 13 | ± | 1 | 6.5 | ± | 0.7 | 13 | ± | 2 |
| | < 0.02 | | | <0.02 | | | 0.02 | |
| 27.5 | ± | 0.2 | 5.8 | ± | 0.1 | 13.6 | ± | 0.2 |
| 197 | ± | 4 | 282 | ± | 1 | 5070 | ± | 30 |
| 4.8 | ± | 0.2 | 3.8 | ± | 0.3 | 6.6 | ± | 0.2 |
| | <0.6 | | 14 | ± | 1 | | <3 | |
| 0.4 | ± | 0.1 | 0.3 | ± | 0.1 | | <0.1 | |
| 3.5 | ± | 0.2 | 10.5 | ± | 0.4 | 22 | ± | 2 |
| 2 | ± | 1 | 2 | ± | 1 | | < 0.1 | |

Table II.    Concentrations of Rare Earth Elements Yttrium and Thorium in the Intestinal Contents of Striped Bass, Filtered Water, and Particulate Matter from Lake Mead, USA, as Determined by ICP-MS.

| Element | Intestinal Contents (ppb) ± SD* | | Filtered Water ng/L of Water ± SD | | Particulates ng/L of water ± SD | |
|---|---|---|---|---|---|---|
| Y | 14.7 | ± 0.6 | 13.9 ± | 0.6 | 91 ± | 2 |
| La | 23.1 | ± 0.3 | 3.70 ± | 0.07 | 198 ± | 1 |
| Ce | 47.7 | ± 0.3 | 2.95 ± | 0.08 | 350 ± | 2 |
| Pr | 5.8 | ± 0.2 | 0.27 ± | 0.02 | 41.1 ± | 0.4 |
| Nd | 21.4 | ± 0.7 | 2.0 ± | 0.2 | 144 ± | 22 |
| Sm | 4.2 | ± 0.2 | 0.39 ± | 0.05 | 23.6 ± | 0.3 |
| Gu | 0.9 | ± 0.1 | 0.11 ± | 0.4 | 7.5 ± | 0.2 |
| Gd | 3.9 | ± 0.3 | 0.53 ± | 0.06 | 36 ± | 1 |
| Tb | 0.49 | ± 0.04 | 0.038 ± | 0.006 | 3.2 ± | 0.1 |
| Dy | 2.5 | ± 0.1 | 0.20 ± | 0.02 | 13.1 ± | 0.4 |
| Ho | 0.50 | ± 0.03 | 0.048 ± | 0.008 | 2.5 ± | 0.1 |
| Er | 13 | ± 1 | 0.17 ± | 0.03 | 7.2 ± | 0.2 |
| Tm | 0.18 | ± 0.02 | 0.039 ± | 0.006 | 0.86 ± | 0.04 |
| Yb | 1.0 | ± 0.1 | 0.24 ± | 0.03 | 5.3 ± | 0.1 |
| Lu | 0.16 | ± 0.03 | 0.043 ± | 0.009 | 0.77 ± | 0.06 |
| Th | 9.0 | ± 0.5 | 0.11 ± | 0.05 | 31 ± | 1 |

* Standard deviation (n = 3)per wet weight for intestinal contents.

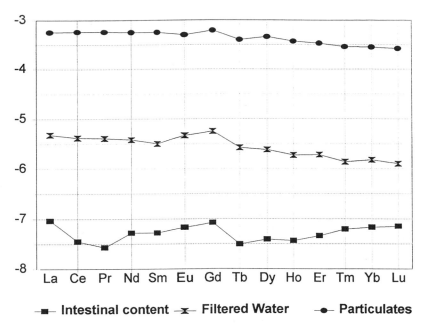

Figure 1. Logarithm of shale-normalized REE concentrations in the intestinal contents of striped bass, filtered water, and particulate matter from Lake Mead, USA.

source of these metals in organs, such as the liver and kidney, which are not in intimate contact with the fish's diet.

## Conclusions

The shale-normalized values of the REEs in the intestinal contents and the lake water particulates show a trend toward depletion of the HREEs. Normalized values for the concentrations of the REEs in the filtered water shows a different pattern, a "w-shaped" bimodal pattern, with enrichment in the HREEs. This difference suggests that the particulate matter may be a significant source for the REEs in the fish's intestinal contents and its diet. The REEs have bioaccumulation factors of about 10,000 in the intestinal contents, when compared to the "dissolved" concentrations of these elements in the lake water, but the results of this study suggest that an accumulation factor of about 150, calculated from the concentrations of REEs found in the particulate matter filtered, is more realistic. Moreover, the ICP-MS was able to determine many elements simultaneously in the digests of the striped bass tissues. It also was able to determine the long-lived radioisotopes, $^{238}U$ and $^{232}Th$, in most of the samples, without additional effort.

## Acknowledgments

The authors thank Megumi Amano for assistance in the ICP-MS measurements and Leslie Gorr for catching the fish and typing the manuscript.

## Literature Cited

1.  Folsom, T.R.; Wong, K.M.; Hodge, V.F. In The Natural Radiation Environment II, CONF-720805; Adams, J.A.S., Lowder, W.M., Gesell, T.F., Eds.; U.S. Energy Research Development Administration, 1975. Chapter 50, pp 863-882.
2.  Rosal, C. "Po-210 Radioactivity in Lake Mead Fish." Master of Science Thesis, Chemistry Department, University of Nevada-Las Vegas, Las Vegas NV 89121-4003, 34 pp (1992).
3.  Stetzenbach, K.J.; Amano, M.; Kreamer, D.K.; Hodge, V.H. *Groundwater* **1994**, *32*, 976-985.
4.  Sholkovitz, E.R. *Am. J.Sci.* **1988**, *288*, 236-281.
5.  Smedley, P.L. Geochim. Cosmochim. Acta **1991**, *55*, 2767-2779.

## Chapter 15

# Aquatic Toxicity of Chemical Agent Simulants as Determined by Quantitative Structure–Activity Relationships and Acute Bioassays

**N. A. Chester[1], G. R. Famini[1], M. V. Haley[1],
C. W. Kurnas[1], P. A. Sterling[1], and L. Y. Wilson[2]**

**[1]U.S. Army Edgewood Research, Development and Engineering Center,
Aberdeen Proving Ground, MD 21010
[2]Department of Chemistry, La Sierra University, Riverside, CA 92515**

Chemical agent simulants (simulants) have been used over the last 30 years to resolve obvious safety issues and provide sources for the development and testing of chemical defense equipment and procedures. In response to increasingly more stringent requirements being placed on the use of the simulants, restrictions involving hazardous waste disposal, and the necessity of performing outdoor tests, a major emphasis has been placed on efforts for determining environmental toxicity and damaging effects. Aquatic bioassays including the Microtox Assay (*Photobacterium phosphoreum*) and 48 hr *Daphnia magna* bioassay, were employed to provide preliminary acute toxicity data. A *D magna* $EC_{50}$ QSAR model from the TOPKAT software package and a QSAR model for estimating *P. phosphoreum* $EC_{50}$s based on the Theoretical Linear Solvation Energy Relationship (TLSER), were used to compare estimates with actual bioassay data. Chemicals tested were chosen from a list of the 70 "most used" simulants which include glycols, ethers, acetates, sulfides, organophosphates, and various other chemicals.

Chemical agent simulants have been used over the last 30 years to resolve obvious safety issues and provide sources for the development and testing of chemical defense equipment and procedures (*1*). Open air testing is often required in order to effectively evaluate chemical defense materiel. Due to stringent environmental regulations for use of simulants in these conditions, and restrictions involving disposal of hazardous waste, it has become necessary to choose wisely the simulants available which not only fit the requirements for usage, but also are the least toxic and damaging to the environment (Duncan, D.P.; Famini, F.R.; Chester, N.A. ERDEC-Technical Report, *in press*). The result has been placement of major emphasis on determining environmental toxicity and harmful effects of simulants.

With the list of possible simulants numbering over 1000 and the "most used" simulants at about 70, there has been great interest in Quantitative Structure Activity Relationship (QSAR) methods that may generate, for a bulk of structures, environmental results for missing data. Data gaps have been found to be extensive

0097–6156/96/0643–0191$15.00/0

for many simulants and generally, both toxicity information for aquatic, terrestrial and wildlife populations, and information on fate of the materials is necessary in order to assess potential environmental impact. Typically, a battery of toxicity assays is conducted for each chemical, representing both aquatic and terrestrial organisms, and also different levels of biological organization. The availability of QSAR models which have been designed to predict toxicity for environmental species are available and may offer a toxicity screening tool for a large quantity of chemicals that can assist in highlighting the compounds that warrant more complete environmental testing with additional in depth, expensive and time consuming toxicity tests.

In order to make a judgment on the applicability of two QSAR models available in-house for use in the simulant program at ERDEC, empirical data was generated for comparison in a post hoc fashion. Bioassays were conducted that represent different trophic levels of aquatic species in order to provide an initial assessment of the aquatic aspect of environmental impact. In this case, the five minute Microtox assay employing the representative marine bacterium *Photobacterium phosphoreum*, and the 48-hour *Daphnia magna* (algae-feeding freshwater crustacean) toxicity assay, were performed to provide acute aquatic toxicity data.

Two different models that generate estimates of acute aquatic toxicities were used to examine the predictive power of QSAR equations. An equation developed by Health Designs, Incorporated (HDi) in collaboration with this laboratory (2) for estimating toxicities for *D. magna,* was chosen from the TOPKAT software package which also contains other aquatic models, and toxicity calculations were done using this equation for ten simulants. A model for estimating *P. phosphoreum* $EC_{50}$s that was developed at this laboratory (3) was selected to generate fourteen model-specific toxicity estimates. Predictions were compared with actual bioassay data to investigate the efficacy of the assay alternatives.

**TOPKAT Model.** TOPKAT is an integrated software package for the estimation of toxic, environmental, and biological effects of chemical structures using statistical quantitative structure-activity relationship techniques (4). The software at Edgewood Research Development, and Engineering Center (ERDEC) contains 13 predictive equations along with the data bases from which they were derived. Several of these mathematical models have been developed for environmental species (5), including a *D. magna* $EC_{50}$ equation (2), which was chosen to provide toxicity estimates. This model was constructed using molecular connectivity indices (MCI - molecular shape descriptors) and substructural keys as the parameters descriptive of the compounds in the model's data base which contribute most to the explanation of the *D. magna* $EC_{50}$. At the time of development of the equation (1989), the data set consisted of 167 chemicals. The data base continues to be enlarged and updated.

Health Designs' MOLSTAC system was used to generate substructural keys as parameters. They fall into the following classes of descriptors (2):

1. Identification of the longest continuous chain of atoms (excluding hydrogen) in the molecule.

2. Identification of carbon chain fragments.

3. Identification of ring systems, including combinations such as the rings forming the bay region in certain carcinogens.

4. Identification of chemically or biologically or both functional substructural fragments.

5. Identification of electron-donating and electron-withdrawing substructural keys.

A total of seventeen parameters were used for the model including primary amine bound to an aromatic ring atom; oxygen-substituted aryl ester; three carbon fragments between two functional groups (electron withdrawing, electron releasing, or combination); valence path-cluster, order 4; benzene; aliphatic alcohol; and valence path MCI, order 2, as examples. The methodology for calculation of the predictive equation uses multiple regression. Equation 1 shows the form of the final regression equation incorporating the seventeen parameters and the coefficient resulting from the regression of each parameter with the corresponding toxicity endpoint (2):

$$Daphnia\ EC_{50} = \text{constant} + (\text{parameter 1 x coefficient 1}) \quad (1)$$
$$+ (\text{parameter 2 x coefficient 2}) + ...$$
$$+ (\text{parameter } n \text{ x coefficient } n)$$

Initial performance tests which validate the model's regression equation used the cross-validation method. Briefly, one chemical is removed from the data set, the equation is recalculated, and the omitted compound is estimated with the equation. The process is repeated for all compounds and the statistics calculated on the estimates. These results can be expressed as the fraction of compounds predicted within certain factors. This model predicted the toxicity ($EC_{50}$) of 85.6% of the compounds from the data set within a factor of ten; however, the TOPKAT program includes important specifications for the validation of an estimate which, along with User's toxicological knowledge of and/or experience with the compound, will provide confidence in the prediction.

**Theoretical Linear Solvation Energy Relationship (TLSER) Model.** Linear free energy relationships (LFER) and quantitative structure activity relationships (QSAR) have been used extensively to correlate and characterize observed macroscopic properties in terms of microscopic structural features. The basic tenet that structure affects a property in a systematic manner is fundamental to chemistry. In the late 1970s to mid 1980s Kamlet and Taft (6, 7) extended the LFER/QSAR methodology dramatically with the development of the linear solvation energy relationship (LSER). The generalized LSER states that all solute/solvent related properties consist of the sum of four effects as described in equation (2); a) a cavity/bulk term, b) a polarizability/dipolarity term c) a hydrogen bonding acidity, and d) a hydrogen bond basicity term.

property = bulk/cavity term(s) + dipolarity/polarizability term(s)
$$+ \text{hydrogen bonding term(s)} + \text{constant} \quad (2)$$

Correlations between properties and structure, which currently number in excess of 300, range in diversity from general toxicity to solubility to nuclear magnetic resonance (NMR) and ultraviolet (UV)-Visible (Vis) spectral shifts (8).

One of the difficulties in using the LSER is the need for empirical descriptors. In order to circumvent this, there have been two independent efforts to substitute molecular orbital based parameters for the empirically derived ones. The effort by Politzer and Murray (9) has focused on using *ab initio* methods to generate molecular electrostatic potential maps (MEPs). These maps provide maxima, minima and deviations which are input as descriptors into an LSER-like equation.

The second approach, has been to use descriptors directly generated from the molecular orbital calculation to substitute directly on a one for one basis for the solvatochromic descriptors. Based on the LSER philosophy and general structure, a new, theoretical set of parameters for correlating a wide variety of properties has been developed (10). These theoretical linear solvation energy relationship (TLSER) descriptors have shown good correlations and physical interpretations for a wide range of properties. These include the following: five nonspecific toxicities (3) including Microtox Test toxicity (*P. phosphoreum* $EC_{50}$) using 42 compounds; activities of some local anesthetics and the molecular transform (11); opiate receptor activity of some fentanyl-like compounds (12); and six physicochemical properties: charcoal absorption, HPLC retention index, octanol-water partition coefficient, phosphonothiolate hydrolysis rate constant, aqueous acid equilibrium constant, and electronic absorption of some ylides (13).

These TLSER parameters are determined solely from computational methods thus permitting nearly *a priori* prediction of properties. Modeled after the LSER parameters, TLSER descriptors were developed to correlate closely with those descriptors; to give equations with correlation coefficients, R, and standard deviations, SD, close to those for LSER; and to be as widely applicable to solute-solvent interactions as the LSER set. Table I gives a summary of these TLSER descriptors as used in this paper and described in the next paragraphs (14).

The TLSER bulk/steric term is described by the molecular van der Waals volume, Vmc, in units of 100 cubic angstroms. The dipolarity/polarizability term uses the polarizability index, $\pi_I$, obtained by dividing the polarizability volume by the molecular volume to produce a unitless, size independent quantity which indicates the ease with which the electron cloud may be moved or polarized. For example, aromatics and chlorine rank high while alkanes and fluorine rank low on the scale.

### Table I. TLSER Descriptors

| Symbol | Name | Definition | Units | Meaning |
|--------|------|------------|-------|---------|
| $V_{mc}$ | molecular volume | molecular volume | $100A^3$ | cavity/steric |
| $\pi_I$ | polarizability index | polarizability/Vmc | none | polarizability |
| $\varepsilon_B$ | covalent basicity | 0.30- $\lvert \Delta E(h,lw) \rvert$ /100 | hev | acceptor HBB |
| q- | electrostatic basicity | maximum $\lvert$ (-) charge | acu | acceptor HBB |
| $\varepsilon_A$ | covalent acidity | 0.30- $\lvert \Delta E(l,hw) \rvert$ /100 | hev | donor HBA |
| q+ | electrostatic acidity | maximum H(+) charge | acu | donor HBA |

A = Angstrom; hev = hecto-electronvolt; acu = atomic charge unit; HBB or HBA = hydrogen bond basicity or acidity; $\Delta E(h,lw) = E(h)-E(lw)$; E(h) = HOMO energy; E(l) = LUMO energy; E(lw) and E(hw) refer to the E(LUMO) and E(HOMO) for water, respectively; $\lvert \ \rvert$ indicate absolute magnitudes.
SOURCE: Reprinted with permission from ref. 14. Copyright 1994.

The acceptor hydrogen bond basicity (HBB) is composed of covalent, $\varepsilon_B$, and electrostatic, q-, basicity terms. Analogously, the donor hydrogen bond acidity (HBA) is made up of covalent, $\varepsilon_A$ and electrostatic, q+, acidity terms. The covalent HBB parameter, $\varepsilon_B$, is the magnitude of the difference between the energy of the highest occupied molecular orbital (HOMO) of the solute and the lowest unoccupied molecular orbital (LUMO) of water. The result is divided by 100 for convenience in presentation and comparison of coefficients; the units are in hectoelectronvolts (heV). Analogously, the covalent HBA parameter, $\varepsilon_A$ is the magnitude of the difference between the energies of the LUMO of the solute and the HOMO of water, again scaled like the covalent HBB with the same units. The water energies are included for aesthetic reasons; the smaller these differences the greater is the ability to form a hydrogen bond with water.

So far no adequate descriptor has been found analogous to $\delta_{H1}^2$, the Hildebrand solubility parameter (15), to provide a theoretical model for the solvent cohesive energy. The first thing that comes to mind is to use $1/V_{mc1}$; this amounts to assuming the enthalpy of vaporization does not change much over the set of solvents. Since the Hildebrand solubility parameters are readily available for a large number of solvents it has been suggested that it be used along with the theoretical descriptors.

Equation (3) is the result of applying the TLSER descriptors to the LSER model, equation (2) (*14*), where logSSP represents the logarithm of some property affected by solvent-solute interactions, and subscripts 1 and 2 refer to solvent and solute respectively:

$$logSSP = a\ \delta_{H1}{}^2 V_{mc2} + b\ \pi_{I1}\pi_{I2} + c\ \epsilon_{B1}\epsilon_{A2} + d\ \epsilon_{A1}\ \epsilon_{B2}$$
$$+ e\ q_{-1}\ q_{+2} + f\ q_{+1}\ q_{-2} + logSSP_0 \qquad (3)$$

This relation can be simplified in the case of multiple solvents for a single solute to give equation (4).

$$logSSP = a_2\ \delta_{H1}{}^2 + b_2\ \pi_{I1} + c_2\ \epsilon_{B1} + d_2\ \epsilon_{A1} + e_2\ q_{-1} + f_2\ q_{+1} + logSSP_0 \qquad (4)$$

The TLSER equation, analogous to equation (3) for multiple solutes in a given solvent, is equation (5).

$$logSSP = a_1\ V_{mc2} + b_1\pi_{I2} + c_1\ \epsilon_{B2} + d_1\ q_{-2} + e_1\ \epsilon_{A2}$$
$$+ f_1\ q_{+2} + logSSP_0 \qquad (5)$$

It is important to note that the theoretical descriptors are from calculations that apply to an isolated molecule in the gas phase. However, one might expect that there could be some relationship between the gas phase structure and the structure in a condensed phase. In TLSERs involving toxicity, the substrate can be considered the "solute", and the medium/organism/receptor, the "solvent". Except in rare cases, therefore, equation 5 can be used as the generalized starting point for developing specific TLSERs. In most cases, some of the six descriptors are not significant in the regression analysis, and can be removed from the final TLSER.

Based on the TLSER descriptors having shown good correlations and physical interpretations for a wide range of properties including nonspecific toxicities, the equation correlating the Microtox test (*Photobacterium phosphoreum*) for a series of 42 compounds was chosen in addition to TOPKAT's *D. magna* EC$_{50}$ model for investigation of the predictive power of structure activity relationships.

This paper addresses the aquatic ecological toxicity of selected simulants and compares the TOPKAT-generated *D. magna* toxicity estimates and TLSER-generated *P. phosphoreum* toxicity estimates to experimental results.

## Materials and Methods

**Chemicals.** Simulants were obtained from chemical suppliers, and purchased at a minimum of 97 percent purity. Stock solutions for tests were prepared fresh at the start of each assay. Water soluble/miscible acetoacetates, glycols, organophosphates and organophosphonates, along with miscellaneous other chemicals were among simulants tested. Chosen chemicals were investigated for safety considerations and known human toxicities prior to testing and required standard operating procedures incorporating fume hoods, laboratory coats, and non powdered polyvinyl gloves.

**Daphnid Toxicity Tests.** The 48 hour *D. magna* acute toxicity test conformed to applicable ASTM and U.S. EPA standards (16, 17), and has been described previously (*18*). Daphnia cultures were originally obtained from Dr. Frieda Taub at the University of Washington (Seattle, WA), and were reared in the laboratory as described by Goulden et al. (*19*) using treated well water. Forty-eight hour tests were conducted in a temperature-controlled room at 20±1°C, using a 16:8 light/dark (315 ft. candles) cycle. Typically, controls and five to six concentrations of simulant, in replicate, were prepared as 100 mls of daphnia media in each of 250 mls beakers. Ten daphnia ($\leq$ 24 hours old) were added to each test vessel, and immobility/ mortality was observed at 24 and 48 hours. The $EC_{50}$s (effective concentration at which 50% of the organisms are immobilized) were calculated using the PROBIT analysis (20), and if needed, verified with graphically tabulated $EC_{50}$s derived from mortality and concentration values.

**Bacterial Toxicity Tests.** Bacteria (reagent) and test solutions for the *Photobacterium phosphoreum* bioassays were purchased by Microbics Corporation, the test's manufacturer. *P. phosphoreum* is a luminescent marine bacterium used in the Microtox aquatic assay as the test organism. Exposure of the organisms to toxicants typically lowers light output in proportion to toxicity. The assay is conducted within the temperature controlled (15°C) wells of a photometer (21) (Microtox Analyzer) and consists of sample solutions, serially diluted by factors of two, and controls (containing no toxicant). A corresponding set of tubes holds diluted Reagent (bacteria) ("Reagent-Tubes"), and following a 15 minute temperature stabilization period, are read at time zero for initial light output ($I_0$). Following this, aliquots from the sample and control tubes are added to the "Reagent-Tubes" and mixed; light output is then measured at predetermined times (t), usually five and 15 minutes, to give final light readings ($I_t$).

Due to the natural decay of light output over time, the timed readings are normalized using the "Blank Ratio" (BR), which is the ratio of the light output of the controls at time t to light output of controls at time 0. This BR is applied to $I_0$s to correct for drift and effects of dilution of organisms, and allows for the measurement of a true baseline. "Light lost" to "light remaining" is calculated, and further data reduction produces an $EC_{50}$ - the effective concentration at which there is a 50% reduction in light output.

**Scoring Criteria.** Scoring criteria for aquatic toxicity results (Table II.) was derived from O'Bryan and Ross' "Chemical Scoring System for Hazard and Exposure Identification" (22) where aquatic refers to animal (except mammalian) and plant life in the water environment, including algae and other lower life forms. Although the above authors define $EC_{50}$ as the concentration at which the effect under observation occurs in 50% of the test group under conditions of study, we have applied this ranking also to $EC_{50}$s generated by Microtox tests where $EC_{50}$ is the concentration at which light output is decreased by 50%. The description of ranking nomenclature by levels of toxicity (from neglible to high toxicity) has been designated by the authors of this paper, based on O'Bryan and Ross' numerical scores given to subgroups of $EC_{50}$s and authors' knowledge and experience. Rankings are used to present toxicity data categorically for easy comparisons.

**TOPKAT Estimation and Validation.** Structures of simulants were entered into TOPKAT using the Simplified Molecular Input Line Entry System (SMILES) (23). SMILES is a linear notation scheme for entry of chemical structures into computer data bases. The *Daphnia magna* $EC_{50}$ model was selected to obtain an $EC_{50}$ estimation for the 48 hour bioassay. A two-step validation process was then employed to ascertain confidence in the toxicity estimate (24). First, a determination was made as to whether the major substructures of the compound being estimated were present in the data base of the model selected, and second, the data base was examined for structurally related compounds whose assay results and also TOPKAT-predicted estimates may either support or refute the estimate in question. This information, together with the user's own experience and knowledge of the compound, provides a measure of confidence that can be placed in the prediction. The resulting estimates were compared to the actual $EC_{50}$s.

TABLE II. SCORING CRITERIA FOR AQUATIC[a] TOXICITY

| SCORE | ACUTE (EC$_{50}$)[b] (mg/L) | CATEGORY[c] |
|-------|---------------------------|-------------|
| 8 | <1 | HIGH |
| 6-7 | 1-10 | HIGH |
| 4-5 | >10-100 | MODERATE |
| 1-3 | >100-1000 | LOW |
| 0 | >1000 | NEGLIGIBLE TOXICITY |

[a] Aquatic refers to animal (except mammalian) and plant life in the water environment, including algae and other lower life forms.

[b] Acute refers to an exposure less than or equal to 96 hours. EC$_{50}$ is the concentration at which the effect under observation occurs in 50% of the test group under conditions of study.

[c] Description of ranking nomenclature designated by authors; ranking applies also to EC$_{50}$s generated by Microtox tests where EC$_{50}$ = concentration at which light output is decreased by 50%.

**TLSER Microtox Equation.** For determination of TLSER parameters for the *P. phosphoreum* EC$_{50}$ model, PCMODEL (Serena Software, Bloomington, IN) and the in-house-developed molecular modeling package, MMADS, were used to construct and view all molecular structures (*25*). Viewing the molecular model helps indicate that the structure is reasonable and might be close to a global minimum. Further molecular geometry optimization was performed using the MNDO algorithm contained in MOPAC (*26, 27*). The orbital energies, partial charges and polarizability volumes are results of this optimization. The molecular volume for the optimized geometry was determined using the algorithm of Hopfinger (*28*). Multilinear regression analysis using MYSTAT (Systat, Evanston, IL.) or MINITAB (Minitab, Inc. State College, PA) was used to obtain the coefficients in the correlation equation given in equation (6):

$$\text{LOG EC}_{50} = -3.63 \frac{V_{mc}}{100} - 45.8 \, \pi_i + 3.71 q_- - 2.92 q_+ + 11.4 \tag{6}$$

$$\text{for } N = 38 \quad \text{where } R = 0.977 \quad sd = 0.37 \quad F = 176$$

The correlation equation was selected based on the following considerations: 1) Most important is the need for an adequate sample size. A good rule of thumb is that there be at least three points for each descriptor. 2) The equation coefficients must be significant at the 0.95 level ("large" t-statistic) or higher. The correlation coefficient, R, should be as close to 1 as possible. In the physical sciences it is desirable for the equation to account for 80% or more of the variance which is given by $R^2$; consequently, a "good" correlation coefficient might have R > 0.90. Depending on the complexity of the system, R values in the range of 0.8 and up can be considered acceptable. 3) There should be minimal cross correlation amongst the parameters; the variance inflation factor (VIF) is a good measure of the cross correlation. The VIF is defined as $1/(1-R^2)$ where R is the correlation coefficient of one variable against the others; small (closer to one) values imply small cross correlation (*29*). Values in the range $1 < VIF < 5$ are considered acceptable. 4) Another concern is that the number of outliers is minimal. Outliers are compounds whose calculated values were three or more standard deviations from the mean. Outlying compounds can be physiochemically explained in several ways. The dominating mechanism or process may be different from the other compounds. The model may be inadequate; TLSER calculations refer to the very dilute (gaseous) phase. The conformation responsible for the dominant reaction characteristic may be considerably different from what is calculated for the gaseous ground state. Lastly, it is possible that an experimental value may be in error. 5) The standard deviation, SD, should be small. An SD of 0.3 when the property involves a base ten logarithm corresponds to an uncertainty factor of two. Another measure of the goodness of fit for the SD is the fraction of the range covered. For example an SD of 0.5 over the interval $0 < logSSP < 8$ would imply that the SD is about 6% of the whole range. Another consideration for the SD is that it should be larger than the experimental uncertainties in the measured properties. If it is not it suggests that the results are artifacts. 6) The Fisher index, F, of the reliability of the correlation equation should also be as large as possible.

## Results

Results of aquatic assays appear in Table III, and reflect categorizations by ranking of $EC_{50}$s as described in Table II. Simulants are listed from most toxic to least toxic, with a few data listed from use of the TOPKAT program, either as an estimate, or bioassay result taken from the *D. magna* $EC_{50}$ model's data base. All other data are from in-house test results.

Table IV reports the TOPKAT-generated *D. magna* $EC_{50}$ estimates with results of validation methodology reported as a measure of confidence. Determinations of the presence of compounds' major substructures in the data base of the model (validation step one) are listed as a degree of "coverage". In a few cases compounds are fully represented in the data base, but for the majority of compounds, parts or all of the structure are not well represented. Conducting validation step two produced searches for similar compounds in the data base. Comparisons of data sets of actual and TOPKAT-predicted $EC_{50}$s for similar data base compounds and the simulants provided a basis for determining how well the model functioned for the simulant being estimated. In many cases, validation methodology resulted in confidence levels based on a range of $EC_{50}$s within which the actual bioassay result was determined to fall.

Experimental versus predicted results for simulants entered into the TLSER equation are shown in Table V. Comparisons against parameters of regression statistics of the equation are highlighted with asterisks (*3*). Fifty-seven percent (eight) fall within three standard deviations, with outliers presented in plain type. All glycols (four) are among the outliers with predictions falling outside of three standard deviations, as well as the two organophosphates, trimethyl phosphate and dimethyl methyl phosphonate.

## Discussion

It is apparent from the aquatic toxicity table that methyl salicylate (MS) is the most toxic of the simulants tested (low/moderate rankings) with each $EC_{50}$ bordering the next most toxic ranking category. MS and dimethyl methyl phosphonate (DMMP) each have at least one moderate ranking for toxicity whereas other combinations for listed simulants consist of low and negligible toxicity rankings only.

## TABLE III. ACUTE AQUATIC TOXICITY FOR SELECTED SIMULANTS

| SIMULANT | MICROTOX x (mg/L) | RANKING | D. magna EC$_{50}$ (mg/L) | RANKING |
|---|---|---|---|---|
| Methyl salicylate | 13 | *moderate* | 113 | *low* |
| Dimethyl methyl phosphonate | 887 | *low* | x = 78.1 | *moderate* |
| Diethyl malonate | 253 | *low* | 324 | *low* |
| Ethyl acetoacetate | 997 | *low* | 646 | *low* |
| Methyl acetoacetate | 2180 | *neg. tox* | 500 *(prelim.)* | *low* |
| Diethyl methyl phosphonate | 2089 | *neg. tox* | >923[††] | *low* |
| Trimethyl phosphate | 170 | *low* | NT@1000 | *neg. tox* |
| Triethyl phosphate | 1169 | *neg. tox* | 553 | *low* |
| Dipropylene glycol monomethyl ether | 3840 | *neg. tox* | est. 233-10 g/L | *neg. tox -low* |
| Polyethylene glycol (200) | 8880 | *neg. tox* | NT@1000. | *neg. tox* |
| Dimethyl sulfoxide | 16036 | *neg. tox* | NT@1000. | *neg. tox* |
| Diethylene glycol | 134288 | *neg. tox* | 10 g/L[†] | *neg. tox* |

x = average of two EC$_{50}$ results
**est.** = estimated toxicity by TOPKAT model
**prelim.** = preliminary test, not definitive
[†] = in TOPKAT database
[††] = in TOPKAT database as EC$_{50}$ = 13.2 mg/L
**NT@1000** = not toxic at 1000 mg/L; when tested at concentrations up to 1000 mg/L, no organisms died or were immobilized

**TABLE IV. TOPKAT ESTIMATES VERSUS *DAPHNIA MAGNA* EXPERIMENTAL RESULTS**

| SIMULANT | TOPKAT ESTIMATE FOR *D. MAGNA* MODEL ($EC_{50}$ mg/L) | *D.MAGNA* $EC_{50}$ (mg/L) |
|---|---|---|
| Methyl salicylate (MS) | nfc; 4.2; lc | 113 |
| Trimethyl phosphate (TMP) | nfc; .958; lc | NT @ 1000 |
| Diethyl malonate (DEM) | fc; 76.9 -769; mc | > 600 |
| Dimethyl methyl phosphonate (DMMP) | nfc; 6.7 - 13.2; mc | x = 78.1 |
| Ethyl acetoacetate | nfc; 100.3-1003; mc | 646 |
| Triethyl phosphate (TEP) | nfc; .513; lc | 553 |
| Diethyl methyl phosphonate (DEMP) | nfc; 3.6; caution, lc in data base @ 13.2 | > 700 |
| Methyl acetoacetate | nfc; 133.1-790.; hc | 400 |
| Polyethylene glycol (200) (PEG 200) | fc; .509 - 10.g/L; mc | NT @ 1000 |
| Dimethyl sulfoxide (DMSO) | nc; 98.6; caution, lc | NT @ 1000 |

nc = not covered; fc = fully covered; nfc = not fully covered (validation step 1)
lc = low confidence; mc = moderate confidence; hc = high confidence (validation step 2)
NT@1000 = not toxic at 1000 mg/L; when tested at concentrations up to 1000 mg/L, no organisms died or were immobilized
x = average of two $EC_{50}$ results

TABLE V. TLSER ESTIMATES VERSUS *PHOTOBACTERIUM PHOSPHOREUM* EXPERIMENTAL RESULTS

| COMPOUND | *P. PHOSPHOREUM* 5 " $EC_{50}$ (mg/L) (MEAN) | TLSER EQUATION PREDICTED $EC_{50}$ |
|---|---|---|
| **Methyl salicylate** | 13 | **45*** |
| Trimethyl phosphate | 170 | 38800*** |
| **Diethyl malonate** | 253 | **255** |
| Dimethyl methyl phosphonate | 887 | 64883*** |
| **Ethyl acetoacetate** | **997** | **1870** |
| **Triethyl phosphate** | **1169** | **183**** |
| **Ethyl-l-lactate** | **2038** | **2200** |
| **Diethyl methyl phosphonate** | **2089** | **3790** |
| **Methyl acetoacetate** | **2180** | **8454**** |
| Dipropylene glycol monomethyl ether | 3840 | 55*** |
| Polyethylene glycol (200) | 8880 | 45*** |
| **Dimethyl sulfoxide** | **16036** | **>100,000** |
| Ethylene glycol | 11287 | 300,000*** |
| Diethylene glycol | 134288 | 6505*** |

\* **between 1 and 2 standard deviations**     Plain Type: Outliers
Plain Type: Outliers
\*\* **between 2 and 3 standard deviations**
\*\*\* **greater than 3 standard deviations**

As classes of compounds, organophosphorus (OP) compounds produced little toxicity with three out of four rankings as negligible/low toxicity combinations; DMMP is the exception. It is interesting to note that our in-house test result of $EC_{50}$ > 923 mg/L for diethyl methyl phosphonate (DEMP) does not agree with the reported $EC_{50}$ of 13.2 mg/L found in the Daphnia model's data base. According to a data base review (Wentsel, R. S.; Checkai, R. T.; Sadusky, M. C.; Chester, N. A.; Guelta, M. A.; Lawhorne, S.; Forster, W. M.; Trafton, T. M. Army Edgewood Research Development, and Engineering Center, APG, MD, unpublished data) and the above in-house data, this particular set of OPs appears to represent some of the less toxic of the OP compounds, which, in many cases are otherwise strong cholinesterase inhibitors and can have highly toxic effects on some environmental species.

Acetoacetates tested exhibit low toxicity, and the miscellaneous classed diethyl malonate and dimethyl sulfoxide exhibit low and negligible toxicity, respectively. Glycols, however, appear as the least toxic class of compounds tested. Although dipropylene glycol monomethyl ether was not able to be tested with the daphnia assay, confidence is high in the TOPKAT-generated estimate to account for the neglible to low toxicity ranking.

Upon running the TOPKAT program, it was found that the *D. magna* $EC_{50}$ model overestimated toxicity for the simulants (compounds) entered by generating smaller $EC_{50}$s than studies from laboratory tests. Following step one of the two-step validation process, the question of full representation of simulants by structures in the data base was addressed, and is termed "coverage". This step helps to identify any major substructural features of the simulants that are outside of the descriptor-design set on which the equation is based. The model's data base undergoes a search for structures containing features of the compound-in-question. Determination of poor or no coverage for part or all of a compound alerts the user to the possibility that a substructure of the compound-in-question, which may have an affect on the toxic endpoint, may not be given consideration by the model.

This condition arose during the toxicity estimation of dimethyl sulfoxide (DMSO), where it was found that no part of the compound was represented in the data base (no structures in the data base contained features of DMSO), thereby producing a poor TOPKAT performance. In the cases where a structure is not fully covered, caution is used, as the estimate through step one of the validation process, may not be valid.

To carry out the second step of validation, the model's data base was searched for similar compounds whose actual bioassay results and TOPKAT-generated $EC_{50}$s either support or refute each estimate. Low confidence in estimates results from poor data base representation and/or comparisons of toxicity data which discredit the prediction.

During several validation procedures, estimates were determined to be valid with moderate to high confidence when reported as a range of $EC_{50}$s. The ranges are bracketed by the estimate, which is the lowest $EC_{50}$ value, and the highest assay-derived $EC_{50}$ available from the most similar data base structure. By comparing estimates and their confidence levels to actual *D. magna* $EC_{50}$s, it was found that in all cases but that of DMMP, if confidence was moderate or high, experimental results fell within estimate-ranges. In cases where confidence was low or caution was advised, experimental results did not concur with estimates. For DMMP, a moderate confidence level was determined, yet the experimental $EC_{50}$ fell outside of the predicted range. Although a determination of moderate for a confidence level allows room for the chance that actual values may be outside of the estimated range, a potential problem in the way the model predicts toxicity for DMMP and the other OPs is suggested.

DEMP is found in the model's data base with an actual experimental result of 13.2 mg/L and is indicated to have been used in development of the equation. ERDEC's preliminary test result for DEMP is reported to be $EC_{50} > 923$ mg/L. Should the low TOPKAT data be invalid, poor results may be partially explained by a model whose few organophosphorus data points may have included one that is dramatically incorrect.

The *P. phosphoreum* TLSER model predicted $EC_{50}$s for the two OPs (trimethyl phosphate (TMP) and dimethyl methyl phosphonate (DMMP)) that are greater than three standard deviations, thereby performing poorly. However, the model performed moderately well for triethyl phosphate (TEP), and well for diethyl methyl phosphonate (DEMP). It is interesting that the equation does better with the di- and tri*ethyl*-containing OPs than the those with di- and tri*methyls*. The TLSER model performed poorly for the glycols with all four estimates considered to be bad outliers. However, if glycols and OPs are removed from consideration then five out of six of the remaining simulants are predicted within an acceptable two standard deviations. Since the DMSO experimental value is so far above the negligible toxicity cutoff point (> 1000 mg/L), the difference between the experimental and predicted results is not considered significant.

As new models are developed, it continues to be necessary to investigate their abilities with comparisons of estimates and experimentally derived values. These kind of data help define the limitations and boundaries of the equations so that they may be more successfully used when experimental avenues are not available. Results of TOPKAT runs and experimental daphnia assays confirm the utility of the predicted toxicities. In fact, for daphnia tests, the number of range-finding assays conducted was effectively reduced in number by five.

## References

1.  Famini, George R. In *Proceedings of the Sixth International Simulant Workshop*; *CRDEC-SP-055*; Clark, D.A.; Famini, G.R., Eds.; U.S. Army Edgewood Research Development and Engineering Center: Aberdeen Proving Ground, Maryland, 1992, pp 13-18.
2.  Enslein, K.; Tuzzeo, T.M.; Blake, B.W.; Hart, J.B.; Landis, W.G. In *Aquatic Toxicology and Environmental Fate; ASTM Special Technical Publication 1007*; Suter III, G.W.; Lewis, M.A., Eds.; American Society for Testing and Materials: Philadelphia, Pennsylvania, 1989, Vol. 11: pp 397-409.
3.  Wilson, L. Y.; Famini, G. R. *J. Med. Chem.*. **1991**, *34*, 1668.
4.  Health Designs, Incorporated. *Topkat User Guide*; Health Designs, Inc.: Rochester, NY, 1987.
5.  Health Designs, Incorporated. *Toxicology Newsletter* **1991**, *14*.
6.  Kamlet, M. J.; Taft, R. W. ; Abboud, J.-L. M. *J. Am. Chem. Soc.* **1977**, *91*, 8325.
7.  Kamlet, M. J.; Doherty, R. M.; Abraham, M. H.; Taft, R. W. *Quant. Struct.-Act. Relat.* **1988**, *7*, 71.
8.  Kamlet, M. J.; Taft, R. W.; Famini, G. R.; Doherty, R. M. *Acta Chem. Scand.* **1987**, *41*, 589.
9.  Murray J. S.; Politzer, P. *J. Chem. Research* **1992**, *5*, 110.
10. Famini, G. R. *Using Theoretical Descriptors in Quantitative Structure Activity Relationships;* CRDEC-TR-085, US Army Chemical, Research, Development and Engineering Center: Aberdeen Proving Ground, MD, 1989, Vol. 5.
11. Famini, G. R.; Kassel, R. J.; King, J. W.; Wilson, L. Y., *Quant. Struct.-Act. Relat.* **1991**, *10*, 344.
12. Famini, G.R.; Ashman, W.P.; Mickiewicz, A.P.; Wilson, L.Y. *Quant. Struct.-Act. Relat.* **1992**, *11*, 162.
13. Famini, G. R.; Penski, C. E.; Wilson, L. Y. *J. Phys. Org. Chem.* **1992**, *5*, 395.

14. Famini, G.R.; Wilson, L.Y. In *Quantitative Treatments of Solute/Solvent Interactions*; Politzer, P.; Murray, J.S., Eds.; Theoretical and Computational Chemistry; Elsevier Science B.V.: Amsterdam, The Netherlands, 1994, Vol. I; pp 213-241.
15. Hildebrand, J. H.; Prausnitz, J. M.; Scott, R. L. *Regular and Related Solutions*; Van Nostrand-Reinhold: Princeton, New Jersey, 1970.
16. American Society for Testing and Materials. *Annual Book of ASTM Standards, Water and Environmental Technology, Section 11*; American Society for Testing and Materials: Philadelphia, PA, 1994; Vol. 11.04; pp 480-499.
17. U.S. Environmental Protection Agency. *User's Guide: Procedures for Conducting Daphnia magna Toxicity Bioassays;* EPA-600/8-87/011. U.S. Environmental Protection Agency: Las Vegas, NV, 1987.
18. Johnson, D.W.; Haley, M.V.; Hart, G.S.; Muse, W.T.; Landis, W.G. *J. Appl. Tox.* **1986,** *6* , 225.
19. Goulden, C.E.; Comotto, R.M.; Hendrickson, Jr., J.A.; Hornig, L.L.; Johnson, K.L. In *Aquatic Toxicology and Hazard Assessment: Fifth Conference;* ASTM STP-766; Pearson, J.G.; Foster, R.B.; Bishop, W.E., Eds; American Society for Testing and Materials: Philadelphia, Pennsylvania, 1982, pp 139-160.
20. Finney, D.J.; *Statistical Method in Biological Assay;* Charles Griffin and Company: London, England, 1978.
21. Microbics Corporation. *Microtox Manual, Preliminary Release*; Microbics Corporation, Carlsbad, CA, 1991.
22. O'Bryan, T.R.; Ross, R.H. *J. of Toxicol. and Env. Health* **1988,** *1,* 119.
23. Weininger, D. *J. Chem. Inf. Comput. Sci.* **1988,** *28*, 31.
24. Health Designs, Incorporated. *TOPKAT Users Workshop;* Health Designs, Incorporated: Rochester, N.Y, 1989.
25. Leornard, J. M.; Famini, G. R. *A User's Guide to the Molecular Modeling Analysis and Display System*; CRDEC-R-030; US Army Chemical Research, Development and Engineering Center: Aberdeen Proving Ground, MD, 1989.
26. Dewar, M. J. K.; Thiel, W. *J. Am. Chem. Soc.* **1977,** *99* , 4899.
27. Stewart, J. J. P. *Mopac Manual*; FJSRL-TR-88-007; Frank J. Seiler Research Laboratory, US Air Force Academy: Colorado Springs, CO, 1988.
28. Hopfinger, A. J. *J. Am. Chem. Soc.* **1980,** *102*, 7126.
29. Belesley, D.A.; Kuh, E.; Welsh, R.E. *Regression Diagnostics*; Wiley: New York, NY, 1980.

# APPLICATION OF MODELING SOLUTIONS IN RISK ASSESSMENT

# Chapter 16

# Use of a Multiple Pathway and Multiroute Physiologically Based Pharmacokinetic Model for Predicting Organophosphorus Pesticide Toxicity

**James B. Knaak[1,3], Mohammed A. Al-Bayati[2], Otto G. Raabe[1,2], and Jerry N. Blancato[4]**

[1]Department of Molecular Biosciences and [2]Institute of Toxicology and Environmental Health, University of California, Davis, CA 95616
[3]Occidental Chemical Corporation, 360 Rainbow Boulevard, Niagara Falls, NY 14302
[4]Characterization Research Division, National Exposure Research Laboratory, U.S. Environmental Protection Agency, 944 East Harmon, Las Vegas, NV 89193–3478

The importance of routes of exposure and metabolic pathways on the outcome of organophosphorus pesticide toxicity studies is often overlooked. A PBPK/PBPD model was developed for isofenphos (1-methylethyl 2-[[ethoxy(1-methylethyl).amino]phosphinothioyl]-oxy]benzoate) using absorption data from whole animal studies, in vitro and in vivo metabolism and enzyme inhibition data. The model shows that at low dose rates the rat, guinea pig, and dog are capable of metabolizing and eliminating isofenphos without producing large decreases in acetylcholinesterase activity by DNIO (des N-isopropyl isofenphos oxon). At high dose rates metabolism by liver microsomal P-450 enzymes result in the continued production of DNIO, and 100% inhibition of the enzyme. Studies with the dog and guinea pig show that these animals are capable of directly hydrolyzing the isopropyl benzoic acid ester of isofenphos thereby reducing the amount of isofenphos available for conversion to IO and DNIO, while OP-hydrolases in the rat, guinea pig and dog remove IO and DNIO from tissues. The dog is also capable of hydrolyzing the benzoic acid esters of des N-isopropyl isofenphos (DNI) and isofenphos oxon (IO), thereby reducing the amount of DNIO formed. The acids of isofenphos, DNI and IO are eliminated in urine.

A rat PBPK/PBPD model describing the percutaneous absorption and metabolism, of isofenphos (1-methylethyl 2-[[ethoxy(1- methylethyl)amino]phosphinothioyl]-oxy]benzoate) and inhibition of acetylcholinesterase (AChE), butyrylcholinesterase (BuChE), and carboxylesterase (CaE) by des N-isopropyl isofenphos oxon (DNIO) was presented at the 1992 ACS Symposium on Biomarkers (1). The model was recently expanded to include the results of in vivo and in vitro metabolism studies by

0097–6156/96/0643–0206$15.75/0

the intravenous route (single bolus and infusion) in the rat, intravenous and dermal routes in the dog, and intraperitoneal route (single bolus and infusion) in the guinea pig. Studies in the dog showed that isofenphos and the P-450 metabolites of isofenphos are extensively hydrolyzed by tissue carboxylesterases to yield carboxylic acids. The P-450 metabolites are also hydrolyzed in the dog by tissue OP-hydrolases to give alkyl phosphates and isopropyl salicylate. The metabolic profile in the guinea pig was intermediate between that of the rat and dog. Several of the P-450 metabolites may also under go deamination prior to being hydrolyzed in these animals. The revised PBPK/PBPD model and output from the model (single bolus intravenous and dermal studies) are presented in this paper.

## MULTIPATHWAY, MULTIROUTE PBPK/PBPD MODEL FOR ISOFENPHOS

**Time Course Data, Topical Application, Intravenous Administration (Single Bolus, Infusion), and Metabolism in the Rat, Guinea Pig, and Dog.** The fate of $^{14}$C-ring labeled isofenphos and $^{3}$H-ethyl labeled isofenphos topically administered to back skin, or intravenously administered as a single bolus or infused dose was studied in the rat, guinea pig and dog. Table I lists the pharmacokinetic studies and gives a brief outline of the protocol used in conducting each study (i.e., animals, number of animals, dose, sacrifice times, tissues taken, route of administration, etc.). During the course of each study, the animals were housed in glass or stainless steel metabolism cages to allow for the complete collection of urine and feces. Urine, feces, and blood taken during the study and tissues at sacrifice (blood, heart, liver, kidney, brain, fat, skin, G.I. tract, and carcass) were analyzed for $^{14}$C and $^{3}$H isofenphos equivalents using scintillation counting procedures. The tissue concentrations (ug equivalents/g) were converted to pmoles $L^{-1}$ of tissue for use in PBPK modeling.

**PBPK/PBPD Model.** The multipathway, multiroute isofenphos PBPK/PBPD model shown in Figure 1 consists of two routes of administration, dermal and intravenous, 17 metabolic pathways (RAM1-17), 6 metabolite concentrations in tissues and the phosphorylation of blood, brain, and liver enzymes (AChE, BuChE and CaE) by DNIO. Tissue compartments involve mixed blood (venous, capillary, and arterial), skin, liver, brain, fat, kidney, slowly perfused organs, and rapidly perfused organs. Flows, Q, are total cardiac (QC), and fractional flows to the tissue compartments. Concentrations, C, are those in arterial/mixed blood and in venous blood leaving the tissue compartments. The venous blood concentrations are dependent on the isofenphos/metabolite concentrations in the tissues, and tissue/blood partition coefficents ($P_i$). Tissue volumes ($V_i$) are expressed in $kg^{-1}$ $kg^{-1}$ of bw and scaled to bw. Metabolic removal of isofenphos/metabolites from liver is described by $V_{max}$ (pmoles $hr^{-1}$ $kg^{-1}$ of bw) and their $K_m$ (umol $L^{-1}$) values. The phosphorylation of AChE, BuChE and CaE by IO was not included in this model consisting of over 1000 lines of code. Equations for the regeneration and aging of AChE were included in the model, but were not used to regenerate or remove AChE (pmol $hr^{-1}$) in tissues. According to Figure 1, six passes through blood, liver, brain, and other tissues were required to model the fate of isofenphos. The model was run using SimuSolv/ACSL (Dow Chemical/Mitchell and Gauthier Associates, Inc.) on a DEC VAX/VMS system.

**Mass Balance Equations.** The PBPK/PBPD model in Figure 1 was used to write the mass balance equations describing the rate of change of the concentration of isofenphos and metabolites (IO, DNI, DNIO, DAI and DAIO) in tissues through time. Mass balance equations for the metabolism of DNIO in blood, brain and liver were included in the model along with equations for the inhibition of AChE, BuChE,

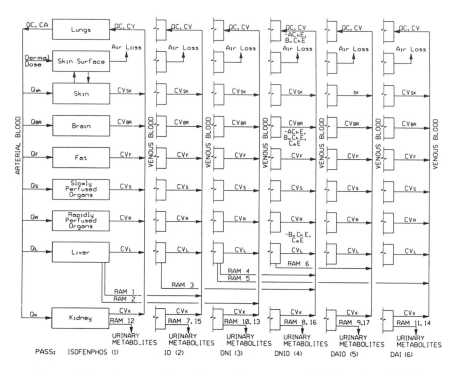

Figure 1. Percutaneous Absorption and Intravenous Administration: Model Physiologically based pharmacokinetic model. Pass 1, IF to DNI, IO and IFA; Pass 2, IO to DNIO, IOA, isopropylaminothiophosphate (AP1) and isopropyl salicylate (IPS); Pass 3, DNI to DNIO, DIA, DNIA, O-ethylaminothiophosphate (AP4) and isopropyl salicylate (IPS); Pass 4, DNIO to phosphorylated enzymes (AChE, BuChE, and CaE), DNIOA, DAIO, aminoethylphosphate (AP2) and isopropyl salicylate; Pass 5, DAIO to DAIOA, ethylphosphate (AP3) and isopropyl salicylate (IPS); Pass 6, DAI to DAIA, O-ethylthiophosphate (AP5)and isopropyl salicylate.(IPS)

**Table I. Pharmacokinetic Studies Conducted with Isofenphos**

| Study | Animal | Dose | Number of animals | Sacrifice/Sampling times, hours |
|---|---|---|---|---|
| [a]/iv | rat, 303 g | 9.0 mg/kg | 3 rats/group, 9 groups | [b]/0.5, 1, 2, 4, 8, 12, 24, 48, 72 |
| [b]/iv infusion | rat, 322 g | 8.1 mg/kg | 4 rats/group, 7 groups | [h]/24, 48, 72, 96, 120, 144, 168 |
| [c]/dermal | rats, 300 g | 3.73 mg/kg | 4 rats/group, 12 groups | [b]/0.5, 1, 4, 8, 12, 24, 48, 72, 96, 120, 144, 168 |
| [d]/ip | guinea pig, 446 g | 22.0 mg/kg | 3 gpigs/group, 12 groups | [b]/0.5, 2, 4, 8, 12, 24, 48, 72, 96, 120, 144, 168 |
| [e]/ip infusion | guinea pig, 513 g | 22.0 mg/kg | 3 gpig/group, 3 groups | [b]/24, 48, 72 |
| [f]/iv | dog, 14.8 kg | 8.0 mg/kg | 3 beagle dogs | [i]/0.25, 0.5, 1, 2 4 6, 8, 12, 24, 48, 72 |
| [g]/dermal | dog, 14.8 kg | 17.4 mg/kg | 1 beagle dog | [i]/1.5, 3, 4.5, 6, 7.5, 24, 48, 72, 96, 120, 144, 168, 192 |

[a]/0.0963 mg of $^{14}C$-ring isofenphos/mg of PEG 400, 25 ul or 28.4 mg of dosing solution administered

[b]/0.01186 mg of $^{14}C$-ring isofenphos/mg of PEG 400, 227.2 mg in osmotic pump, pump rate 1.14 mg/hr

[c]/1.12 mg of $^{14}C$-ring isofenphos per 0.2 ml of acetone, 0.2 ml topically applied to 12 cm$^2$ of skin

[d]/19.6 mg of $^{14}C$-ring isofenphos/mg of PEG 400, 0.5 ml of dosing solution administered

[e]/ 0.049 mg of $^{14}C$-ring isofenphos/mg of PEG 400, 227.2 mg in osmotic pump, pump rate 1.14 mg/hr

[f]/121 mg of $^{14}C$-ring isofenphos/ml of PEG 400, 1.0 ml of dosing solution administered

[g]/257 mg of $^{3}H$-ethyl isofenphos/ 7.44 ml of acetone (34.5 mg/ml) applied to 400 cm$^2$ of back skin

[h]/Tissues harvested: heart, kidney, liver, brain, fat, G.I. tract and carcass

[i]/Blood taken at indicated time intervals for enzyme and radioassay work

and CaE by DNIO in these tissues based on those developed by Gearhart et al.([2]) for DFP. The mass balance equations for percutaneous absorption and loss to air were published earlier by Knaak et al.([3], [4]) for isofenphos, while the equations for tissue/blood exchange, and metabolism were developed by Ramsey and Andersen ([5]). The intravenous dose was introduced as a single bolus dose directly into mixed blood. The mass balance equations for this route of administration are also presented in the following equations.

**Mass Balance for percutaneous absorption**

$dA_{surf}/dt = K_p*A*(C_{sk}/P_{a/sk}-C_{surf})-K_a A_{surf}$, pmol hr$^{-1}$

$A_{surf} = $ Integ (dAsurf/dt, SKO)

$C_{surf} = A_{surf}/V_{sk}$, pmol cm$^{-3}$

$C_{sk} = A_{sk}/V_{sk}$, pmol cm$^{-3}$

$dA_{air}/dt = K_a*A_{surf}$, pmol hr$^{-1}$

where:

$A_{surf}$ = Amount of isofenphos on skin surface, pmol (applied dose)

$C_{surf}$ = Concentration on skin surface, pmol cm$^{-3}$

$K_p$ = permeability constant, cm hr$^{-1}$

$A_{sk}$ = Amount in skin, pmol

$C_{sk}$ = Concentration within skin, pmol cm$^{-3}$

$V_{sk}$ = Volume of skin, pmol cm$^{-3}$

$A_{air}$ = Amount in air, pmol

$K_a$ = rate of loss to air, hr$^{-1}$

$P_{a/sk}$ = partition coefficient, air/skin

$A$ = Area of treated skin, cm$^2$

SKO = Topical dose, pmol

**Mass Balance for Skin**

$dA_{sk}/dt = K_p*A*(C_{surf}-C_{sk}/P_{a/sk}) + Q_{sk}*(CA_1-C_{sk}/P_{sk/b})$

$A_{sk} = $ Integ (dAsk/dt, 0)

$CV_{sk} = (A_{sk}/Vsk*P_{sk/b})$

$C_{sk} = A_{sk}/V_{sk}$

$P_{sk/b}$ = partition coefficient, skin/blood

$CA_1$ = Concentration of isofenphos in arterial/mixed blood

$R_{sk} = Q_{sk}(CV_{sk}*1000)*SSK$

where:

$Q_{sk}$ = Blood flow to skin, L hr$^{-1}$

$CV_{sk}$ = Concentration in venous blood leaving skin

$R_{sk}$ = Rate topical dose is delivered to arterial/mixed blood, pmol hr$^{-1}$

SSK = on, off switch in Discrete Section of Model

## Mass Balance for Intravenous Administration

$$dAB/dt = (Q_i * CV_i + R_{sk} + R_{iv} + R_{ivf} - RAM12B - QC + CA_i)$$

where:

$Q_i$ = Blood flow to tissue i, L hr$^{-1}$

$CA_i$ = Conc of IF/metabolites in arterial blood entering tissue i, pmol L$^{-1}$

$CV_i$ = Conc of IF/metabolites in venous blood leaving tissue i, pmol L$^{-1}$

$QC$ = Total cardiac blood flow, L hr$^{-1}$

$R_{sk}$ = Rate topical dose is delivered to arterial/mixed blood, pmol hr$^{-1}$

$R_{iv}$ = Rate iv dose is administered, pmol hr$^{-1}$

$R_{ivf}$ = Rate iv infusion dose is administered. pmol hr$^{-1}$

$RAM12B$ = Rate of hydrolysis of IF by CaE in blood

## Mass Balance for Tissue/Blood Exchange

$$Q_i CA_i dt = dA_i = Q_i CV_i dt, \text{ pmol hr}^{-1}$$
$$dA_i/dt = Q_i(CA - CV_i), \text{ pmol hr}^{-1}$$
$$CV_i = A_i/(V_i * P_i), \text{ pmol L}^{-1}$$
$$C_i = A_i/V_i, \text{ pmol L}^{-1}$$

where:

$A_i$ = Amount of IF/metabolites in tissue i, pmol

$C_i$ = Conc of IF/metabolites in tissue i, pmol L$^{-1}$

$P_i$ = Tissue i blood/tissue partition coefficient for parent/metabolites

## Mass Balance Equation for Metabolism

$$dA_m/dt = V_{max} * CV_l/(K_m + CV_l), \text{ pmol hr}^{-1}$$
$$dA_l/dt = Q_l(CA_l - CV_l) - dA_m/dt, \text{ pmol hr}^{-1}$$

where:

$A_m$ = Amount metabolized in liver, pmol

$V_{max}$ = Michaelis Menten rate constant, pmol hr$^{-1}$ kg$^{-1}$ of bw

$K_m$ = Substrate Conc, pmol L$^{-1}$, at half maximum velocity

$CV_l$ = Conc venous blood in liver, pmol L$^{-1}$

$A_l$ = Amount in liver, pmol

$Q_l$ = Blood flow to liver, L hr$^{-1}$

$CA_l$ = Concentration in arterial blood entering liver, pmol L$^{-1}$

## Inhibition of AChE and BuChE in Blood

$$dAAB/dt = (K_{AChE1} * C_{AEB} * C_{b4} * V_b), \text{ pmol hr}^{-1}$$
$$dABB/dt = (K_{BChE3} * C_{BEB} * C_{b4} * V_b), \text{ pmol hr}^{-1}$$

where:
$V_b$ = Volume of blood, L
AAB = Inhibited AChE, pmol
ABB = Inhibited BuChE, pmol
$K_{AChE1}$ = AChE bimolecular inhibition rate constant, $(pmol\ L^{-1})^{-1}\ hr^{-1}$
$K_{BChE3}$ = BuChE bimolecular inhibition rate constant, $(pmol\ L^{-1})^{-1}\ hr^{-1}$
$C_{AEB}$ = Concentration of Free AChE in blood, $pmol\ L^{-1}$
$C_{BEB}$ = Concentration of Free BuChE in blood, $pmol\ L^{-1}$
$C_{b4}$ = Concentration of DNIO in the blood, $pmol\ L^{-1}$

## Mass Balance Equation for DNIO in the Blood

$dC_{b4}/dt = Q_b*(CA_4 - CV_4)$
$-(V_{max6b}*CA_4/(K_{m6b}+CA_4))$, deamination
$-(V_{max8b}*CA_4/(K_{m8b}+CA_4))$, hydrolysis by OP-hydrolase
$-(V_{max16b}*CA_4/(K_{m16b}+CA_4))$, hydrolysis by carboxylesterase
$-(K_{AChE1}*C_{AEB}*C_{b4}*V_b)$, inhibition of AChE
$-(K_{BChE3}*C_{BEB}*C_{b4}*V_b)$, inhibition of BuChE, pmol $hr^{-1}$

where:
$Q_b$ = Blood flow, L $hr^{-1}$
$CA_4$= DNIO conc in arterial blood, $pmol\ L^{-1}$
$CV_4$ = DNIO conc in venous blood, $pmol\ L^{-1}$
$V_{max6b}$ = DNIO deaminated to DAIO in blood, $pmol\ L^{-1}$
$K_{m6b}$ = DNIO conc in blood giving half maximum rate of hydrolysis, $pmol\ L^{-1}$
$V_{max8b}$ = DNIO hydrolyzed to AP2 and IPS in blood, $pmol\ L^{-1}$
$K_{m8b}$ = DNIO conc in blood giving half maximum rate of hydrolysis, $pmol\ L^{-1}$
$V_{max16b}$ = DNIO hydrolyzed to DNIOA in blood, $pmol\ L^{-1}$
$K_{m16b}$ = DNIO conc in blood giving half maximum rate of hydrolysis, $pmol\ L^{-1}$

## Inhibition of AChE, BuChE, and CaE in the Brain

d AABr/dt = $(K_{AChE2}*C_{AEBr}*C_{br4}*V_{br})$, pmol $hr^{-1}$
dABBr/dt = $(K_{BChE4}*C_{BEBr}*C_{br4}*V_{br})$, pmol $hr^{-1}$
dACBr/dt = $(K_{CaE6}*C_{CEBr}*C_{br4}*V_{br})$, pmol $hr^{-1}$

where:
$V_{br}$ = Volume of brain, L
AABr = Inhibited AChE in brain, pmol
ABBr = Inhibited BuChE in brain, pmol
ACBr = Inhibited CaE in brain, pmol
$K_{AChE2}$ = AChE bimolecular inhibition rate constant, $(pmol\ L^{-1})^{-1}\ hr^{-1}$
$K_{BChE4}$ = BuChE bimolecular inhibition rate constant, $(pmol\ L^{-1})^{-1}\ hr^{-1}$

$K_{CaE6}$ = CaE bimolecular inhibition rate constant, (pmol L$^{-1}$)$^{-1}$ hr$^{-1}$
$C_{AEBr}$ = Free AChE in brain, pmol L$^{-1}$
$C_{BEBr}$ = Free BuChE in brain, pmol L$^{-1}$
$C_{CEBr}$ = Free CaE in brain, pmol L$^{-1}$
$C_{br4}$ = Concentration of DNIO in the brain, pmol L$^{-1}$

**Mass Balance Equation for DNIO in the Brain**

$dC_{br4}/dt = Q_{br}*(CA_4 - CV_{br4})$
$-(V_{max6br}*CA_4/(K_{m6br}+CA_4))$, deamination, pmol hr$^{-1}$
$-(V_{max8br}*CA_4/(K_{m8br}+CA_4))$, hydrolysis by OP-hydrolase, pmol hr$^{-1}$
$-(V_{max16br}*CA_4/(K_{m16br}+CA_4))$, hydrolysis by carboxylesterase, pmol hr$^{-1}$
$-(K_{AChE2}*C_{AEBr}*C_{br4}*V_{br})$, inhibition of AChE, pmol hr$^{-1}$
$-(K_{BChE4}*C_{BEBr}*C_{br4}*V_{br})$, inhibition of BuChE, pmol hr$^{-1}$
$-(K_{CaE6}*C_{CEBr}*C_{br4}*V_{br})$, inhibition of CaE, pmol hr$^{-1}$

where:
$Q_{br}$ = Blood flow to brain, L hr$^{-1}$
$CA_4$ = DNIO conc in arterial blood in brain, pmol L$^{-1}$
$CV_{br4}$ = DNIO conc in venous blood in brain, pmol L$^{-1}$

$V_{max6br}$ = DNIO deaminated to DAIO in brain, pmol hr$^{-1}$
$K_{m6br}$ = DNIO conc in brain giving half maximum rate of hydrolysis, pmol L$^{-1}$
$V_{max8br}$ = DNIO hydrolyzed to AP2 and IPS in brain, pmol hr$^{-1}$
$K_{m8br}$ = DNIO conc in brain giving half maximum rate of hydrolysis, pmol L$^{-1}$
$V_{max16br}$ = DNIO hydrolyzed to DNIOA in brain, pmol hr$^{-1}$
$K_{m16br}$ = DNIO conc in brain giving half maximum rate of hydrolysis, pmol L$^{-1}$

**Inhibition of BuChE, and CaE in the Liver**

$dABL/dt = (K_{BChE5}*C_{BEL}*C_{L4}*V_L)$, pmol hr$^{-1}$
$dACL/dt = (K_{CaE7}*C_{CEL}*C_{L4}*V_L)$, pmol hr$^{-1}$

where:
$V_L$ = Volume of liver, L
ABL = Inhibited BuChE, pmol
ACL = Inhibited CaE in liver, pmol
$K_{BChE5}$ = BuChE bimolecular inhibition rate constant, (pmol L$^{-1}$)$^{-1}$ hr$^{-1}$
$K_{CaE7}$ = CaE bimolecular inhibition rate constant, (pmol L$^{-1}$)$^{-1}$ hr$^{-1}$
$C_{BEL}$ = Free BuChE in liver, pmol L$^{-1}$
$C_{CEL}$ = Free CaE in liver, pmol L$^{-1}$
$C_{L4}$ = Conc of DNIO in the liver, pmol L$^{-1}$

**Mass Balance Equation for DNIO in the Liver**

$dC_{Br4}/dt = Q_l*(CA_4 - CV_{L4})$
$-(V_{max6}*CA_4/(K_{m6}+CA_4))$, deamination, pmol hr$^{-1}$
$-(V_{max8}*CA_4/(K_{m8}+CA_4))$, hydrolysis by OP-hydrolase, pmol hr$^{-1}$
$-(V_{max16}*CA_4/(K_{m16}+CA_4))$, hydrolysis by CaE, pmol hr$^{-1}$
$-(K_{BChE5}*C_{BEL}*C_{l4}*V_L)$, inhibition of BuChE, pmol hr$^{-1}$
$-(K_{CaE7}*C_{CEL}*C_{l4}*V_L)$, inhibition of CaE, pmol hr$^{-1}$

where:
$Q_L$ = Blood flow to liver, L hr$^{-1}$
$CA_4$ = DNIO conc in arterial blood in liver, pmol L$^{-1}$
$CV_{L4}$ = DNIO conc in venous blood in liver, pmol L$^{-1}$
$V_{max6}$ = DNIO deaminated to DAIO in liver, pmol hr$^{-1}$
$K_{m6}$ = DNIO conc in liver giving half maximum rate of hydrolysis, pmol L$^{-1}$
$V_{max8}$ = DNIO hydrolyzed to AP2 and IPS in liver, pmol hr$^{-1}$
$K_{m8}$ = DNIO conc in liver giving half maximum rate of hydrolysis, pmol L$^{-1}$
$V_{max16}$ = DNIO hydrolyzed to DNIOA in liver, pmol hr$^{-1}$
$K_{m16}$ = DNIO conc in liver giving half maximum rate of hydrolysis, pmol L$^{-1}$

**Physiological Parameters.** Physiological parameters compiled by Arms and Travis and published by EPA were used in the model (6).

**Partition Coefficients.** The partition coefficients for isofenphos were determined according to the procedure of Knaak et al. (7), while the coefficients for metabolites were estimated based on their octanol/water partition coefficients relative to isofenphos. See Table II.

**Table II. Partition Coefficients used in the Model**

| Chemical | L/B | F/B | K/B | Br/B | A/Sk | Sk/B |
|----------|-----|-----|-----|------|------|------|
| IF | 21.0 | 48.0 | 27.0 | 17.0 | 1.6 | 2.0 |
| IO | 2.0 | 5.0 | 2.0 | 2.0 | | |
| DNI | 2.0 | 5.0 | 2.0 | 1.0 | | |
| DNIO | 1.2 | 2.0 | 1.2 | 1.0 | | |
| DIA | 1.2 | 2.0 | 1.2 | 1.0 | | |
| DIAO | 1.2 | 2.0 | 1.2 | 1.0 | | |

Chemicals identified in Figure 2.
L/B, Liver/Blood; F/B, Fat/Blood; K/B. Kidney/Blood; Br/B, Brain/Blood; A/Sk, Air/Skin; Sk/B, Skin/Blood

**Metabolic Pathways.** Figure 2 gives the combined metabolic pathway of isofenphos in the rat, guinea pig, and dog. The metabolites, IO, DNI, and DNIO, produced in vitro by P-450 isozymes fortified with NADPH were previously described by Knaak et al. (8), while the deaminated metabolites, DAI and DAIO, have not been identified, but are possible products of metabolism (9). The carboxylic acids of isofenphos, DNI, IO, and DNIO, were partially identified as urinary metabolites of isofenphos in dog urine by anion exchange chromatography and negative ion

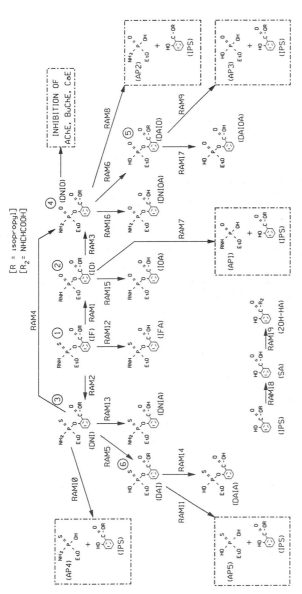

Figure 2. Combined metabolic pathway for Isofenphos in the Rat, Guinea Pig and Dog. IF, isofenphos; IO, isofenphos oxon; DNI, des N-isopropyl isofenphos; DNIO, des N-isopropyl isofenphos oxon; DAIO, des amino-isofenphos oxon; DAI, desaminoisofenphos; AP1, O-ethylisopropylamino-phosphate; IPS, isopropyl salicylate; AP2, O-ethylaminophosphate; AP3, O-ethylphosphate; AP4, O-ethylaminothiophosphate; AP5, O-ethylthiophosphate; IFA, carboxylic acid of isofenphos; DNIA, carboxylic acid of DNI; DAIA, carboxylic acid of DAI; IOA, carboxylic acid of IO; DNIOA, carboxylic acid of DNIO; DAIOA, carboxylic acid of DAIO; SA, 2-hydroxy salicylic acid; 2-OH-HA, 2-hydroxy hippuric acid; DIA, des isopropylamino isofenphos; DIAO, des isopropylamino isofenphos oxon.

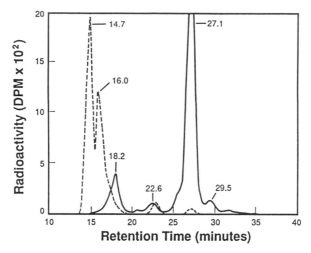

Figure 3. Anion Exchange Chromatogram of Rat Urinary Metabolites of [3]H-ethyl and [14]C-ring labeled Isofenphos. Dashed line, [3]H-ethyl labeled metabolites; solid line, [14]C-ring labeled metabolites. Urines (100 ul volumes) were chromatographed on a 250 x 10 mm SynChropak AX100 HPLC column using 0.5M Tris/HCl buffer, pH 8.0. Metabolites were detected using a Radiomatic Flo-one\Beta detector. Flow rate, 1.0 ml per minute. Metabolites identified in Table III on the basis of their retention times (RT) in minutes.

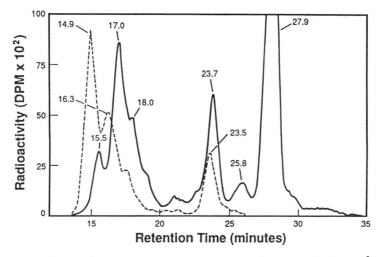

Figure 4. Anion Exchange Chromatogram of Dog Urinary Metabolites of [3]H-ethyl and [14]C-ring labeled Isofenphos. Dashed line, [3]H-ethyl labeled metabolites; solid line, [14]C-ring labeled metabolites. See Figure 3 for conditions. Metabolites identified in Table III on the basis of their retention times (RT) in minutes.

electrospray mass spectroscopy (10). Figures 3 and 4 give the anion exchange column chromatographic profiles of the $^{14}$C and $^{3}$H-labeled urinary metabolites present in 24 hr rat and dog urines, respectively. Major peaks indicated by their retention times in minutes are identified in Table III. Figure 5 gives the chromatographic profiles of the rat and guinea pig 24 hr $^{3}$H-ethyl labeled urinary metabolites. The metabolites chromatographed as three major peaks with retention times of 7.8, 8.3, and 10.8 min. A flow rate of 2.0 ml/min was used to elute the $^{3}$H-ethyl labeled alkyl phosphates (first two peaks) chromatographing in a manner similar to the alkyl phosphates present in rat urine. The carboxylic acids of DNIO, IO, and DNI are believed to partially co-elute with the alkyl phosphates. The metabolite chromatographing at 10.8 min is believed to be the carboxylic acid of isofenphos. This metabolite elutes at 22.6 minutes when a flow rate of 1.0 ml/min was used as shown in Figures 3 and 4.

Table III. Metabolites Identified in Urines

| Metabolites | Formula Weight | Molecular Weight | [d]Electro-Spray mass, m/z | [e]Urines, RT Fig. 3,4 |
|---|---|---|---|---|
| [a]**Carboxylic acids:** | | | | |
| IFA | $C_{12}H_{18}O_4N_1P_1S_1$ | 303.26 | | 22.6, 23.7 |
| DNIA | $C_9H_{12}O_4N_1P_1S_1$ | 261.24 | 260 | 18.2, 18.0 |
| IOA | $C_{12}H_{18}O_5N_1P_1$ | 287.26 | | 17.0 |
| DNIOA | $C_9H_{12}O_5N_1P_1$ | 245.18 | 244 | 15.5 |
| DAIOA | $C_9H_{11}O_6P_1$ | 246.16 | 245 | |
| DAIA | $C_9H_{11}O_5P_1S_1$ | 262.14 | 261 | |
| | | | | |
| [b]**Alkyl Phosphates:** | | | | |
| AP1 | $C_5H_{14}O_3N_1P_1$ | 167.15 | | 16.0, 16.3 |
| AP2 | $C_2H_8O_3P_1$ | 110.01 | | 14.7, 14.9 |
| AP3 | $C_2H_7O_4P_1$ | 126.01 | 125 (std) | |
| AP4 | $C_2H_8O_2N_1P_1S_1$ | 140.16 | | |
| AP5 | $C_2H_7O_3P_1S_1$ | 142.07 | | |
| | | | | |
| [c]**Ring Metabolites:** | | | | |
| 2-OH HA | $C_9H_9O_4N_1$ | 195.18 | 194 | 27.1, 27.9 |
| 2-OH BA | $C_7H_6O_3$ | 138.12 | 137 | |

[a]Carboxylic acids of: isofenphos, IFA; des N-isopropyl isofenphos, DNIA; isofenphos oxon, IOA; des N-isopropylisofenphos oxon, DNIOA; des amino isofenphos oxon, DAIOA; des amino isofenphos, DAIA.
[b]Alkyl phosphates of: isofenphos oxon, AP1; des N-isopropyl isofenphos oxon, AP2; des amino isofenphos oxon, AP3; des N-isopropyl isofenphos, AP4; des amino isofenphos, AP5
[c]Ring Metabolites: 2-hydroxy hippuric acid, 2-OH HA; 2-hydroxy benzoic acid, 2-OH BA
[d] VG Trio MS (Fison Inc.), Ionization: electrospray (-), dog urinary metabolites
[e]Rat, guinea pig and dog urinary metabolites chromatographing on anion exchange resin, see Figures 3,4 and 5

The metabolites in dog urine were prepared for mass spectroscopy using the method Weisskopf and Seiber (11). Briefly, 1.8 g of granular ammonium sulfate and 4 drops of 70% acetic acid were added to 4 ml of urine in a 15 ml test tube and mixed for 30 seconds. The urine was then passed through a cyclohexyl solid phase extraction

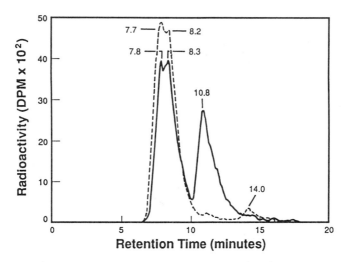

Figure 5. Anion Exchange Chromatogram of Rat and Guinea Pig Urinary Metabolites of $^3$H-ethyl labeled Isofenphos. Dashed line, rat metabolites; Solid line, Guinea Pig metabolites. See Figure 3 for conditions. Flow rate 2.0 ml per minute.

column (Varian Bond Elut) prewashed with one column volume (~6.0 ml) of acetone, water, and ammonium sulfate saturated water. The cartridge was then washed with 2 ml of 20% acetone in hexane and vacuum dried to remove water. The metabolites were eluted with 1.4 ml of acetone and analyzed by negative ion electrospray-mass spectrometry. A methylating reagent was not used. Table III gives the electrospray ions found in the dog urine. The ions (M-H)⁻ were one mass unit less than the molecular weights of the metabolites (carboxylic acids of DNI, DNIO, DAIO, DAI, 2-hydroxy hippuric acid and 2-hydroxy benzoic acid). Ions were not obtained for the alkyl phosphates (AP1-AP5), because they were not retained by the cyclohexyl solid phase extraction column. Analytical standards of AP3, 2-hydroxy hippuric acid and 2-hydroxy benzoic acid were separately analyzed and gave (M-H)⁻ ions in the negative ion mode at m/z 125, 194, and 137, respectively.

**Metabolic Rate Constants, P-450** In the presence of NADPH, liver P-450 isozymes catalyze the metabolism of isofenphos to IO, DNI and DNIO. The rat, guinea pig, and dog $V_{max}$ (pmoles hr$^{-1}$ kg$^{-1}$ of bw)and $K_m$ (umol L$^{-1}$)values for these conversions (8) were used in the model. The $V_{max}$, $K_m$ values used in the PBPK model are given in Tables IV and V for the reactions carried out by P-450 isozymes and hydrolytic enzymes in liver microsomes or cytosol. During fitting, the in vitro P-450 $V_{max}$ values (8) were reduced to fit the model to $^{14}$C-tissue concentrations (liver, fat, kidney, etc.). The smaller extraction ratios ($V_{max}/K_m$) improved the overall fit, but did not completely eliminate discrepancies between in vivo data and the model.. The reason for this adjustment may be related to the distribution (location) of the metabolic enzymes involved, and competitive, noncompetitive or uncompetitive inhibition by isofenphos and metabolites. According to the in vitro P-450 studies conducted by Knaak et al. (8) involving the rat, isofenphos has a greater affinity for the P-450 isozymes than either IO or DNI, while guinea pig P-450 isozymes have a greater affinity for IO than for isofenphos. In the case of the dog, isofenphos and DNI have higher affinities than IO or DNIO. On the basis of their octanol:water partition coefficients the order of their affinity for P-450 isozymes should be isofenphos>DNI>IO>DNIO.

**Metabolic Rate Constants, OP-Hydrolases.** Tissue OP-hydrolases catalyze the hydrolysis of IO, DNIO, and DNI to their respective alkyl phosphates and 2-OH isopropylsalicylate. The kinetics of these hydrolytic reactions ($V_{max}$ and $K_m$) are unknown for IO, DNI, and DNIO (12). According to Maxwell (12), the thiono esters are less water soluble than their respective oxons and have smaller $K_m$ values (higher affinity) than the oxons. Liver and blood OP-hydrolases rapidly hydrolyze both the thiono (P=S) and phosphate (P=O) esters to their respective alkyl phosphates. Hydrolysis products (alkyl phosphates) of isofenphos, IO, DNIO, or DNI were not detected in the NADPH fortified P-450 studies (8). $K_m$ values for OP-hydrolases catalyzing the hydrolysis of P=O (182 uM) and P=S (15 uM) compounds were taken from published studies and used in the model(13). The $V_{max}$ values for these reactions were estimated by the model using urinary metabolite data.

**Metabolic Rate Constants, Carboxylesterases.** In the absence of NADPH, liver microsomal carboxylesterases catalyze the hydrolysis of the isopropylester of isofenphos (14), while in the presence of NADPH, newly formed IO and DNIO per se inhibit microsomal carboxylesterases and the hydrolysis of the isopropyl esters of isofenphos, IO, DNI, or DNIO. Carboxylesterases, however, are capable of hydrolyzing isopropyl esters when very low concentrations of these oxons are present in the incubation mixture. A published $K_m$ value of 62 uM was used for the carboxylesterase catalyzed hydrolysis of isofenphos, DNI, DNIO, DAI, and DAIO (15). $V_{max}$ values were estimated by the model using urinary metabolite data.

**Inhibition (phosphorylation) of Liver, Brain and Plasma CaE by Oxons.** Equations for the in vivo inhibition (phosphorylation) of liver and brain CaE by DNIO were included in the model. These equations were not used in this modeling exercise because CaE inhibition was measure only in the dog studies. The in vitro inhibition (phosphorylation) of liver, brain, and plasma CaEs by isofenphos, DNI, IO, and DNIO were investigated by Pierce (16). The oxons, IO and DNIO, were strong inhibitors of CaE, while neither isofenphos or DNI per se inhibited CaE. The $I_{50}$, $pI_{50}$, and $k_i$ (bimolecular rate constants) values for these reactions are given in the Table VI. The $k_i$ value for the inhibition of CaE was calculated using the equation of O'Brien (17), where $k_i = 0.685/I_{50}$*time. The nM $I_{50}$ concentrations are 32 and 418 fold less than the uM $K_m$ values for the P-450 catalyzed formation of IO and hydrolysis by OP-hydrolases, respectively (see Tables IV and V).

**Inhibition (phosphorylation) of Brain and Blood AChE, BuChE by Oxons.** The in vivo inhibition of brain AChE and BuChE were measured in the dog studies. Brain AChE and BuChE inhibition equations involving DNIO were included in the PBPK/PBPD model, but were not used in this modeling exercise. The in vivo inhibition of red cell and plasma cholinesterases were measured in the rat and dog iv single bolus, rat iv-infusion, guinea pig ip and ip-infusion, and dog dermal studies. In practice, blood samples were taken after the iv administrations or topical applications of isofenphos and analyzed for cholinesterase activity using the automated colorimetric method of Knaak et al. (18). In several studies, the dose producing 50% inhibition was calculated using a logprobit regression program (SAS, Cary, NC). The enzyme inhibition data (percentage inhibition) from these studies were placed in the Command File for estimating bimolecular rate constants. To facilitate the fitting process the bimolecular rate constants ($k_i = $ (pmol $L^{-1}$)$^{-1}$ hr$^{-1}$) were mathematically represented in the PBPK/PBPD model by $k_p/K_a$ (phosphorylation constant, hr$^{-1}$, divided by affinity constant, pmol $L^{-1}$). Table VII gives the $K_a$, $k_p$, and $k_i$ values for a number of toxic oxons formed in the agricultural work place from active OP ingredients (malathion, parathion, and guthion) (19, 20). The $K_a$ value for the inhibition of acetylcholinesterase by diethyl paraoxon (21.0 x $10^6$ pmol $L^{-1}$) was used in the model for the inhibition of BuChE and AChE by DNIO (20).The $k_p$ (hr$^{-1}$) values were estimated using in vivo AChE and BuChE inhibition data in the Command File and the SimuSolv "optimize" procedure. The $k_i$ (AChE) modeling values (~3.3 x $10^4$ (mol $L^{-1}$)$^{-1}$ hr$^{-1}$) involving DNIO were in agreement with the in

Table IV. $V_{max}$, $K_m$ values used in Intravenous Route

| Enzymes | Metabolism | Rat $V_{max}$/a | Rat $K_m$/a | Guinea Pig $V_{max}$/a | Guinea Pig $K_m$/a | Dog $V_{max}$/a | Dog $K_m$/a |
|---|---|---|---|---|---|---|---|
| **P-450** | | $(x10^6)$ | $(x10^6)$ | $(x10^6)$ | $(x10^6)$ | $(x10^6)$ | $(x10^6)$ |
| Vmax1 | IF to IO | 68.6 | 14.1 | 90.5 | 7.4 | 50.0 | 23.3 |
| Vmax2 | IF to DNI | 19.0 | 12.5 | 0.0 | 1.0 | 11.2 | 9.3 |
| Vmax3 | IO to DNIO | 36.3 | 9.5 | 37.6 | 11.9 | 44.8 | 2.2 |
| Vmax4 | DNI to DNIO | 12.4 | 7.9 | 19.7 | 8.0 | 23.9 | 14.4 |
| **Deaminases** | | | | | | | |
| Vmax5 | DNI to DAI | 0.0 | 1.0 | 0.0 | 1.0 | 0.0 | 1.0 |
| Vmax6 | DNIO to DAIO | 0.0 | 1.0 | 0.0 | 1.0 | 0.0 | 1.0 |
| **OP Hydrolases**/b | | $(x10^8)$ | $(x10^6)$ | $(x10^8)$ | $(x10^6)$ | $(x10^8)$ | $(x10^6)$ |
| Vmax7 | IO to Ring/AP1 | 3.9 | 182 | 3.9 | 182 | 7.1 | 182 |
| Vmax8 | DNIO to Ring/AP2 | 5.2 | 182 | 6.0 | 182 | 0.8 | 182 |
| Vmax9 | DAIO to Ring/AP3 | 0.0 | 182 | 0.0 | 182 | 0.0 | 182 |
| Vmax10 | DNI to Ring/AP4 | 0.20 | 15.0 | 0.0 | 15.0 | 0.0 | 15 |
| Vmax11 | DAI to Ring/AP5 | 0.0 | 182 | 0.0 | 182 | 0.0 | 182 |
| **CaE**/c | | $(x10^8)$ | $(x10^6)$ | $(x10^8)$ | $(x10^6)$ | $(x10^8)$ | $(x10^6)$ |
| Vmax12 | IF to IFA | 0.176 | 62 | 5.8 | 62.0 | 0.22 | 62 |
| Vmax13 | DNI to DNIA | 2.7 | 62 | 0.0 | 62.0 | 0.83 | 62 |
| Vmax14 | DAI to DAIA | 0.0 | 62 | 0.0 | 62.0 | 0.0 | 62 |
| Vmax15 | IO to IOA | 0.0 | 62 | 0.0 | 62.0 | 2.4 | 62 |
| Vmax16 | DNIO to DNIOA | 0.0 | 62 | 0.0 | 62.0 | 0.20 | 62 |
| Vmax17 | DAIO to DAIOA | 0.0 | 62 | 0.0 | 62.0 | 0.0 | 62.0 |

a $V_{max}$, pmoles hr$^{-1}$ kg$^{-1}$ of bw; $K_m$, pmoles L$^{-1}$
b Wallace and Dargan (13).
c Talcott (15).

**Table V. $V_{max}$, $K_m$ Values used in Percutaneous Absorption Route**

| Enzymes | Metabolism | Rat | | Dog | |
|---|---|---|---|---|---|
| | | $Vmax^{/a}$ | $Km^{/a}$ | $Vmax^{/a}$ | $Km^{/a}$ |
| **P-450** | | $(x10^6)$ | $(x10^6)$ | $(x10^6)$ | $(x10^6)$ |
| Vmax1 | IF to IO | 36.15 | 14.1 | 25.97 | 23.3 |
| Vmax2 | IF to DNI | 10.1 | 12.5 | 5.84 | 9.3 |
| Vmax3 | IO to DNIO | 15.9 | 9.5 | 23.26 | 2.2 |
| Vmax4 | DNI to DNIO | 5.48 | 7.9 | 12.41 | 14.4 |
| Deaminases | | | | | |
| Vmax5 | DNI to DAI | 0.0 | 1.0 | 0.0 | 1.0 |
| Vmax6 | DNIO to DAIO | 0.0 | 1.0 | 0.0 | 1.0 |
| **OP Hydrolases**[b] | | $(x\ 10^8)$ | $(x\ 10^6)$ | $(x\ 10^8)$ | $(x\ 10^6)$ |
| Vmax7 | IO to Ring/AP1 | 3.9 | 182 | 7.1 | 182 |
| Vmax8 | DNIO to Ring/AP2 | 5.2 | 182 | 0.84 | 182 |
| Vmax9 | DAIO to Ring/AP3 | 0.0 | 182 | 0.0 | 182 |
| Vmax10 | DNI to Ring/AP4 | 0.20 | 15.0 | 0.2 | 15 |
| Vmax11 | DAI to Ring/AP5 | 0.0 | 182 | 0.0 | 182 |
| **CaE**[/c] | | $(x\ 10^8)$ | $(x\ 10^6)$ | $(x\ 10^8)$ | $(x\ 10^6)$ |
| Vmax12 | IF to IFA | 0.176 | 62 | 0.22 | 62 |
| Vmax13 | DNI to DNIA | 2.7 | 62 | 0.83 | 62 |
| Vmax14 | DAI to DAIA | 0.0 | 62 | 0.0 | 62 |
| Vmax15 | IO to IOA | 0.0 | 62 | 2.4 | 62 |
| Vmax16 | DNIO to DNIOA | 0.0 | 62 | 0.2 | 62 |
| Vmax17 | DAIO to DAIOA | 0.0 | 62 | 0.0 | 62.0 |

[a] $V_{max}$, pmoles hr$^{-1}$ kg$^{-1}$ of bw; $K_m$, pmoles L$^{-1}$
[b] Wallace and Dargan (13).
[c] Talcott (15).

**Table VI. Inhibition of Carboxylesterase in Tissues of Male and Female Guinea Pigs and Female Rats by Isofenphos-Oxon[a].**

| Constant | Male Gpig Liver | Male Gpig Brain | Male Gpig Plasma | Female Gpig Liver | Female Rat Liver |
|---|---|---|---|---|---|
| $I_{50}$[b] | 435 nmol L$^{-1}$ | 709 nmol L$^{-1}$ | 371 nmol L$^{-1}$ | 7343 nmol L$^{-1}$ | 927 nmol L$^{-1}$ |
| $pI_{50}$[c] | 6.36 | 6.15 | 6.43 | 5.13 | 6.033 |
| $k_i$[d] | $2.66 \times 10^4$ | $1.63 \times 10^4$ | $3.12 \times 10^4$ | $1.57 \times 10^4$ | $1.25 \times 10^4$ |

[a] Data from Pierce (16).
[b] $I_{50}$ = molar concentration producing 50% inhibition
[c] $pI_{50}$ = negative log of molar $I_{50}$ concentration
[d] $k_i = 0.695/I_{50}*t$; $t = 1/60$ hrs, units = (mol L$^{-1}$)$^{-1}$ hr$^{-1}$

**Table VII. Affinity Constants ($K_a$) and phosphorylation constants ($k_p$) for a series of oxons for the inhibition of acetylcholinesterase**

| Compound | $K_a$ (mmol L$^{-1}$) | $k_p$ (min$^{-1}$) | $k_i$ ($k_p/K_a$) |
|---|---|---|---|
| Diethyl malaoxon[a] | 3.6 | 52 | 240 (mol L$^{-1}$)$^{-1}$hr$^{-1}$ |
| Diethyl paraoxon[a] | 0.36 | 42.7 | $1.935 \times 10^3$ (mol L$^{-1}$)$^{-1}$hr$^{-1}$) |
| Ethyl gutoxon[b] | 0.020 | 72.90 | $6.075 \times 10^4$ (mol L$^{-1}$)$^{-1}$hr$^{-1}$) |
| Diethyl paraoxon[b] | 0.02169 | 38.17 | $2.93 \times 10^4$ (mol L$^{-1}$)$^{-1}$ hr$^{-1}$) |

[a] Values from Chiu, Main and Dauterman (19).
[b] Values from Wang and Murphy (20).

vitro diethyl paraoxon data and the DNIO $k_i$ value ($3.86 \times 10^4$ (mol $L^{-1})^{-1}$ $hr^{-1}$) calculated from in vitro electric eel AChE $I_{50}$ data published by Gorder et al. (21). The electric eel $k_i$ (ACHE) value was also similar to the $k_i$ (CaE) values obtained by Pierce (16) shown in Table VI involving IO.

## MULTIPATHWAY, MULTIROUTE PBPK/PBPD MODEL OUTPUT

**Model Output, Absorption, Metabolism and Elimination.** The fate of intravenously administered bolus doses of $^{14}$C-ring labeled isofenphos in the rat, guinea pig and dog are given in Table VIII along with values obtained from the PBPK/PBPD model. The output from the model closely agreed with in vivo intravenous study data. Table IX gives comparable data for the fate of topically applied $^{14}$C-ring and $^{3}$H-ethyl labeled isofenphos to the back skin of the rat and dog, respectively. Here again, output from the model closely agreed with in vivo data.

**Table VIII. Fate of $^{14}$C-ring Isofenphos Intravenously Administered to the Rat, Guinea Pig, and Dog in Percent of Dose**

| | Rat | | Guinea Pig | | Dog | |
|---|---|---|---|---|---|---|
| | Bolus | Model | Bolus | Model | Bolus | Model |
| Urine & Feces | 95.76 | 99.9 | 99.7 | 99.98 | na | 98.5 |
| Carcass | 3.84 | 0.1 | 0.2 | 0.0 | na | 1.5 |
| Total | 99.6 | 100 | 99.9 | 100 | na | 100 |
| | | | | | | |
| Excreted in Urine as: | | | | | | |
| 2-OH Hippuric Acid | 73.7 | 81.73 | 53.9 | 51.0 | 39.13 | 42.12 |
| 2-OH Benzoic Acid | 1.74 | na | na | na | na | na |
| Alkyl phosphates | 75.4 | 81.73 | 53.9 | 51.0 | 43.99 | 42.12 |
| Carboxylic Acids | 19.2 | 18.17 | 46.0 | 49.0 | 55.37 | 56.4 |
| Inhibited enzymes | na | 0.074 | na | 0.015 | na | 0.047 |

Alkyl phosphates: AP1 and AP2; na = not available
Carboxylic Acids: IFA, IOA, DNIA and DNIOA
See Table II for identification of metabolites
iv rat dose = 8.0 mg/kg ($7.018 \times 10^6$ pmol), iv guinea pig dose = 22.19 mg/kg ($2.86 \times 10^7$ pmol)
iv dog dose = 8.0 mg/kg ($3.427 \times 10^8$ pmol)

**Phosphorylation (Inhibition) of AChE and BuChE in Rat, Guinea Pig and Dog Blood by DNIO.** Table X gives the concentration of AChE and BuChE in blood (pmol $L^{-1}$) and total amount of uninhibited enzyme (pmol) present in the blood of the rat, guinea pig and dog prior to the administration of isofenphos and after administration along with the percentage of inhibited enzyme. The molar bimolecular rate constants obtained in the intravenous and dermal studies were of similar magnitude ($\times 10^4$), except for the values obtained for the dermal rat study ($\times 10^5$). The maximum inhibition rates (nonlinear, peak values from the model) for rat AChE and BuChE in the dermal study were very small compared to the rates obtained in the other studies involving the rat, guinea pig and dog. $k_i$ values for liver and brain enzymes were not included in Table X, because sufficient in vivo inhibition data for these tissues and enzymes were not available for modeling. Figures 6 and 7 give the percentage of rat blood AChE and BuChE inhibited (model output and in vivo data) by an 8.0 and 3.73 mg/kg intravenous and topical dose, respectively, as a function of time.

**Table IX.** Fate of [14]C-ring and [3]H-ethyl-Isofenphos Topically Administered to the Rat and Dog in Percent of Dose.

|  | Rat | | Dog | |
|---|---|---|---|---|
|  | Dermal | Model | Dermal | Model |
| Loss to air | 35.0 | 44.0 | na | 44.4 |
| Retained on skin | 15.3 | 10.0 | na | 9.2 |
| Urine & Feces. | 43.2 | 45.0 | na | 44.9 |
| Carcass | 6.0 | 1.0 | na | 1.5 |
| Total | 99.5 | 100.0 | na | 100.0 |
|  |  |  |  |  |
| Excreted in Urine as: |  |  |  |  |
| 2-OH Hippuric Acid | 31.5 | 29.5 | 17.56 | 15.52 |
| 2-OH Benzoic Acid | na | na | 0.0 | 0.0 |
| Alkyl phosphates | 31.5 | 29.5 | 17.56 | 15.52 |
| Carboxylic Acids | 11.7 | 15.5 | 26.44 | 28.9 |
| Inhibited Enzymes | na | 0.2 | na | 0.013 |

Alkyl phosphates: AP1 and AP2
Carboxylic Acids: IFA, IOA, DNIA and DNIOA
See Table II for identification of metabolites.
Dermal rat dose = 1.12 mg/kg ($3.2 \times 10^6$ pmol); Dermal dog dose = 17.36 mg/kg ($7.44 \times 10^6$ pmol)
na: not available

**Table X.** Rate of Inhibition and Amount of AChE and BuChE Inhibited by Des N-Isofenphos Oxon

| Enz/Source | Enzyme[a] pmol L$^{-1}$ (x 10$^3$)/pmol | k$_i$, (mol L$^{-1}$)$^{-1}$ hr$^{-1}$ | Enzyme inhibited by DNIO pmol | Percent of enzyme inhibited Model | Study | Maximum Inhibition Rates[b] pmol hr$^{-1}$ |
|---|---|---|---|---|---|---|
| **Rat iv** |  |  |  |  |  |  |
| AChE/Blood | 1.1/19.9 | $2.88 \times 10^4$ | 11.77 | 58.87 | 50.0 | 1.4 |
| BuChE/Blood | 4.8/87.3 | $3.47 \times 10^4$ | 61.85 | 70.87 | - | 7.0 |
| **Guinea Pig ip** |  |  |  |  |  |  |
| AChE/Blood | 1.1/29.43 | $3.719 \times 10^4$ | 12.32 | 41.9 | 44.7 | 2.8 |
| BuChE/Blood | 4.8/128.4 | $7.314 \times 10^4$ | 105.76 | 82.36 | 86.6 | 24.0 |
| **Dog iv** |  |  |  |  |  |  |
| AChE/Blood | 1.1/976 | $4.719 \times 10^4$ | 800.8 | 81.9 | 75.1 | 180 |
| BuChE/Blood | 4.8/4262 | $5.714 \times 10^4$ | 4231 | 99.2 | 82.3 | 900 |
| **Rat skin** |  |  |  |  |  |  |
| AChE/Blood | 1.1/19.99 | $6.014 \times 10^5$ | 5.69 | 28.48 | 25.0 | 0.065 |
| BuChE/Blood | 4.8/87.26 | $8.873 \times 10^5$ | 36.67 | 42.02 | - | 0.4 |
| **Dog skin** |  |  |  |  |  |  |
| AChE/Blood | 1.1/976 | $4.671 \times 10^4$ | 293.8 | 30.1 | 30.3 | 2.5 |
| BuChE/Blood | 4.8/4262 | $9.608 \times 10^4$ | 2634.5 | 61.8 | 64.3 | 25.0 |

[a] Enzyme concentrations from Maxwell et al. (22)
[b] Output from model

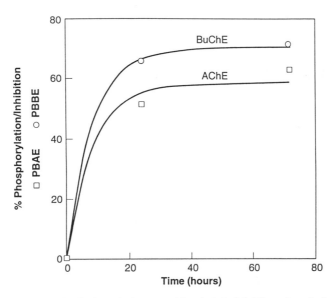

Figure 6. Physiologically based pharmacokinetic Model: Phosphorylation of Rat blood AChE and BuChE by des N-Isopropyl Isofenphos Oxon (DNIO) after administration of 8.0 mg/kg intravenous dose. Solid line model output. Squares and Circles in vivo data. PBAE, % blood AChE phosphorylated; PBBE, % blood BuChE phosphorylated

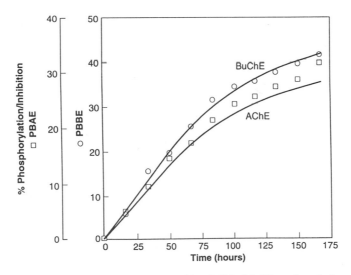

Figure 7. Physiologically based pharmacokinetic Model: Phosphorylation of Rat blood AChE and BuChE by des N-Isopropyl Isofenphos Oxon (DNIO) after administration of 3.73 mg/kg topical dose. Solid line model output. Squares and Circles in vivo data. PBAE, % blood AChE phosphorylated; PBBE, % blood BuChE phosphorylated.

## DISCUSSION

The modeling results presented here consist of our first look at the pharmacokinetics and pharmacodynamics of isofenphos across routes (single bolus intravenous and dermal) and experimental animals (pathways). Results of intravenous infusion studies were not included in this presentation because of limited space. Isofenphos was initially metabolized by P-450 isozymes to two oxons (IO and DNIO), which are strong inhibitors of peripheral and nervous system AChE and neuropathy target esterase (NTE), tissue and blood BuChE, and tissue CaE. The oxons and possibly DNI are further metabolized by OP-hydrolases to alkyl phosphates and isopropyl salicylate. Deamination does not appear to be a major pathway in the metabolism of isofenphos. In vitro laboratory studies are needed to determine the rate constants ($V_{max}$, $K_m$) for the metabolism of IO, DNI, and DNIO by OP-hydrolases.

The $V_{max}$ rates (pmoles $hr^{-1}$ $kg^{-1}$ of bw) for the P-450 metabolism of isofenphos to IO, DNI and DNIO developed by Knaak et al. ([8]) were initially used in the model.The $V_{max}$ rates were calculated based on total P-450 isozymes in liver homogenates. These $V_{max}$ rates were reduced (0.26 x $V_{max}$ for dermal and 0.5 x $V_{max}$ for intravenous administration) to prevent the to rapid removal of isofenphos and metabolites from liver and other tissues. The reason for this adjustment was unclear at first until we compared these rates against $V_{max}$ rates based on P-450 isozymes harvested in microsomes. The microsomes contained 22, 19, and 15% of total P-450 isozymes (activity) present in rat, guinea pig and dog liver homogenates, respectively. The activity in rat microsomes was almost identical to that used for the percutaneous route given in Table V. This activity, however, was not sufficient to model the elimination of the 8 mg/kg intravenous dose administered to the rat. In the case of the guinea pig, use of the microsomal rates reduced the rate of conversion of isofenphos to IO and resulted in an abnormal amount of isofenphos (72%) being metabolized to IFA by carboxylesterases. Microsomal $V_{max}$ rates reduced P-450 metabolism in the dog model below in vivo pharmacokinetic data.These observations suggest that $V_{max}$ estimates, based on harvested P-450 microsomal isozymes, should be carefully evaluated before and during use. Analytical data on each metabolite (tissue time-metabolite concentration data) should be obtained when ever possible. Modeling is possible using [14]C equivalency data, but it does not provide information on the nature and concentration of each metabolite under the [14]C-tissue concentration curve.

In modeling it was also unclear as to why only very small amounts of isofenphos were metabolized in the rat by liver carboxylesterases compared to the amounts metabolized by the guinea pig or dog. An intravenous dose of 8 mg/kg in the dog inhibited 40-75% red cell AChE, 71-94% plasma BuChE and only 5% liver CaE after 1, 24 and 72 hrs post administration ([14]). According to Chambers et al. ([23]) and Pierce ([16]), rat liver CaE are strongly inhibited by oxons. We are currently examining the inhibition of CaE by DNIO and the effect of this inhibitor on the hydrolysis of IO using the equation ($V_{max}$ + CVL/$K_m$ (1 + CVL/$K_a$) + CVL) for competitive inhibition ([24]) in our PBPK/PBPD model. Nanomolar concentrations ($K_a$ = 435 nM) of DNIO were found to effectively reduce the amount of IO ($K_m$ = 182 uM) hydrolyzed to IOA. Large differences in affinity between inhibitor and substrate for CaE may be responsible for the differences in the metabolic pathway between the rat, guinea pig and dog. No information is currently available concerning the status of CaE activity in either the rat or guinea pig after administration of isofenphos. In vitro laboratory studies involving carboxylesterases from the rat, dog and guinea pig are needed to determine the rate constants ($V_{max}$, $K_m$) for the metabolism of isofenphos, DNI and perhaps IO in the presence of DNIO. DNIO appears to be a strong inhibitor of its own metabolism and that of IO, DNI, and isofenphos.

Equations describing the inhibition of AChE, BuChE, and CaE by DNIO are in the current model. We plan to add equations for IO in the next planned revision of the model. In vivo inhibition data on dog brain AChE, blood BuChE and AChE and liver CaE are available for modeling. Equations involving the resynthesis of AChE, BuChE and CaE wll be used along with equations describing the aging of phosphorylated enzymes.

## CONCLUSIONS

A PBPK/PBPD model, as presented for isofenphos, is a powerful tool for describing and testing the effects of in vitro and in vivo derived experimental values such as skin permeability constants, evaporation rates and $V_{max}$, and $K_m$ values on oxon tissue concentrations, enzyme inhibition and elimination rates. A properly validated model may be used to extrapolate exposure, pharmacokinetics and effects data from one route to another and from animals to humans. The relationship between dose (i.e., acute, subchronic and chronic), AChE inhibition (red cells, peripheral and central nervous system), resynthesis of AChE and clinical manifestations (e.g., muscle weakness, tremor, blurred vision, etc.) maybe critically examined through time using the model.

## ACKNOWLEDGMENTS

We wish to thank D. Uyeminami and F. Gielow, University of California, and Jerry Elig, Computer Sciences Corporation, Las Vegas, NV for their assistance. This project was sponsored by the United States Environmental Protection Agency and Cooperative Agreement No. CR-816332 with the University of California, Davis.

## LITERATURE CITED

1.  Knaak, J.B., Al-Bayati, M.A., Raabe, O.G., and Blancato, J.N.,  In "Biomakers of Human Exposure to Pesticides", Saleh, M.A., Blancato, J.N. and Nauman, C.H., Eds., ACS Symposium Series 542, ACS, Washington, DC 1994, p284.
2.  Gearhart, J.M.; Jepson, G.W.; Clewell, H.J. II; Andersen, M.E.; and Conolly, R.B., Toxicol. Appl. Pharmacol., 1990, 106, 295.
3.  Knaak, J.B.; Al-Bayati, M.A.; Raabe, O.G.; and Blancato, J.N.  In "Prediction of Percutaneous Penetration: Methods, Measurements, Modelling"; Scott, R.C.; Guy, R.H.; and Hadgraft, J., Eds., IBC Technical Service Ltd. London, 1990; pp. 1-18.
4.  Knaak, J.B.; Al-Bayati, M.A.; Raabe, O.G.,  In "Health Risk Assessment through Dermal and Inhalation Exposure and Absorption of Toxicants";

Wang, R.G.; Knaak, J.B.; and Maibach, H.I., Eds., CRC Press, Boca Raton, Fl., 1992, Chapter 1.

5. Ramsey, J.C.; and Andersen, M.E. Toxicol. Appl. Pharmacol., 1984, 73, 159.
6. Arms, A.D.; and Travis, C.C., Reference Physiological Parameters in Pharmacokinetic Modeling, Report No. EPA/600/6-88/004, National Technical Information Service, Springfield, VA 22161.
7. Knaak, J.B., Al-Bayati, M.A. and Raabe, O.G., Toxicology Letters 1995, 79, 87.
8. Knaak, J.B.; Al-Bayati, M.A.; Raabe, O.G.; and Blancato, J.N., Toxicol. Appl. Pharmacol., 1993, 120, 106.
9. Ueji, M.; and Tomizawa, C., J. Pesticide Sci., 1985, 10, 691.
10. Hugener, M., Tinke, A.P., Niessen, W.M.A, Tjaden, U.R., van der Greef, J., J. Chromatogr., 1993, 647(2), 375..
11. Weisskopf, C.P. and Seiber, J.N., In "Biological Monitoring for Pesticide Exposure: Measurement, Estimation, and Risk Reduction, R.G.M. Wang, C.A. Franklin, R.C. Honeycutt and J.C. Reinert, Eds., ACS Symposium Series 382, ACS, Washington, DC 1989, p 206.
12. Maxwell, D.M. and Brecht, K.M., Chem. Res. Toxicol. 1992, 5(1), 66.
13. Wallace, K.B. and Dargan, J.E., Toxicol. Appl. Pharmacol. 1987, 90, 235.
14. Al-Bayati, M.A., Pierce, J.P., Knaak, J.B. and Raabe, O.G. University of California, unpublished report.
15. Talcott, R.E., Toxicol. Appl. Pharmacol. 1979, 47, 145.
16. Pierce, J.P., MS Thesis, Sept 1995, Sacramento State University, Physiological Science.
17. Toxic Phosphorus Esters: Chemistry, Metabolism, and Biological Effects, O'Brien, R.D., Ed., Academic Press, New York and London, 1960, Chapter 3.
18. Knaak, J.B., Maddy, K.T., Jackson, T., Fredrickson, A.S.. Peoples, S.A., and Love, R., Toxciol. Appl. Pharmacol. 1978, 45, 755.
19. Chiu, Y.C., Main, A.R. and Dauterman, W.C. Biochem. Pharmacol. 1969, 18, 2171.
20. Wang, C.; and Murphy, S.D., Toxicol. Appl. Pharmacol., 1982, 66, 409.
21. Gorder, G.W., Kirino, O., Hirashima, A., Casida, J.E., J. Agric. Food Chem. 1986, 34, 941.
22. Maxwell, D.M., Lenz, D.E., Groff, W.A., Kaminski, and Froehlich, H.L. Toxicol. Appl. Pharmacol., 1987, 88,66.
23. Chambers, H., Brown, B., Chambers, J.E., Pestic. Biochem. Physiol. 1990, 36, 308.
24. Tardif, R., Lapare, S., Krishman, K., Brodeur, J., Toxicol. Appl. Pharmacol. 1993, 120, 266.

Chapter 17

# Use of Spot Urine Sample Results in Physiologically Based Pharmacokinetic Modeling of Absorbed Malathion Doses in Humans

Michael H. Dong[1], John H. Ross[1], Thomas Thongsinthusak[1], and Robert I. Krieger[2]

[1]Worker Health and Safety Branch, California Department of Pesticide Regulation, 1020 N Street, Sacramento, CA 95814
[2]Personal Exposure Program, Department of Entomology, University of California, Riverside, CA 92507

Recently state health and regulatory agencies in California used a PB-PK model to estimate the absorbed doses of malathion in individuals allegedly exposed to aerial sprays during an urban pesticide application. Dose simulation in that study was performed on the results of single urine samples collected within 48 h of a potential exposure, and was based on a model validated with observed values from only a single volunteer. As a continuing effort another case study is presented in this chapter to validate the model with more human literature data. Results from this validation study showed that the time courses of the serial urinary malathion (metabolites) excretion presented in the literature were consistent with those simulated by the PB-PK model. When urine results collected from the literature cases at 8 - 12 h, 12 - 24 h, and 24 - 36 h or 24 - 48 h after initial exposure were postulated as spot samples, the majority (64/85) of the individual total absorbed doses simulated were within two-fold of their measured values, with no simulation doses exceeding three-fold. This validation study further showed that the accuracy would be improved considerably, if the simulation for each literature case were performed on two or more spot samples collected at different time points preferably within the first 24 or 36 h of exposure.

Biomarkers may be defined as properties that are capable of indicating that an individual has had exposure to a particular chemical (*1*). These indicators are measurable in biological media such as human cells or fluids, and are referred to as markers of internal dose. There are biological markers of other categories, such as markers of biologically effective dose and markers of disease (*2*). However, it is markers of internal dose that have given biological monitoring its long-standing place in the assessment of human exposure to pesticides. Of all the methods available for biological monitoring of exposure, the test most commonly used to assess the internal dose is the measurement of the chemical or its metabolite(s) in urine.

Despite its popularity in biological monitoring, urine analysis is not a test without limitations. Routine collection of 24-hour samples in human subjects is usually impractical. Urine analyses for pesticide exposure assessment are accordingly often necessarily performed on spot specimens. The excretion and other related disposition kinetics must then be used to interpret these analytical results. In addition, these results normally must be corrected for the total urine volume anticipated or excreted. One approach to a more effective interpretation of biomarker data based on spot samples is to simulate the internal dose through use of a physiologically based pharmacokinetic (PB-PK) model. This type of dose simulation can also simplify or even eliminate the task required for the correction of urine volume.

PB-PK models are designed to simulate the body as a series of tissue compartments, across which a chemical is absorbed, distributed, metabolized, and excreted in accord with pharmacokinetic rate laws and constants. Well-constructed PB-PK models can be powerful tools for interpretation of internal dose biomarker data, even of those data based on analysis of spot specimens. Such use was demonstrated in a 1994 study by Dong et al. (3), who used a PB-PK model to estimate the absorbed malathion doses in 11 adults and children allegedly exposed to aerial sprays during an urban pesticide application. Dose simulation in that study was performed on single urine samples collected within 48 h of a potential exposure, and was based on a PB-PK model validated with observed values from only a single volunteer. As a continuing effort, another case study is presented in this chapter to validate the PB-PK model with more human data now available in the literature.

In addition to validating the PB-PK model, the purpose of this chapter is to discuss the utility of this type of simulation in interpreting the data on internal dose markers that are analyzed especially on spot samples. It is also hoped that by presenting the results of this validation study in the literature, there will be a fuller appreciation of the absorbed doses of malathion simulated in the earlier 1994 case study.

## Materials and Methods

**PB-PK Simulation.** In recent years numerous investigators have used PB-PK models to simulate tissue dose in the animal or human body, especially in those exposed to organic chemical vapors via inhalation. Over 15 references have been cited for work done in this subject area by Dong et al. (3), in which the basic structure of a PB-PK model for dermal exposure is included. The work done in that paper represents in effect a first collaborative effort by several state health and regulatory agencies in California to simulate absorbed doses of pesticides on spot urine samples from human volunteers. The PB-PK model used in the 1994 collaborative study was first constructed by the California Office of Environmental Health Hazard Assessment (4,5) to predict primarily urinary excretion of dermally applied malathion. The model was later modified by Dong et al. (6) to account more closely for the actual disposition and metabolic fate of malathion in humans. That modified model, together with the physiological and biochemical parameter values that were provided in the collaborative study, was used throughout this validation study. As in the collaborative study, all PB-PK simulations performed in this validation study were implemented with a microcomputer program written in BASICA (7).

**Literature Data.** Human data published on urinary excretion of malathion metabolites from dermal exposure are quite limited. All urine samples should be collected serially over a period of several days, if they are to be useful for dose simulation performed on spot samples. In addition, the dermal dose used should be at a level typical of human exposure and prepared in some aqueous or organic-based formulation. A review of the literature indicated that thus far only one study has been published to present data fulfilling these criteria for malathion. This is the work conducted in 1994 by Dary et al. (8). In their study, dermal absorption of neat malathion, a 50% emulsifiable concentrate (EC), and a 1% and 10% aqueous mixture of the 50% EC formulation was examined in 12 adult volunteers. The urine samples from these human volunteers were collected over a 3-day period for recovery of radiolabeled malathion (mainly its metabolites). The total absorbed doses of malathion accumulated from these urinary radiorecoveries over the 72-hour period are listed in Table I below. These cumulative absorbed doses (in percent of their applied doses) were later used to validate the doses simulated from PB-PK modeling performed on spot urine samples.

In addition to the literature data published in 1994, there is one other earlier dermal absorption study available which provides serial urinary excretion of malathion over two consecutive 7-day periods. That study was published in 1983 by Wester *et al. (9)*, whose interest was to contrast the percutaneous absorption of malathion from single dose applications with that after repeated administration. The data in the 1983 study were considered to be less desirable for the purpose of this validation study, in that the topical dose given to each of the 5 human volunteers was neat malathion at a level (5 mg/cm$^2$) which is much higher than those seen with typical exposure. Another drawback with the 1983 literature data is that there were no urine samples collected for 24 - 36 h after initial exposure. This interval is one of the crucial ones for the collection of spot samples (since from this time onward the excreted dose would be very low). Yet for completeness, the total absorbed doses calculated over the first 7 days (i.e., from single doses) for each of the 5 subjects in the 1983 study were also used in this validation study, and are included in Table I.

The third (and final) study of this type available in the literature is the classic work published in 1974 by Feldmann and Maibach *(10)*. Their data were not used in this validation study since only the *group* mean values of urinary radiorecoveries taken over 6 volunteers were presented. The topical dose used in that classic work was also neat malathion, though at a much lower and hence a more typical dermal exposure level (4 µg/cm$^2$). Note that the intent of this validation study was to make use of the spot urine samples in estimating the total exposure for individuals rather than for groups of people. In general, it is relatively easier to more accurately estimate the total exposure for a group of people than for an individual.

There is still one other work not considered in this validation study. This is an earlier pilot study by Dong *et al. (6)*, in which a comparison was made between dermal absorption of malathion simulated with PB-PK modeling for a single volunteer and those absorption values determined primarily in the 1983 and the 1974 studies cited above. Since the PB-PK model had been modified with much of the simulation results from that pilot study, it did not seem appropriate to *re*validate the simulation here with the same urinary data collected from that single volunteer. Another concern with the use of that set of urinary data is that the volunteer was an obese person with a body weight (~ 150 kg) twice that of an average adult. Body weight is related to blood flow, fat content, and some other physiological parameters included in the PB-PK model. Although there is a preferred formula that has been used to adjust for this body weight difference *(3-5)*, the adjustment is far from perfect especially when an obese (*vs.* a lean heavyweight) person is considered. The rest of the human literature data included for comparison of malathion absorption in that pilot study were not presented in a time series (and hence could not be used here as spot samples for validating the PB-PK model).

**Validation Procedures.** For the purposes of this validation study, the PB-PK model was first applied to simulate the *serial* urinary excretion of malathion metabolites (including the small amount of unchanged malathion) observed in the 1983 and the 1994 literature cases. The endpoints used for simulation were the absorbed doses (in percent of the applied doses) accumulated to 72 h for cases in groups A through D and to the first 168 h for those in group E. The objective of this first phase was to assure that the time course of urinary excretion of malathion metabolites could be reproduced by the PB-PK model. The model was then used to simulate the total absorbed doses of malathion in these same literature cases under the assumption that only their spot urine samples collected at 8 - 12 h, 12 - 24 h, and 24 - 36 h or 24 - 48 h

Table I.  Percent of the Applied Dose of $^{14}$C-Malathion Recovered in the Urine of Human Volunteers at Selected Spot Intervals Following Topical Administration

| Subject[a,b] | Dermal Dose mg/cm$^2$ | 8-12 h | 12-24 h | 24-36 h | 24-48 h | Cumulative Abs. Dose[c] |
|---|---|---|---|---|---|---|
| | | Percent of the Applied Dose Recovered | | | | |
| A1 | 0.35 | 3.86 | 11.54 | 2.50 | | 18.36 |
| A2 | 0.92 | 0.55 | 1.06 | 0.58 | | 2.45 |
| A3 | 0.31 | 0.80 | 6.26 | 1.34 | | 9.53 |
| A4 | 2.03 | 1.35[d] | 2.28 | 1.03 | | 5.05 |
| A5 | 0.43 | 0.41 | 1.72 | 0.50 | | 3.10 |
| A6 | 0.80 | | 4.78[e] | 0.55 | | 5.62 |
| B1 | 1.32 | 0.20 | 1.37 | 0.53 | | 2.33 |
| B2 | 1.20 | 0.74 | 2.14 | 1.12 | | 4.24 |
| B3 | 1.56 | 2.16 | 3.53 | 0.94 | | 7.23 |
| B4 | 0.57 | 2.19[d] | 5.45 | 1.15 | | 9.01 |
| B5 | 0.39 | 1.30 | 5.42 | 1.62 | | 9.04 |
| B6 | 0.54 | 0.81[d] | 0.47 | 1.11 | | 2.57 |
| C1 | 0.02 | 4.03 | 7.53 | 1.70 | | 16.86 |
| C2 | 0.03 | 4.22 | 10.36 | 3.10 | | 19.59 |
| C3 | 0.04 | 1.54 | 3.72 | 2.39 | | 9.40 |
| C4 | 0.03 | 20.50[d] | 6.97 | 0.30 | | 28.60 |
| C5 | 0.03 | 4.67[d] | 4.55 | 0.83 | | 10.57 |
| C6 | 0.03 | 0.53 | 8.94 | 1.02 | | 11.93 |
| D1 | 1.19 | 0.33 | 1.39 | 0.54 | | 3.13 |
| D2 | 1.20 | 1.18[d] | 3.74 | 2.08 | | 7.48 |
| D3 | 1.18 | | 2.83[e] | 1.13 | | 4.40 |
| D4 | 1.19 | 1.25[d] | 1.37 | 1.13 | | 4.76 |
| D5 | 1.17 | 0.33 | 2.25 | 1.95 | | 5.27 |
| D6 | 0.87 | 3.25[d] | 4.85 | 2.28 | | 10.99 |
| E1 | 5.00 | 0.35 | 1.41 | | 0.87 | 4.00 |
| E2 | 5.00 | 0.56 | 1.88 | | 0.34 | 3.97 |
| E3 | 5.00 | 1.03 | 2.14 | | 0.51 | 5.30 |
| E4 | 5.00 | 0.25 | 0.37 | | 0.34 | 2.07 |
| E5 | 5.00 | 0.83 | 0.46 | | 0.13 | 3.07 |

[a] subjects A1 through D6 were from the 1994 study by Dary *et al.* (*8*); after resting for two weeks, individuals in group A that were dosed with neat malathion were given again a 1.0% aqueous mixture as members of group C; individuals in group B receiving the 50% EC formulation were dosed again with the 10.0% aqueous mixture as members of group D after resting for two weeks.

[b] subjects E1 through E5 were from the 1983 study by Wester *et al.* (*9*); absorbed doses (from neat malathion) presented here for group E were *not* corrected for excretion by other routes (e.g., feces).

[c] cumulative absorbed dose in percent of the applied dose; accumulated to 72 h for groups A through D, and to the first (as from single doses only) 168 h for group E.

[d] at spot interval of 4 - 12 h, since no dose recovery was measured for 4 - 8 h and dose recovery for 8 - 12 h in this study subject was reported as dose accumulated to this interval.

[e] at spot interval of 4 - 24 h, since no dose recovery was measured for 4 - 8 h or for 8 - 12 h and dose recovery for 12 - 24 h in this study subject was reported as dose accumulated to this interval.

where applicable) after initial exposure were available.   These time points were selected because they were considered to be reasonable for biological monitoring of exposure to malathion or to similar organophosphates.   For example, for all the 11 adults and children in the 1994 collaborative study (*3*), the time lapse from first dermal contact until urine collection was reported to be within 12 h, 12 - 36 h, 24 - 36h, or 36 - 48 h.

In its simplest form, dose simulation with a PB-PK model would involve predictions of the *total* absorbed dose based upon a small (cumulative, temporal) amount of dose estimated from biomarker data.   This task objective, together with the simulation procedures and the kinetic equations involved, was explicitly provided in the 1994 collaborative study.   However, despite this provision it is important to note that urine contents measured in *spot* samples are normally not predictive of the amount of urinary excretion accumulated *to* that spot time point, unless there are no earlier voids given by the individual between that time and initial exposure.   The cumulative percent of the applied dose recovered to that time point as quantified in spot urine samples, which is to be used as an endpoint for simulation of the total cumulative absorbed dose, is hence necessarily determined by some type of extrapolation.

One extrapolation method is to measure the urine concentration in the spot sample in some mass unit per volume of urine, and then multiply this concentration by the total urine volume that is expected to be voided between that time and initial exposure.   This was the extrapolation method used in the 1994 collaborative study since its malathion metabolites were reported only in μg per liter of urine.

In the present validation study, a more accurate as well as more direct endpoint was used for dose simulation.   Here urinary excretion *during* a spot interval was modeled by iterative simulation to the amount reported in the literature for that spot interval.   The total cumulative absorbed dose of interest would then be read off the *entire* simulated serial urinary excretion.   Note that even though only a portion of the urine content in a series is to be used as an endpoint for dose simulation, each run of the PB-PK modeling will simulate the urine and other tissue contents from time 0 to whatever time (e.g., at 72 h or 168 h) the analyst wishes the simulation to stop.   The spot intervals used for validation with the 1994 literature data from Dary *et al.* (*8*) were 8 - 12 h, 12 - 24 h, and 24 - 36 h.   As only cumulative percent of the applied dose recovered in urine was provided, the absorbed dose from the 1994 study for each postulated spot interval was determined by taking the difference between the absorbed dose of malathion accumulated to a given spot interval and the absorbed dose accumulated to the preceding interval in the series.   Where the absorbed dose of malathion accumulated to the preceding interval was not available, the absorbed dose accumulated to the earlier interval was used instead.   The spot interval targeted for validation would then be extended accordingly.   These irregular, extended spot intervals are footnoted in Table I.   The spot intervals used for validation with the 1983 literature data from Wester *et al.* (*9*) were 8 - 12 h, 12 - 24 h, and 24 - 48 h, since there were no urine samples collected specifically for the 24 - 36 h interval in that study.   The absorbed dose presented in the 1983 study for each interval in the series, while also in percent of the applied dose, was *non*cumulative.

## Results and Discussion

Results of this validation study showed that the time courses of the serial malathion metabolites (including the very small amount of unchanged malathion) present in human urine reported in the literature were reproducible by the PB-PK model.   These findings were also found to be quite consistent with those observed earlier in a pilot study on urine results from a single volunteer (*6*).   The time courses of malathion metabolites simulated in this validation study for subjects A1, B1, C1, D1, and E1, together with their observed serial urinary excretion, are summarized in Figures 1 through 5, respectively.

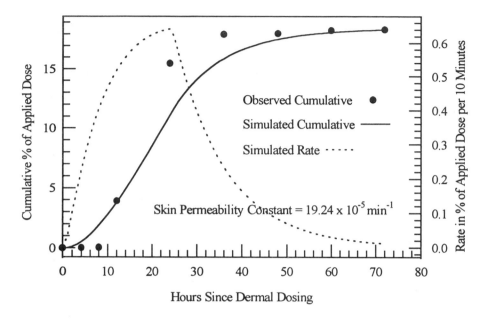

Figure 1.  Excretion of urinary malathion (mainly its metabolites) observed *versus* simulated for subject A1 in Dary *et al.* (*8*).

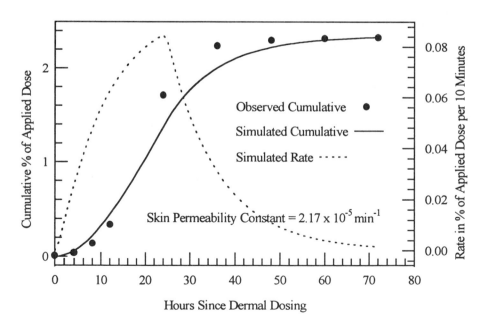

Figure 2.  Excretion of urinary malathion (mainly its metabolites) observed *versus* simulated for subject B1 in Dary *et al.* (*8*).

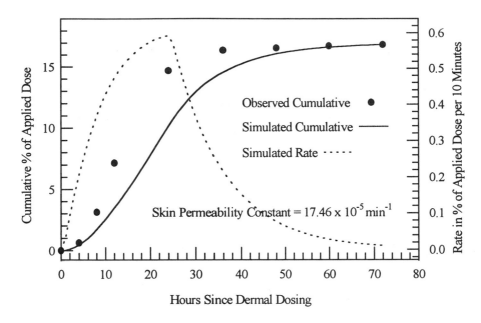

Figure 3. Excretion of urinary malathion (mainly its metabolites) observed *versus* simulated for subject C1 in Dary *et al.* (*8*).

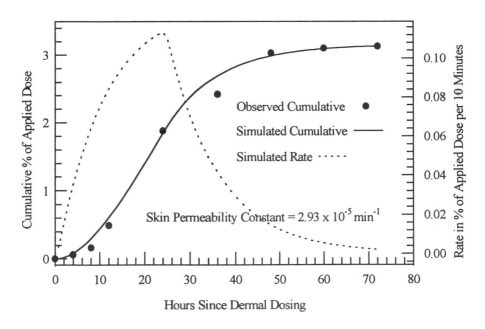

Figure 4. Excretion of urinary malathion (mainly its metabolites) observed *versus* simulated for subject D1 in Dary *et al.* (*8*).

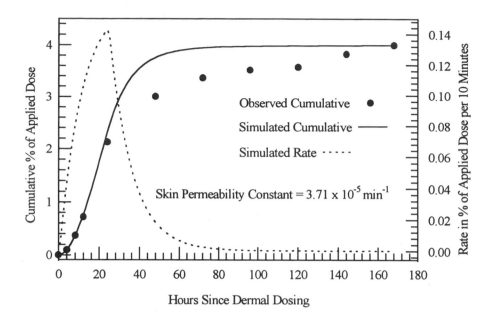

Figure 5. Excretion of urinary malathion (mainly its metabolites) observed *versus* simulated for subject E1 in Wester *et al*. (*9*).

Figures 1 through 5 exemplify the high degree of reproducibility achieved by the PB-PK model, especially in light of the fact that a physiological and not a physical model is considered here. The observed values presented in these figures were from the human studies by Dary *et al*. (*8*) and by Wester *et al*. (*9*), who like other investigators did not have full control of the human experimental situation (e.g., subjects were not cannulated, their dietary intake not assessed, their fluid intake not controlled, etc.). Insofar as there was no assurance for the *absolute* accuracy of *each* of these observed values, it did not seem worthwhile to perform any goodness-of-fit test for the simulated time series.

Also included in Figures 1 through 5 are the excretion rates (per 10 minutes) simulated at various time intervals for these same individuals. As shown (dotted lines), the excretion rates for these subjects (and for the rest of those not shown) peaked at approximately 18 to 24 h after initial exposure. The peak hours of excretion rates simulated in this validation study were found to be in good agreement with those observed in the 1983 study by Wester *et al*. (*9*) and in the 1994 study by Dary *et al*. (*8*). These peak hours seen in the two literature studies (and in this simulation study as well) were approximately 2 to 3 times greater than those simulated earlier in the 1994 collaborative study by Dong *et al*. (*3*). The reason for this difference in peak hours is that the dermal dose applied to the volunteers in the literature studies was left on their skin for a period of 24 h, whereas the subjects in the collaborative study were assumed to be exposed for approximately 8 h before they took a bath or shower. Insofar as duration is concerned, the 8-hour exposure scenario seems to be more realistic (or more frequently encountered in real life situations).

Figure 1 indicates that there might be a few hours of short lag before the dermal absorption of malathion could take place in subject A1. Similar indication was seen earlier in the single volunteer in the pilot study (*6*), but not particularly in the other literature cases considered in this validation study.   As footnoted in Table I, subjects in study group A that were dosed with neat malathion were later given again a 1.0% aqueous mixture as members of study group C.   However, it is not clear from the information available whether subject A1 and subject C1 were the same individual. If they were, then the short lag seen in Figure 1 (i.e., subject A1) and not in Figure 3 (i.e., subject C1) might have been due at least in part to the dilution of the malathion used in that experiment.   The emotional, nutritional, and physiological state, etc. at the time the study subject was exposed to the malathion dilution might also have played an important role.

The variable skin permeability constants shown in Figures 1 through 5 were the only model input parameter that needed to be adjusted until the model prediction of malathion (and its metabolites) matched the amount estimated from the biomarker data (for the spot interval in question).   This adjustment should not have violated the underlying principles of PB-PK modeling averred by McDougal *et al.* (*11*) and by Knaak *et al.* (*12*).   According to these investigators, the total absorbed dose from dermal exposure is simply a function of three input parameters.   These input parameters are the skin permeability constant $P$, the total applied (topical) dose $D$, and the permeation time $T$.   In this validation study, the input parameters were further reduced to $T$ (which was fixed at 24 h) and $P$ because the absorbed dose was calculated in percent of the applied dose.   (In the 1994 collaborative study, the parameter $D$ was written as the product of the total exposed surface area $A$ and the exposure concentration $C$.)   As described elsewhere (*3-7,13*), the PB-PK model consists of a series of differential equations which collectively contain numerous biochemical and physiological variables and constants, such as blood flow rate to a particular organ, the organ volume, and Michaelis constant.   However, for purposes of this dose simulation, these (other) variables and constants are not considered as the true *input* parameters in that they are known or at least fixed (regardless of the amount of dermal dose to be absorbed).

The total internal doses of malathion simulated from the various spot urine samples are summarized in Table II.   This table shows that none of the total absorbed malathion doses simulated for the 17 volunteers was more than 3 times its measured value.   The majority (64/85 = 75%) of the simulation doses were within two-fold of their measured values.   A few (6/85 = 7%) simulation doses were found to be unacceptably ($\geq$ three-fold) lower than their measured values, however; these were the absorbed doses simulated from spot samples collected at 24 - 36 h or 24 - 48 h.   The averages taken over the three absorbed doses simulated (from the three spot samples) for each study subject were found to be highly comparable to the measured values.   As shown in Table II, the ratios of these averages to the observed (measured) values ranged from 0.7 to 1.4, with an arithmetic mean of 1.1.

By and large, the absorbed doses simulated on spot samples collected between 12 and 24 h (after initial exposure) appeared to be most comparable to their literature values.   This group of simulation doses had an arithmetic mean ratio (to the measured values) of 1.2.   This finding is not surprising at all, given that the excretion rates for these subjects were seen to peak at this interval.   The dose simulation theoretically should be most accurate on spot samples collected at this interval, in that sensitivity and specificity would become less an issue at this interval in quantifying the urine content from the samples collected.   Despite the scarcity of data, there appears to be some indication from Table II that neither chemical formulation nor exposure level will have a significant effect on the accuracy of dose simulation.

As pointed out in the collaborative study by Dong *et al.* (*3*), the PB-PK model used here specified that slightly over 75% of an absorbed dose of malathion would be recovered in urine with another 20% recovered in feces.   The urinary recovery simulated with this PB-PK model was thus approximately 15 to 20% lower than those dose recoveries observed experimentally by Feldmann and Maibach (*10*) and by Ross *et al.* (*14*).   This difference in urinary excretion for malathion metabolites (including

Table II. Total Percent of Absorbed Dose of Malathion Observed in the Urine of Human Volunteers *versus* Simulated, based on Selected Spot Intervals Following Topical Administration

| Subject[a] | Observed Dose[b] | 8-12 h | 12-24 h | 24-36 h | 24-48 h | Average[c] |
|---|---|---|---|---|---|---|
| | | Total Percent of Absorbed Dose Simulated from Spot Sample[d] | | | | |
| A1 | 18.4 | 34.3 (1.9) | 29.1 (1.6) | 9.4 (0.5) | | 24.3 (1.3) |
| A2 | 2.5 | 5.5 (2.3) | 2.7 (1.1) | 2.1 (0.9) | | 3.5 (1.4) |
| A3 | 9.5 | 8.0 (0.8) | 15.8 (1.7) | 5.0 (0.5) | | 9.6 (1.0) |
| A4 | 5.1 | 7.9 (1.6)[e] | 5.8 (1.2) | 3.8 (0.8) | | 5.8 (1.2) |
| A5 | 3.1 | 4.2 (1.3) | 4.4 (1.4) | 1.8 (0.6) | | 3.5 (1.1) |
| A6 | 5.6 | | 8.4 (1.5)[f] | 2.0 (0.4) | | 5.2 (0.9) |
| B1 | 2.3 | 2.0 (0.9) | 3.5 (1.5) | 1.9 (0.8) | | 2.5 (1.1) |
| B2 | 4.2 | 7.4 (1.7) | 5.4 (1.3) | 4.2 (1.0) | | 5.7 (1.3) |
| B3 | 7.2 | 20.5 (2.8) | 8.9 (1.2) | 3.5 (0.5) | | 11.0 (1.5) |
| B4 | 9.0 | 12.5 (1.4)[e] | 13.8 (1.5) | 4.3 (0.5) | | 10.2 (1.1) |
| B5 | 9.0 | 12.7 (1.4) | 13.7 (1.5) | 6.0 (0.7) | | 10.8 (1.2) |
| B6 | 2.6 | 4.8 (1.9)[e] | 1.2 (0.5) | 4.1 (1.6) | | 3.4 (1.3) |
| C1 | 16.9 | 35.5 (2.1) | 19.0 (1.1) | 6.3 (0.4) | | 20.3 (1.2) |
| C2 | 19.6 | 36.9 (1.9) | 26.2 (1.3) | 11.7 (0.6) | | 24.9 (1.3) |
| C3 | 9.4 | 14.9 (1.6) | 9.4 (1.0) | 9.0 (1.0) | | 11.1 (1.2) |
| C4 | 28.6 | 70.5 (2.5)[e] | 17.6 (0.6) | 1.1 (0.04) | | 29.8 (1.0) |
| C5 | 10.6 | 25.1 (2.4)[e] | 11.5 (1.1) | 3.1 (0.3) | | 13.2 (1.3) |
| C6 | 11.9 | 5.3 (0.5) | 22.6 (1.9) | 3.8 (0.3) | | 10.6 (0.9) |
| D1 | 3.1 | 3.3 (1.1) | 3.5 (1.1) | 2.0 (0.6) | | 3.0 (0.9) |
| D2 | 7.5 | 6.9 (0.9)[e] | 9.5 (1.3) | 7.8 (1.0) | | 8.1 (1.1) |
| D3 | 4.4 | | 5.0 (1.1)[f] | 4.2 (1.0) | | 4.6 (1.1) |
| D4 | 4.8 | 7.3 (1.5)[e] | 3.5 (0.7) | 4.2 (0.9) | | 5.0 (1.1) |
| D5 | 5.3 | 3.3 (0.6) | 5.7 (1.1) | 7.3 (1.4) | | 5.4 (1.0) |
| D6 | 11.0 | 18.1 (1.7)[e] | 12.3 (1.1) | 8.5 (0.8) | | 13.0 (1.2) |
| E1 | 4.0 | 3.6 (0.9) | 3.6 (0.9) | | 2.4 (0.6) | 3.2 (0.8) |
| E2 | 4.0 | 5.7 (1.4) | 4.8 (1.2) | | 0.9 (0.2) | 3.8 (1.0) |
| E3 | 5.3 | 10.3 (1.9) | 5.5 (1.0) | | 1.4 (0.3) | 5.7 (1.1) |
| E4 | 2.1 | 2.6 (1.3) | 1.0 (0.5) | | 0.9 (0.5) | 1.5 (0.7) |
| E5 | 3.1 | 8.4 (2.7) | 1.2 (0.4) | | 0.4 (0.1) | 3.3 (1.1) |
| *mean* | 7.9 | *14.0 (1.6)* | *9.5 (1.2)* | *4.9 (0.7)* | *1.2 (0.4)* | *9.0 (1.1)* |

[a] same individuals as those listed in Table I; groups A through D were from Dary *et al.* (*8*) and group E, from Wester *et al.* (*9*).

[b] same (rounded off to one decimal place) as those listed under cumulative absorbed dose in Table I.

[c] arithmetic mean taken over the total absorbed doses of malathion simulated from the three selected spot urine samples.

[d] in parentheses is the ratio of the total absorbed dose simulated from the selected spot sample (or of the average simulated dose) to the observed value.

[e] simulated for spot interval of 4 - 12 h (*see* footnote *d* in Table I for explanation).

[f] simulated for spot interval of 4 - 24 h (*see* footnote *e* in Table I for explanation).

the small amount of unchanged malathion) should have caused a slight oversimulation of the total absorbed doses from the PB-PK modeling. Underlying this dose overestimation is the fact that in order for the same amount of urinary output to be equivalent to the amount estimated from biomarker data, a greater absorbed dose has to be simulated or called for if the presumed ratio of urinary excretion to excretion by other routes (e.g., feces) is smaller.

For the ratio of urinary excretion to excretion by other routes in the model to be consistent with that observed experimentally, an adjustment should be made in the model for the physiological and biochemical parameter values, including the fecal and the urinary constants, that were assumed especially for the gastrointestinal tract and the kidney compartments. Thus far the PB-PK model has not been revised to increase this excretion ratio. The main reason for not revising the PB-PK model in this regard is that this excretion ratio could very well be sex-, body weight (fat)-, or age-related. More experimental data are required in order to confirm this speculation. After all, the greater excretion ratio observed by Feldmann and Maibach (*10*) was based on 6 human volunteers whose individual physiological characteristics were not given. Also, the urinary excretion of these 6 subjects from intravenous administration (to account for excretion via urine versus other routes) was reported in the form of a group mean, with a large variation coefficient of 10.8% (relative to the 15 - 20% difference seen in excretion ratio). The greater excretion ratio observed in Ross *et al.* (*14*) likewise must be interpreted with care, as that study for human clearance of malathion involved only a single volunteer.

The human subjects in this validation study were assumed to have a typical body weight of 70 kg since their individual physiological characteristics were not given. Some of the physiological and biochemical parameter values used in a (this) PB-PK model are related closely to body weight. Accordingly, the accuracy of dose simulation might have been improved somewhat if these parameter values used in the PB-PK model were tailored more closely to those actually governing the excretion and disposition kinetics in the literature cases. Sensitivity analyses performed in the 1994 collaborative study (*3*) indicated that model predictions were most affected (up to as much as two-fold) by the skin-blood partition coefficient, which tends to be highly chemical-specific and is related closely to skin permeability. Although many other input parameters (such as tissue volume and tissue perfusion rate) are also sensitive to body weight, they tend to be affected *proportionally* as well as *unselectively* by body weight and hence as a group have less effect on dose simulation.

Overall, the absorbed doses simulated in this validation study were much more comparable to the literature values than were those doses simulated earlier in a pilot trial published recently (*13*). Only literature data published in the 1994 study by Dary *et al.* (*8*) were attempted in that trial study. The highest doses simulated for the 12 human subjects in that trial were up to 5 or 6 times greater than the measured values. Those comparatively poorer results were mainly due to the use of a different endpoint for dose simulation. In that trial study, model predictions of the urinary excretion of malathion accumulated *up to* (instead of *during*) the spot interval were simulated. In addition, the literature cumulative dose value for each of the spot samples was approximated by *multiplying* the total time lapsed since initial exposure *by* the hourly rate of excretion determined for that interval. (Note that in real life situations where spot specimens are used, there will not be any actual value known for total urinary excretion *up to* the spot interval in question.) This type of approximation should yield poorer simulation results since, as shown in Figures 1 through 5, excretion rates for malathion do vary over time. That method of approximation was nonetheless used in the trial study to offer an alternative for a situation where the urine content in spot samples is reported in some mass unit per volume of urine. An actual case in point is again the 1994 collaborative study by Dong *et al.* (*3*), in which the urine contents of malathion metabolites for the 11 adults and children were available only in µg per liter of urine.

## Conclusion

The results in this validation study suggest that PB-PK simulation can be performed on spot urine samples to predict the total absorbed dose of malathion in humans, with an accuracy likely to be well within a few fold of actual (dermal) exposure. The best spot samples appear to be those collected around the peak hours of urinary excretion. There appears to be some indication that neither chemical formulation nor exposure level will have a significant effect on the accuracy of dose simulation. The data also show that the accuracy could be improved considerably, if the dose simulation for each study subject were performed on two or more spot specimens collected at different time points preferably within the first 24 or 36 h of exposure to malathion.

It was also found that spot urine samples would be most useful in PB-PK modeling of the total absorbed doses (of malathion) if the urine content were measured as the total amount accumulated during a given spot interval, instead of in some mass unit per volume of urine. It is further recommended that for each study subject, the values of some basic physiological parameters such as body weight, sex, and age be supplied along with the spot urine sample(s). Above all, the spot *interval* (especially relative to initial exposure) should be measured or reported as accurately as feasible.

## Disclaimer

No official endorsement of particular computer software by the California Department of Pesticide Regulation or by the University of California is intended or implied; the opinions expressed are those of the authors and not necessarily those of the Department, of any other California State agency, or of the University of California.

## Literature Cited

1.  Mendelsohn, M. L. In *Biomarkers and Occupational Health*; Mendelsohn, M. L.; Peeters, J. P.; Normandy, M. J., Eds.; Joseph Henry Press: Washington, D. C., 1995; pp *iii-iv*.
2.  Vine, M. F. In *Environmental Epidemiology - Effects of Environmental Chemicals on Human Health*; Draper, W. M., Ed.; Advances in Chemistry Series 241; American Chemical Society: Washington, D. C., 1994; pp 105-120.
3.  Dong, M. H; Draper, W. M.; Papanek, P. J., Jr.; Ross, J. H.; Woloshin, K. A.; Stephens, R. D. In *Environmental Epidemiology - Effects of Environmental Chemicals on Human Health*; Draper, W. M., Ed.; Advances in Chemistry Series 241; American Chemical Society: Washington, D. C., 1994; pp 189-208.
4.  *Health Risk Assessment of Aerial Application of Malathion-Bait*; California Environmental Protection Agency, Office of Environmental Health Hazard Assessment: Sacramento, CA (Reprints available from Copies Unlimited, 5904, Sunset Boulevard, Los Angeles, CA 90028.), 1991.
5.  Rabovsky, J.; Brown, J. P. *J. Occup. Med. Toxicol.* **1993**, *2*, 131-168.
6.  Dong, M. H.; Ross, J. H.; Thongsinthusak, T.; Sanborn, J. R.; Wang, R. G. M. *Physiologically-Based Pharmacokinetic (PB-PK) Modeling for Dermal Absorption of Pesticide (Malathion) in Man*; Worker Health and Safety Branch, California Department of Pesticide Regulation: Sacramento, CA, 1993; Technical Report HS-1678 (abstract published in *The Toxicologist*, **1993**, *13*, 355).
7.  Dong, M. H. *Comp. Meth. Prog. Biomed.* **1994**, *45*, 213-221.
8.  Dary, C. C.; Blancato, J. N.; Castles, M.; Reddy, V.; Cannon, M.; Saleh, M. A.; Cash, G. G. In *Biomarkers of Human Exposure to Pesticides*; Saleh, M. A.; Blancato, J. N.; Nauman, C. H., Eds.; American Chemical Society Symposium Series No. 542; American Chemical Society: Washington, D. C., 1994; pp 231-263.

9.   Wester, R. C.; Maibach, H. I.; Bucks, D. A. W.; Guy, R. H.  *Toxicol. Appl. Pharmacol.* **1983**, *68*, 116-119.
10.  Feldmann, R. J.; Maibach, H. I. *Toxicol. Appl. Pharmacol.* **1974**, *28*, 126-132.
11.  McDougal, J. N.; Jepson, G. W.; Clewell, H. J., III; Gargas, M. L.; Andersen, M. E. *Fund. Appl. Toxicol.* **1990**, *14*, 299-308.
12.  Knaak, J. B.; Al-Bayati, M. A.; Raabe, O. G.  In *Health Risk Assessment - Dermal and Inhalation Exposure and Absorption of Toxicants*; Wang, R. G. M.; Knaak, J. B.; Maibach, H. I., Eds.; CRC: Boca Raton, FL, 1993, pp 3-29.
13.  Dong, M. H., Thongsinthusak, T.; Ross, J. H.; Krieger, R. I.  *Validation of a Physiologically Based Pharmacokinetic (PB-PK) Model Used to Simulate Absorbed Malathion Doses in Humans*; Worker Health and Safety Branch, California Department of Pesticide Regulation:  Sacramento, CA, 1995; Technical Report HS-1718 (presented as a poster paper at the 209th American Chemical Society Annual National Meeting in Anaheim, California, April 2 - 6, 1995).
14.  Ross, J. H.; Thongsinthusak, T.; Krieger, R. I.; Frederickson, S.; Fong, H. R.; Taylor, S.; Begum, S.; Dong, M. H. *Human Clearance of Malathion*.  Worker Health and Safety Branch, California Department of Pesticide Regulation: Sacramento, CA, 1991; Technical Report HS-1617 (abstract published in *The Toxicologist*, **1991**, *11*, 160).

Chapter 18

# Symbolic Solutions of Small Linear Physiologically Based Pharmacokinetic Models Useful in Risk Assessment and Experimental Design

Robert N. Brown

Characterization Research Division, National Exposure Research Laboratory, U.S. Environmental Protection Agency, 944 East Harmon, Las Vegas, NV 89193–3478

Symbolic solutions of the ordinary differential equations (ODEs) describing linear physiologically based pharmacokinetic (PBPK) models yield qualitative and quantitative information about target organ dose beyond that provided by general purpose ODE algorithms. Symbolic tools simplify model formulation, manipulation and solution; yield accurate, efficient and robust solutions; and provide improved diagnostic tools for regulatory risk assessment and management. An intuitive terminology, physiologically consistent system matrix factorization and mathematically efficient matrix normalization for a previously studied 4-compartment prototype (P4) is introduced to clarify physiological structure, simplify symbolic solution and facilitate extensions to an 8-compartment (P8) model. Using area under the concentration curve (AUC) for target organ dose, the classical scalar formula ($AUC = CL^{-1}A_0$) is generalized to P4 and P8. Determinant expansions simplify system inversion, and AUC and steady-state concentration formulas for bolus and infusion inputs. Biomonitoring formulas for reconstruction of external exposure from internal dose are given. Comparative toxicology formulas are derived for route-to-route exposure extrapolation of dose, for efficient body burden biomonitoring and parameter estimation, and for identification of dominant subpathways.

To improve its risk assessment models for exposure-dose-response prediction, and to strengthen its risk characterization, model validation and sensitivity-uncertainty analysis processes, the U.S. Environmental Protection Agency (EPA) is investigating new symbolic (analytic) methods for the analysis of physiologically based pharmacokinetic and pharmacodynamic (PBPK/PD) models describing the exposure and disposition of environmental toxicants in humans and animals. Such tools will help to more efficiently set and defend environmental regulatory risk prevention and risk management strategies, guidelines, and standards. Others, such as the U.S. Food and Drug Administration, need new PBPK tools to improve drug safety and efficacy. With the growing globalization of human risk and ecosystem issues, the need for improved system analysis tools increases. The development here of symbolic (analytic) dose formulas should help provide such critical diagnostic tools.

PBPK models examine toxicant dose at all biological scales: systemic, target tissue, and sub-cellular. They refine the formulation, computation and uncertainty analysis of biological processes. PBPK models are commonly described by mixed systems of linear and nonlinear ordinary differential equations (ODEs) that reduce to linear systems at low concentrations typical of human and some test animal exposures. Both toxicant concentrations and integrated concentrations can be computed using general purpose stiff ODE algorithms such as Gear's [4], such as for the analysis of dichloromethane toxicity [1]. For general linear systems, however, symbolic analytic solutions are well known [2]. Their adaptation and refinement for PBPK processes should provide intuitive, accurate, efficient and robust algorithms for quantitative risk assessment (QRA) and experimental design (ED) applications, complementing outputs of general purpose numerical solvers. Furthermore, linear symbolic methods may motivate development of better nonlinear solution methods.

Previous work [3] developed simple integrated concentration area under the curve (AUC) and steady-state concentration ($C_{SS}$) dose formulas for a sequence of 1-4 compartment linear PBPK prototypes (P1-P4) with blood, fat, muscle and liver compartments, under bolus and infusion inputs. The purpose here is to introduce a more intuitive terminology, a physiologically more consistent system matrix factorization, and a mathematically more efficient matrix normalization to better standardize physiological formulation, symbolic solution and extension of application formulas to larger PBPK models such as the 8-compartment (P8) model below. Reformulation helps to properly generalize scalar AUC formulas and sporadically appearing nonphysiological model formulas ([5], [7]) to general multidimensional PBPK models.

## New Matrix Factorizations for a P4 PBPK Prototype

The previous work transformed a 1-compartment classical compartmental model into a 1-compartment pseudo PBPK model (P1) that included bloodflow (Q), volume (V), partition coefficient (R) and systemic elimination clearance ($CL_e$) parameters. This was expanded to 2, 3 and 4-compartment PBPK prototypes. The P4 prototype includes (i) a main compartment ($A \equiv$ composite arterial-venus blood, possibly lumped with the lung, kidney or other rapidly perfused organs); (ii) a slowly perfused strong sink compartment ($F \equiv$ fat, possibly lumped with other slowly perfused strong sink organs such as bone); (iii) a slowly perfused weak sink compartment ($M \equiv$ muscle, possibly lumped with other weak sink slowly perfused compartments such as skin); and (iv) a large flow systemic clearance compartment ($L \equiv$ liver, possibly lumped with GI tract).

**Mass Balance Equations.** The P4 model is given (inputs suppressed) by

$$C_A' = \frac{1}{V_A}\left(\frac{Q_{AF}C_F}{R_{AF}} + \frac{Q_{AM}C_M}{R_{AM}} + \frac{Q_{AL}C_L}{R_{AL}}\right) - \frac{1}{V_A}\left(\frac{Q_A C_A}{R_A} + V_A k_{eA} C_A\right)$$

$$C_F' = \frac{1}{V_F}\frac{Q_{FA}C_A}{R_{FA}} - \frac{1}{V_F}\left(\frac{Q_{AF}C_F}{R_{AF}} + V_F k_{eF} C_F\right)$$

$$C_M' = \frac{1}{V_M}\frac{Q_{MA}C_A}{R_{MA}} - \frac{1}{V_M}\left(\frac{Q_{AM}C_M}{R_{AM}} + V_M k_{eM} C_M\right)$$

$$C_L' = \frac{1}{V_L}\frac{Q_{LA}C_A}{R_{LA}} - \frac{1}{V_L}\left(\frac{Q_{AL}C_L}{R_{AL}} + V_L k_{eL} C_L + \frac{V_{mL}^{(A)}}{K_{mL}}\frac{C_L}{\left(1 + \frac{C_L}{K_{mL}}\right)}\right), \qquad (1)$$

but can be rewritten in factored matrix form (with input vector $i$ added) as

$$C' = -K_c C + i = -V^{-1}(CL)C + i \qquad (2)$$

where $C' = V^{-1}A'$ is the concentration derivative vector in mass/volume-time units; $A'$ the amount vector $A'$; $V^{-1}$ the diagonal matrix of inverse volumes. The $Q_i$ are bloodflows in volume/time units with $Q_A = Q_{FA} + Q_{MA} + Q_{LA}$ and $Q_{FA} = Q_{AF}$, $Q_{MA} = Q_{AM}$, $Q_{LA} = Q_{AL}$ since arterial bloodflows equal venous flows. Linear clearance terms are also given by $(V_A k_{eA})$, where $k_{eA}$ is a kinetic transfer rate (e.g., metabolic elimination from blood) for volume $V_A$; or by $V_{mL}^{(A)}/K_{mL}$ in the linear range of a Michaelis-Menten equation as the denominator perturbation $(1 + C_L/K_{mL}) \to 1$, where $V_{mL}^{(A)}$ is the maximum metabolite formation velocity in the liver in mass/time units and $K_{mL}$ is the Michaelis constant in mass/volume units. The partition coefficient $R_{ij}$ denotes the ratio of toxicant concentration in compartment $j$ vs. $i$. The concentration form system kinetic rate matrix, in time$^{-1}$ units, is $K_c = V^{-1}CL$, with

$$-CL = (-1) \begin{pmatrix} CL_{AA} & -CL_{AF} & -CL_{AM} & -CL_{AL} \\ -CL_{FA} & CL_{FF} & 0 & 0 \\ -CL_{MA} & 0 & CL_{MM} & 0 \\ -CL_{LA} & 0 & 0 & CL_{LL} \end{pmatrix} =$$

$$\begin{pmatrix} -\frac{Q_A}{R_A} - k_{eA}V_A & \frac{Q_{AF}}{R_{AF}} & \frac{Q_{AM}}{R_{AM}} & \frac{Q_{AL}}{R_{AL}} \\ \frac{Q_{FA}}{R_{FA}} & -\frac{Q_{AF}}{R_{AF}} - k_{eF}V_F & 0 & 0 \\ \frac{Q_{MA}}{R_{MA}} & 0 & -\frac{Q_{AM}}{R_{AM}} - k_{eM}V_M & 0 \\ \frac{Q_{LA}}{R_{LA}} & 0 & 0 & -\frac{Q_{AL}}{R_{AL}} - k_{eL}V_L - \frac{V_{mL}^{(A)}}{K_{mL}} \end{pmatrix},$$

$$(3)$$

the off-diagonal elements of the system clearance matrix $CL$ denoting local circulatory clearance to compartment $i$ from $j$. Note that $A' = -(CL)V^{-1}A = -KA$ is the amount form of the mass balance equations and yields the amount form system kinetic rate matrix $(K = (CL)V^{-1})$. Thus, the relationship between concentration and amount forms is given by $K_c = V^{-1}KV$ so that $K_{c,ij} = V_i^{-1}K_{ij}V_j$, and $K_{c,ij} = K_{ij}$ only when $V_i = V_j$. Note also that $CL$ can be rewritten in 't-normalized' form as

$$CL = S_t CL_t \qquad (4)$$

$$= \begin{pmatrix} 1_A & -\pi_{AF} & -\pi_{AM} & -\pi_{AL} \\ -\pi_{FA} & 1_F & 0 & 0 \\ -\pi_{MA} & 0 & 1_M & 0 \\ -\pi_{LA} & 0 & 0 & 1_L \end{pmatrix} \begin{pmatrix} CL_{tA} & 0 & 0 & 0 \\ 0 & CL_{tF} & 0 & 0 \\ 0 & 0 & CL_{tM} & 0 \\ 0 & 0 & 0 & CL_{tL} \end{pmatrix} \qquad (5)$$

where $CL_{ti}$ is the total clearance from $i$ and $S_t$ is $CL$ with $i^{th}$ column divided by $CL_{ti}$ so that diagonal entries are unity $(\pi_{ii} \equiv 1_i)$ and off-diagonal $ij^{th}$ entries represent the fractional clearance to compartment $i$ from $j$ $(\pi_{ij} = CL_{ij}/CL_{tj} = k_{ij}V_j/k_{tj}V_j = k_{ij}/k_{tj})$. $S_t$ might be called the total- or t-normalized unitless structural interconnection matrix for the system. This normalized factorization simplifies symbolic solutions since $S_t$ is easier to invert than $CL$. Use of $S_t$ also emphasizes the intuitive importance of fractional flows between compartments in model formulation, understanding and solution.

For some applications it may be useful to split the diagonal clearance elements $(CL_{ti} \equiv CL_{ii})$ into $CL_{ti} = CL_{ri} + CL_{ei}$, $CL_{ri}$ denoting all clearances from $i$ that circulate to other modeled compartments (e.g., bloodflows) and $CL_{ei}$ denoting total $i^{th}$ compartment noncirculatory systemic elimination clearances (e.g., urinary flow or metabolic elimination). Use of a $CL_{ei}$ for each organ reflects an extension of traditional pharmacokinetic practice where systemic clearance typically referred only to the kidney's clearing of toxicant from blood. In summary, the above terminology, matrix factorization and matrix normalization gives a flexible framework for easy formulation, manipulation, solution and application of linear PBPK models to QRA or ED problems.

Note that all column sums of $S_t$ are nonnegative ($S_{t\bullet j} \geq 0 \; \forall \; j$), equality holding only if $CL_{ej} = 0$. For typical systems, a few nonzero systemic eliminations should guarantee invertibility of $S_t$ and thus finite AUCs. Invertibility is assumed herein, but consult matrix texts like [6] for rigorous invertibility conditions.

**First Integral Concentration Solution.** The integral of 2 is given formally by the matrix exponential convolution integral [2] as

$$C(t) = \int_0^t e^{-K_c(t-s)} i(s)ds + e^{-K_c t} C_0 \qquad (6)$$

for continuous input $i(s)$ plus bolus input $C_0 = V^{-1}A_0$, where $C_0 = (C_{10}, C_{20}, C_{30}, C_{40})^T$ is the vector of initial bolus concentrations at time zero.

**Second Integral AUC Solution For Bolus Dose Input.** The second integral integrated concentration, or area under the curve (AUC) from time zero to infinity, is a useful dose metric for studying some drug efficacy effects and many long latency cancer effects - effects relatively independent of short-term exposure-dose variations - thus providing the focus for QRA applications below. AUC from a bolus input can be derived using matrix exponential tools [2] as

$$AUC = \int_0^\infty C(\tau)d\tau = -K_c^{-1} e^{-K_c t} C_0 \mid_0^\infty = K_c^{-1} C_0$$

$$= CL^{-1}A_0 = (S_t CL_t)^{-1} A_0 = CL_t^{-1} S_t^{-1} A_0, \qquad (7)$$

which generalizes the classical scalar formula $AUC = A_0/CL_e$ [5].

Using classical matrix row and column determinant expansion tools [6], the complete inverse structural matrix $S_t^{-1}$ is given formally by

$$S_t^{-1} \equiv \left(S_t^{-1}\right)_{i,j=1..4} = |S_t|^{-1} S_{ta}, \qquad (8)$$

$|S_t| \equiv \det(S_t)$ being the of determinant of $S_t$ and $S_{ta}$ the adjoint matrix

$$S_{ta} = \begin{pmatrix} 1_F 1_M 1_L & \begin{matrix}\pi_{AF} 1_M 1_L \\ (1_A 1_M 1_L \end{matrix} & \pi_{AM} 1_F 1_L & \pi_{AL} 1_F 1_M \\[1em] \pi_{FA} 1_M 1_L & \begin{matrix} -\pi_{AM}\pi_{MA} 1_L \\ -\pi_{AL}\pi_{LA} 1_M) \end{matrix} & \pi_{FA}\pi_{AM} 1_L & \pi_{FA}\pi_{AL} 1_M \\[1em] \pi_{MA} 1_F 1_L & \pi_{MA}\pi_{AF} 1_L & \begin{matrix} (1_A 1_F 1_L \\ -\pi_{AF}\pi_{FA} 1_L \\ -\pi_{AL}\pi_{LA} 1_F) \end{matrix} & \pi_{MA}\pi_{AL} 1_F \\[1em] \pi_{LA} 1_F 1_M & \pi_{LA}\pi_{AF} 1_M & \pi_{LA}\pi_{AM} 1_F & \begin{matrix} (1_A 1_F 1_M \\ -\pi_{AF}\pi_{FA} 1_M \\ -\pi_{AM}\pi_{MA} 1_F) \end{matrix} \end{pmatrix}$$

$$(9)$$

where the $ij^{th}$ adjoint component (i.e., transposed cofactor $C_{tji}$) is defined as

$$(S_{ta})_{ij} \equiv C_{tji} = (-1)^{i+j} \det(S_t(j', i')) = (-1)^{i+j} M_{tji}, \quad i, j = 1, \ldots 4 \tag{10}$$

with $ji^{th}$ minor given by $M_{tji} = \det(S_t(j', i'))$ - the determinant with $j^{th}$ row and $i^{th}$ column of $S_t$ deleted. The structural determinant is expandable as

$$|S_t| = 1_A 1_F 1_M 1_L - \pi_{AF}\pi_{FA} 1_M 1_L - \pi_{AM}\pi_{MA} 1_F 1_L - \pi_{AL}\pi_{LA} 1_A 1_F 1_M$$

$$= 1 - \pi_{AF}\pi_{FA} - \pi_{AM}\pi_{MA} - \pi_{AL}\pi_{LA}. \tag{11}$$

Thus, AUC has scalar components, summed over all input organs, given by

$$AUC_i = \sum_{j=1}^{4} CL_{ti}^{-1}\left(S_t^{-1}\right)_{ij} A_{j0} = \sum_{j=1}^{4} \frac{(S_{ta})_{ij}}{|S_t|}\frac{A_{j0}}{CL_{ti}}, \quad i = 1, \ldots, 4. \tag{12}$$

When $|S_t| > 0$, $S_t$ is invertible and the AUCs are finite. Also all adjoint components are necessarily nonnegative ($S_{ta} \geq 0$), yielding nonnegative $AUCs$.

Note that each $AUC_i$ is also derived by applying Cramer's Rule to the unnormalized matrix equation $(CL)AUC = A_0$, yielding $AUC_i = \left|CL^{(i)}\right|/|CL|$, the superscript $i$ denoting replacement of the $i^{th}$ column of $CL$ by the initial condition $A_0$. Equivalently, Cramer's Rule, applied to the factored form $(S_t CL_t)AUC = A_0$, yields $AUC_i = \left|S_t^{(i)}\right|/(|S_t| CL_{ti})$. Again, manipulation of $S_t$ is easier than inverting $CL$ because of its many unity elements on and off the diagonal. While Cramer's Rule is compact, the adjoint formulation above yields more intuitive formulas and will thus be preferred in the applications below.

**Simplification of P4 Single Input, Single Output (SISO) AUCs.** P4 AUCs (12) are linear combinations of bolus inputs from all organs. Reduction to single input, single output (SISO) formulas simplifies presentation and yields an $i^{th}$ compartment target organ dose from a $j^{th}$ compartment bolus input of

$$AUC_{ij} = (CL^{-1})_{ij} A_{j0} = CL_{ti}^{-1}(S_t^{-1})_{ij} A_{j0} = \frac{(S_{ta})_{ij}}{|S_t|}\frac{A_{j0}}{CL_{ti}} = \frac{V_j}{V_i}\frac{(S_{ta})_{ij}}{|S_t|}\frac{C_{j0}}{k_{ti}}, \tag{13}$$

where $CL_{ti} = k_{ti}V_i$. Note how use of the amount forms ($A_{j0}$ vs. $C_{j0} = A_{j0}/V_j$) simplifies symbolic AUC results by yielding unity volume ratios.

Manipulation of (13) using the t-normalized adjoint (9) shows that the 16 SISO AUC results for P4 can be reduced to 5 qualitatively different subcases, illustrated for greater clarity using specific organs:

$$AUC_{AF} = \pi_{AF}\frac{1}{|S_t|}\frac{A_{F0}}{CL_{tA}} = \pi_{AF}\frac{|S_t - AF|}{|S_t|}\frac{A_{F0}}{CL_{tA}} \tag{14}$$

$$AUC_{FA} = \pi_{FA}\frac{1}{|S_t|}\frac{A_{A0}}{CL_{tF}} = \pi_{FA}\frac{|S_t - FA|}{|S_t|}\frac{A_{A0}}{CL_{tF}} \tag{15}$$

$$AUC_{MF} = \frac{\pi_{MA}\pi_{AF}}{|S_t|}\frac{A_{F0}}{CL_{tM}} = \pi_{MAF}\frac{|S_t - MAF|}{|S_t|}\frac{A_{F0}}{CL_{tM}} \tag{16}$$

$$AUC_{AA} = \frac{1}{|S_t|}\frac{A_{A0}}{CL_{tA}} = \frac{|S_t - A|}{|S_t|}\frac{A_{A0}}{CL_{tA}} \tag{17}$$

$$AUC_{FF} = \frac{1 - \pi_{AM}\pi_{MA} - \pi_{AL}\pi_{LA}}{|S_t|} \frac{A_{F0}}{CL_{tF}} = \frac{|S_t - F|}{|S_t|} \frac{A_{F0}}{CL_{tF}}. \tag{18}$$

Shorthand notations like $\pi_{MAF} \equiv \pi_{MA}\pi_{AF}$ simplify symbolic formulas by removing redundant path information. The potentially inconsistent notation $|S_t - AF|$ (equivalently $|S_t - \pi_{AF}|$ or $|ML|$ for P4) denotes the $2^{nd}$ order determinant derived from the $4^{th}$ order matrix $S_t^{(4)}$ - after subtracting the columns and rows corresponding to the $AF$ pathway. Interchangable useage of such equivalents helps clarify the nature of and simplify the presentation of subdeterminants. Specifically, the P4 model SISO AUC formulas above, plus P8 model results later, suggest that general model AUCs are decomposable into sums of physiologically meaningful factors as

$$AUC_{ij} = \sum_{\pi_{i...j}} \pi_{i...j} \frac{|S_t - i...j|}{|S_t|} \frac{A_{j0}}{CL_{ti}} \equiv \sum_{\pi_{i...j}} \pi_{i...j} \frac{|S_t - \pi_{i...j}|}{|S_t|} \frac{A_{j0}}{CL_{ti}}, \tag{19}$$

where $\sum_{\pi_{i...j}}$ is the sum of all toxic pathways from input organ $j$ to target organ $i$; where $\pi_{i...j}$ represents the fractional conductance (product of all fractional clearances) along a specific toxic pathway; where the determinant ratio $|S_t - \pi_{i...j}| / |S_t|$ acts like a circulation multiplier along the specific toxic pathway, describing the apparent average number of times each molecule of parent toxicant reenters the toxic pathway at input $j$ for repeat conductance down the toxic pathway to target $i$; and $A_{j0}/CL_{ti}$ is a crude AUC, valid as a total AUC only if all input were to go directly, without additional recirculation, to the target organ for clearance. The use of nonstandard notations, such as $|S_t - i...j| \equiv |S_t - \pi_{i...j}|$ which seem to subtract organs or a scaler from a matrix, can be made consistent by the procedural interpretation given earlier and it helps conceptualize symbolic AUC evaluation as a natural recursive process yielding ever smaller subdeterminants. These symbolic AUC formulas are equivalent to different looking SISO formulas given in [3], but are more intuitive and simpler than previous results.

### Quantitative Risk Assessment Applications

Analytic formulas for $AUC$ dose or steady-state concentration $(C_{ss})$ are useful for clarifying the absolute and relative role of model parameters in QRA and ED applications such as (i) exposure extrapolation of target organ dose from bolus to infusion input; (ii) inverse reconstruction of external environmental exposure from internal organ dose measurements; (iii) interspecies and intraspecies scaling of target organ dose; (iv) route-to-route exposure extrapolation of target organ dose; (v) comparative target organ body burdens from single exposure routes; (vi) comparative subpathway risks for given exposure route, target organ pairs; (vii) efficient biomonitoring experimental designs for model identification and parameter-state estimation; (viii) model expansion-collapse strategies for biological refinement or computational simplification; (ix) parameter sensitivity and statistical uncertainty analysis of dose; (x) high-to-low exposure extrapolation of target dose. Examples of linear model applications of (i), (ii), (iv)-(vii) are explored below. Interspecies scaling (iii) was considered previously [3].

**Steady-State Concentration Formulas.** For constant infusions, steady-state concentrations can be algebraically derived from $0 = -K_c C_{ss} + i$ (2) as

$$C_{ss} = K_c^{-1}i \equiv K_c^{-1}t^{-1}C_0 = (V^{-1}S_tCL_t)^{-1}t^{-1}V^{-1}A_0 = CL_t^{-1}S_t^{-1}t^{-1}A_0, \tag{20}$$

with constant infusion rate $i$ equivalent to a bolus input $A_0$ spread over a volume $V_0$ and an arbitrary time $t$ (i.e., $i = t^{-1}V^{-1}A_0$). Thus, symbolic steady state concentration formulas mimic AUC formulas already developed above (7) after a simple change of input units, reinforcing a similar interpretation for the 1-compartment model case [3]. Note that since linear model AUCs are additive functions of toxicant input, AUC is constant for a fixed amount of input, whether administered as a bolus, a short-term infusion, or any combination thereof.

**Inverse Reconstruction of External Exposure from Internal Dose.** For the inverse problem of reconstructing a bolus exposure from internal biomeasurements, one can invert the time dependent first-integral (6) as

$$A_0 = VC_0 = Ve^{K_c t}C(t), \qquad (21)$$

or the time-independent second-integral (7) as

$$A_0 = (CL)AUC. \qquad (22)$$

Using second-integral reconstructions, note that computation of bolus input for P4 may require AUC observations and parameter information from as few as 2 compartments. For example, liver bolus dose is given using (3) by

$$A_{L0} = -CL_{LA}AUC_A + CL_{tL}AUC_L, \quad (CL_{tL} \equiv CL_{LL}), \qquad (23)$$

whereas the blood bolus dose is given by

$$A_{A0} = CL_{tA}AUC_A - CL_{AF}AUC_F - CL_{AM}AUC_M - CL_{AL}AUC_L. \qquad (24)$$

Thus, to compute bolus input $A_{i0}$, one requires the clearances and AUCs for compartment $i$ plus all compartments inputting to $i$ - a biomonitoring experimental design problem motivating development of batteries of cost-effective multiple compartment cumulative exposure biomarkers. Note that matrix normalization and clearance inversion are unnecessary for reconstruction.

**Route-to-Route Dose Extrapolation.** Symbolic formulas for linear models are useful for comparing target organ dose from two routes of toxicant exposure under bolus or infusion input. Comparative dose formulas were introduced previously [3], but are developed here more simply using mnemonic indexing, physiologically consistent matrix factorizations and mathematically efficient matrix normalizations - facilitating extension to larger PBPK models like P8 below. A general comparative risk (CR) formula for compartment $i$ dose from $j$ and $k$ inputs is given, after cancellation of common factors, by

$$CR_{ij/ik} \equiv \frac{AUC_{ij}}{AUC_{ik}} = \frac{|S_t|^{-1} CL_{ti}^{-1} (S_{ta})_{ij} A_{j0}}{|S_t|^{-1} CL_{ti}^{-1} (S_{ta})_{ik} A_{k0}} = \frac{(S_{ta})_{ij} A_{j0}}{(S_{ta})_{ik} A_{k0}}. \qquad (25)$$

Using (9) and (12), the P4 examples below show the utility of CR formulas.

First, the muscle compartment's comparative toxicant risk from liver and blood bolus inputs (via ingestion to liver and inhalation to blood) is given by

$$CR_{ML/MA}^{(4)} = \frac{AUC_{ML}}{AUC_{MA}} = \frac{(S_{ta})_{ML}}{(S_{ta})_{MA}} \frac{A_{L0}}{A_{A0}} = \frac{\pi_{MA}\pi_{AL}1_F}{\pi_{MA}1_F 1_L} \frac{A_{L0}}{A_{A0}} = \frac{\pi_{AL}A_{L0}}{A_{A0}}. \qquad (26)$$

where $\pi_{AL} = CL_{AL}/(CL_{AL} + CL_{eL})$. For equal bolus inputs ($A_{L0} = A_{A0}$), CR is $\leq 1$. In particular, $CR = 1$ if liver systemic elimination clearance is zero ($CL_{eL} = 0 \Rightarrow \pi_{AL} = 1$), but $CR \to 0$ as metabolic elimination increases ($CL_{eL} \to \infty$) and dominates circulatory (bloodflow) clearance ($CL_{AL}$). This example provides a simple qualitative-quantitative mathematical description of the liver's biologically critical first pass effect in protecting peripheral organs from toxicant ingestion relative to inhalation. The importance of comparative formulas for biomonitoring experimental design derives from the use of dual AUC biomeasures to provide more efficient model parameter estimates (e.g., for $\pi_{AL}$ or its $CL_{eL}$ subcomponent) via the cancelling of common system components (e.g., $|S_t|$ and $\pi_{MA}$) - something not possible using single AUC biomeasures where uncertainties in $|S_t|$ or $\pi_{MA}$ could add unnecessary uncertainty or bias to statistical estimates. By symmetry arguments for P4, $CR^{(4)}_{FL/FA} = \pi_{AL}A_{L0}/A_{A0}$. Also by symmetry, $CR^{(4)}_{MF/MA} = \pi_{AF}A_{F0}/A_{A0}$ and typically reduces to unity for equal bolus inputs since fat elimination clearance is usually zero ($CL_{eF} = 0$).

Second, the blood's comparative risk for liver and blood inputs is given by

$$CR^{(4)}_{AL/AA} = \frac{AUC_{AL}}{AUC_{AA}} = \frac{(S_{ta})_{AL}}{(S_{ta})_{AA}}\frac{A_{L0}}{A_{A0}} = \frac{\pi_{AL}1_F1_M}{1(1_F1_M1_L)}\frac{A_{L0}}{A_{A0}} = \pi_{AL}\frac{A_{L0}}{A_{A0}} \qquad (27)$$

equaling $CR^{(4)}_{ML/MA}$ above. Thus an alternative AUC biomeasure ratio can be used as a secondary check to confirm parameter estimates (i.e., $\pi_{AL}$ or $CL_{eL}$ above) and further emphasizes the potential experimental design importance of developing multiple AUC exposure biomeasures. By symmetry, $CR^{(4)}_{AM/AA} = \pi_{AM}A_{M0}/A_{A0}$ and $CR^{(4)}_{AF/AA} = \pi_{AF}A_{F0}/A_{A0}$.

Third, the blood's comparative risk resulting from liver and fat inputs is

$$CR^{(4)}_{AL/AF} = \frac{AUC_{AL}}{AUC_{AF}} = \frac{(S_{ta})_{AL}}{(S_{ta})_{AF}}\frac{A_{L0}}{A_{F0}} = \frac{\pi_{AL}1_F1_M}{\pi_{AF}1_M1_L}\frac{A_{L0}}{A_{F0}} = \frac{\pi_{AL}}{\pi_{AF}}\frac{A_{L0}}{A_{F0}}, \qquad (28)$$

so that CR $\to 0$ as liver systemic clearance ($CL_{eL}$) increases faster via metabolism than fat systemic clearance ($CL_{eF}$). Else, CR $\to 1$ if $A_{L0} = A_{F0}$ as liver and fat systemic clearances go to zero. For experimental design purposes this comparative risk measure requires one additional AUC biomeasure (the fifth) and yields a tertiary experimental check on potential estimates for $\pi_{AL}$ or $CL_{eL}$ above. By symmetry, $CR^{(4)}_{AF/AM} = (\pi_{AF}A_{F0})/(\pi_{AM}A_{M0})$, which tends to unity (if $A_{F0} = A_{M0}$) since fat and muscle systemic clearances are typically zero.

Fourth, the muscle's comparative risk for liver and fat inputs is given by

$$CR^{(4)}_{ML/MF} = \frac{AUC_{ML}}{AUC_{MF}} = \frac{(S_{ta})_{ML}}{(S_{ta})_{MF}}\frac{A_{L0}}{A_{F0}} = \frac{\pi_{MA}\pi_{AL}1_F}{\pi_{MA}\pi_{AF}1_L}\frac{A_{L0}}{A_{F0}} = \frac{\pi_{AL}}{\pi_{AF}}\frac{A_{L0}}{A_{F0}}, \qquad (29)$$

which not surprisingly equals $CR^{(4)}_{AL/AF}$ above. Thus for P4, changing target organ to the peripheral ($M$) compartment from the main ($A$) does not change comparative risk from two other peripheral inputs ($L, F$). Since muscle AUCs may be more difficult to measure than blood AUCs, $CR^{(4)}_{AL/AF}$ may still be a more convenient comparative biomeasure for parameter estimation than $CR^{(4)}_{ML/MF}$. By symmetry, $CR^{(4)}_{LM/LF} = (\pi_{AM}A_{M0})/(\pi_{AF}A_{F0})$ and $CR^{(4)}_{FL/FM} = (\pi_{AL}A_{L0})/(\pi_{AM}A_{M0})$.

Fifth, the liver's comparative dose for liver and blood inputs is given by

$$CR^{(4)}_{LL/LA} = \frac{(S_{ta})_{LL}\,A_{L0}}{(S_{ta})_{LA}\,A_{A0}} = \frac{1\,(1_A 1_F 1_M - \pi_{AF}\pi_{FA}1_M - \pi_{AM}\pi_{MA}1_F)\,A_{L0}}{\pi_{LA}1_F 1_M}\,\frac{A_{L0}}{A_{A0}}. \quad (30)$$

If systemic clearance occurs only in the liver (i.e., $CL_{eL} > 0$, but $\pi_{AF} = \pi_{AM} = 1$), then $CR^{(4)}_{LL/LA} = 1$ (if $A_{L0} = A_{A0}$) since the numerator, $1 - \pi_{FA} - \pi_{MA} = \pi_{LA}$, cancels the same term in the denominator. But $CR^{(4)}_{LL/LA} \approx \pi_{LA}^{-1} \to \infty$ if blood, fat or muscle systemic elimination strongly increase via metabolism (e.g., if $CL_{eA} \to \infty$). Thus, the relative importance of ingestion versus inhalation inputs ($A_{L0}$ vs. $A_{A0}$) is typically greater when liver is the target organ than when muscle is the target (26) - another variation of the liver's first-pass effect . Due to the form of the numerator cofactor, when systemic clearance occurs outside the liver, many system components do not completely cancel via ratioing - even though $|S_t|$ cancels.

Sixth, the liver CR for liver and fat inputs is given by

$$CR^{(4)}_{LL/LF} = \frac{(S_{ta})_{LL}\,A_{L0}}{(S_{ta})_{LF}\,A_{F0}} = \frac{1\,(1_A 1_F 1_M - \pi_{AF}\pi_{FA}1_M - \pi_{AM}\pi_{MA}1_F)\,A_{L0}}{\pi_{LA}\pi_{AF}\,(1_M)}\,\frac{A_{L0}}{A_{F0}} \quad (31)$$

which is similar to, but larger than, $CR^{(4)}_{LL/LA}$ as fat elimination ($CL_{eF}$) increases. This ratio may also be more difficult to biomonitor.

In summary, these dual entry route comparative risk examples emphasize the potential critical importance to risk assessment applications of experimental development of cost-effective multiple site internal dose biomarker measures. To be sure, biological model misspecifications may introduce some biases, but the cancelling inherent in using single or multiple comparative risk expressions can also be expected to alleviate some, if not all, bias in practice.

**Comparative Body-Burdens.** Single entry route formulas are also useful in developing efficient estimators of difficult to monitor 'in-line' AUCs, e.g., comparative risk for $i$ and $k$ targets from a bolus input to $j$, is given for all linear models by

$$CR_{ij/kj} = \frac{AUC_{ij}}{AUC_{kj}} = \frac{CL_{ti}^{-1}\left(S_t^{-1}\right)_{ij}A_{j0}}{CL_{tk}^{-1}\left(S_t^{-1}\right)_{kj}A_{j0}} = \frac{CL_{ti}^{-1}\left(S_{ta}\right)_{ij}}{CL_{tk}^{-1}\left(S_{ta}\right)_{kj}}. \quad (32)$$

The comparative muscle, blood risk for P4 from a liver input is given by

$$CR^{(4)}_{ML/AL} = \frac{AUC_{ML}}{AUC_{AL}} = \frac{CL_{tM}^{-1}\left(S_{ta}\right)_{ML}}{CL_{tA}^{-1}\left(S_{ta}\right)_{AL}} = \frac{CL_{tM}^{-1}}{CL_{tA}^{-1}}\,\frac{\pi_{MA}\pi_{AL}}{\pi_{AL}}\,\frac{1_F}{1_F 1_M} = \frac{CL_{MA}}{CL_{tM}}. \quad (33)$$

Cross-multiplication for this comparative body burden formula implies that

$$AUC_{ML} = AUC_{AL}\frac{CL_{MA}}{CL_{tM}} \quad (34)$$

so that a difficult to monitor dose ($AUC_{ML}$) for P4 may be estimated from an easier to monitor dose ($AUC_{AL}$) if one knows neighboring clearances.

**Biomonitoring Experimental Design.** Comparative body burden formulas (32) can also be useful in developing efficient biomonitoring experimental designs for parameter estimation using single routes of entry. For example, $CR_{ML/AL}^{(4)} = CL_{MA}/CL_{tM}$ in (33) allows efficient estimation of muscle elimination clearance ($CL_{eM} \leq CL_{tM} = CL_{eM} + CL_{rM}$) using knowledge of muscle and blood circulatory clearances ($CL_{MA} = Q_{MA}/R_{MA}$, $CL_{rM} = Q_{AM}/R_{AM}$) and biomeasurements of the muscle/blood AUC ratio following a bolus liver input (or steady-state concentration ratio following infusion). Parameter misspecifications in peripheral compartments may introduce little or no bias since $|S_t|$ cancels in the ratioing process - further motivation for development of cost-effective biomonitoring pairs. Canceling of common pathway parameters between monitored compartments further simplifies estimation formulas. Thus, development of efficient biomonitors starts with the optimal tactic of taking neighboring compartment biomeasurements at the target site and retreating, as necessary, to more easily monitored compartments nearby along the toxic pathway.

## P8 Model

The P4 model [3] assumes bolus input into the liver or blood compartment and implicitly lumps urinary elimination with blood - misleading assumptions for some QRAs. The P4 mass-balance equations can be extended to a more realistic physiological input, output (IO) model that modulates bolus inputs via a gastrointestinal (GI) tract that slows food absorption, and via a urinary system that clears blood waste products. An 8-compartment P8 extension with GI-lumen (Gl), GI-wall (Gw), Kidney (K) and Urine (U) is given (inputs suppressed) as

$$C_A' = \frac{CL_{AF}C_F}{V_A} + \frac{CL_{AM}C_M}{V_A} + \frac{CL_{AK}C_K}{V_A} + \frac{CL_{AL}C_L}{V_A} - \left(\frac{CL_A C_A}{V_A} + \frac{CL_{eA}C_A}{V_A}\right)$$

$$C_F' = \frac{CL_{FA}C_A}{V_F} - \left(\frac{CL_{AF}C_F}{V_F} + \frac{CL_{eF}C_F}{V_F}\right)$$

$$C_M' = \frac{CL_{MA}C_A}{V_M} - \left(\frac{CL_{AM}C_M}{V_M} + \frac{CL_{eM}C_M}{V_M}\right)$$

$$C_K' = \frac{CL_{KA}C_A}{V_K} - \left(\frac{CL_{AK}C_K}{V_K} + \frac{CL_{UK}C_K}{V_K} + \frac{CL_{eK}C_K}{V_K}\right) \quad (35)$$

$$C_U' = \frac{CL_{UK}C_K}{V_U} - \left(\frac{CL_{eU}C_U}{V_U}\right)$$

$$C_{Gl}' = \frac{CL_{GlL}C_L}{V_{Gl}} - \left(\frac{CL_{GwGl}C_{Gl}}{V_{Gl}} + \frac{CL_{eGl}C_{Gl}}{V_{Gl}}\right)$$

$$C_{Gw}' = \frac{CL_{GwA}C_A}{V_{Gw}} + \frac{CL_{GwGl}C_{Gl}}{V_{Gw}} - \left(\frac{CL_{LGw}C_{Gw}}{V_{Gw}} + \frac{CL_{eGw}C_{Gw}}{V_{Gw}}\right)$$

$$C_L' = \frac{CL_{LA}C_A}{V_L} + \frac{CL_{LGw}C_{Gw}}{V_L} - \left(\frac{CL_{AL}C_L}{V_L} + \frac{CL_{GlL}C_L}{V_L} + \frac{CL_{eL}C_L}{V_L}\right)$$

where $CL_{AL} = Q_{AL}/R_{AL} \equiv (Q_{LA} + Q_{LGw})/R_{AL}$, $Q_{LA}$ is the direct bloodflow to liver and $Q_{AL}$ is the composite liver plus gut-wall bloodflow to $A$; and where $CL_A$ is the total clearance to all organs with bloodflow.

P8 is given in input matrix form by $C' = -V^{-1}(CL)C + i$ where $CL \equiv CL^{(8)}$

$$
= \begin{pmatrix}
CL_{tA} & -CL_{AF} & -CL_{AM} & -CL_{AK} & 0 & 0 & 0 & -CL_{AL} \\
-CL_{FA} & CL_{tF} & 0 & 0 & 0 & 0 & 0 & 0 \\
-CL_{MA} & 0 & CL_{tM} & 0 & 0 & 0 & 0 & 0 \\
-CL_{KA} & 0 & 0 & CL_{tK} & 0 & 0 & 0 & 0 \\
0 & 0 & 0 & -CL_{UK} & CL_{tU} & 0 & 0 & 0 \\
0 & 0 & 0 & 0 & 0 & CL_{tGl} & 0 & -CL_{GlL} \\
-CL_{GwA} & 0 & 0 & 0 & 0 & -CL_{GwGl} & CL_{tGw} & 0 \\
-CL_{LA} & 0 & 0 & 0 & 0 & 0 & -CL_{LGw} & CL_{tL}
\end{pmatrix} ,
$$

$$(36)$$

or in t-normalized factored form by $C' = -V^{-1}(S_t CL_t)C + i$ where

$$
S_t \equiv S_t^{(8)}
$$

$$
= \begin{pmatrix}
1_A & -\pi_{AF} & -\pi_{AM} & -\pi_{AK} & 0 & 0 & 0 & -\pi_{AL} \\
-\pi_{FA} & 1_F & 0 & 0 & 0 & 0 & 0 & 0 \\
-\pi_{MA} & 0 & 1_M & 0 & 0 & 0 & 0 & 0 \\
-\pi_{KA} & 0 & 0 & 1_K & 0 & 0 & 0 & 0 \\
0 & 0 & 0 & -\pi_{UK} & 1_U & 0 & 0 & 0 \\
0 & 0 & 0 & 0 & 0 & 1_{Gl} & 0 & -\pi_{GlL} \\
-\pi_{GwA} & 0 & 0 & 0 & 0 & -\pi_{GwGl} & 1_{Gw} & 0 \\
-\pi_{LA} & 0 & 0 & 0 & 0 & 0 & -\pi_{LGw} & 1_L
\end{pmatrix} .
$$

$$(37)$$

**P8 SISO Examples.** As with P4, SISO expressions are derived by determinant expansion. The liver dose for a GI lumen input is

$$
AUC_{LGl}^{(8)} = (CL^{-1})_{LGl} A_{Gl0} = \frac{\left(S_{ta}^{(8)}\right)_{LGl}}{|S_t|} \frac{A_{Gl0}}{CL_{tL}} = \pi_{LGw}\pi_{GwGl} \frac{|S_t - LGwGl|}{|S_t|} \frac{A_{Gl0}}{CL_{tL}}
$$

$$
= \pi_{LGw}\pi_{GwGl} \frac{(1 - \pi_{AF}\pi_{FA} - \pi_{AM}\pi_{MA} - \pi_{AK}\pi_{KA})}{|S_t|} \frac{A_{Gl0}}{CL_{tL}}
$$

$$(38)$$

where $\pi_{LGwGl} \equiv \pi_{LGw}\pi_{GwGl}$ denotes toxic pathway conductance and $|S_t - LGwGl|$ $\equiv |AFMKU|$ is the adjoint's contribution to the multiplier. The structural determinant's contribution to the multiplier is expandable as

$$
|S_t| \equiv \left|S_t^{(8)}\right| = \begin{array}{l} |LGwGl| - \pi_{AF}\pi_{FA}|LGwGl| - \pi_{AM}\pi_{MA}|LGwGl| \\ -\pi_{AK}\pi_{KA}|LGwGl| - \pi_{AL}\pi_{LA} - \pi_{AL}\pi_{LGw}\pi_{GwA} \end{array}
$$

$$(39)$$

where $|LGwGl| = 1 - \pi_{LGw}\pi_{GwGl}\pi_{GlL}$ is the $3 \times 3$ subdeterminant of $S_t^{(8)}$ generated by the (L,Gw,Gl) row-column intersections.

The double pathway liver dose for a blood (inhalation surrogate) input is

$$
AUC_{LA}^{(8)} = (CL^{-1})_{LA} A_{A0} = \frac{S_{taLA}^{(8)}}{|S_t|} \frac{A_{A0}}{CL_{tL}}
$$

$$
= (\pi_{LA} \frac{|S_t - LA|}{|S_t|} + \pi_{LGw}\pi_{GwA} \frac{|S_t - LGwA|}{|S_t|}) \frac{A_{A0}}{CL_{tL}} = (\pi_{LA} + \pi_{LGwA}) \frac{1}{|S_t|} \frac{A_{A0}}{CL_{tL}}
$$

$$(40)$$

since $|S_t - LA| = |S_t - LGwA| = 1$ for P8 and $\pi_{LGwA} \equiv \pi_{LGw}\pi_{GwA}$.

For later comparisons, the kidney dose from ingestion is given by

$$AUC_{KGl}^{(8)} = \left(CL^{-1}\right)_{KGl} A_{Gl0} = \frac{(S_{ta})_{KGl}}{|S_t|} \frac{A_{Gl0}}{CL_{tK}}$$

$$= \pi_{KA}\pi_{AL}\pi_{LGw}\pi_{GwGl} \frac{|S_t - KALGwGl|}{|S_t|} \frac{A_{Gl0}}{CL_{tK}} = \pi_{KALGwGl} \frac{1}{|S_t|} \frac{A_{Gl0}}{CL_{tK}} \quad (41)$$

since $|S_t - KALGwGl| \equiv |FMU| = 1$ and $\pi_{KALGwGl} \equiv \pi_{KA}\pi_{AL}\pi_{LGw}\pi_{GwGl}$.
Also, the corresponding kidney dose from a blood input is given by

$$AUC_{KA}^{(8)} = \frac{(S_{ta})_{KA}}{|S_t|} \frac{A_{A0}}{CL_{tK}} = \pi_{KA} \frac{|S_t - KA|}{|S_t|} \frac{A_{A0}}{CL_{tK}} = \pi_{KA} \frac{|LGwGl|}{|S_t|} \frac{A_{A0}}{CL_{tK}}, \quad (42)$$

where $|S_t - KA| \equiv |LGwGlFMU| = |LGwGl|$ by determinant expansion.

**P8 Inverse Reconstruction.** Extending the P4 example (23) to P8, liver bolus dose is reconstructed from $A_0 = (CL)AUC$ and (36) as

$$A_{L0}^{(8)} = -CL_{LA}AUC_A - CL_{LGw}AUC_{Gw} + CL_{tL}AUC_L. \quad (43)$$

Similarly, blood bolus dose is given by

$$A_{A0}^{(8)} = CL_{tA}AUC_A - CL_{AF}AUC_F - CL_{AM}AUC_M - CL_{AK}AUC_K - CL_{AL}AUC_L, \quad (44)$$

and thus larger models necessarily tend to require extra AUC biomeasures.

**P8 Route-to-Route Extrapolation.** Using (38) and (40), comparative liver risk from stomach (GI-lumen) versus blood inputs is given by

$$CR_{LGl/LA}^{(8)} = \frac{AUC_{LGl}}{AUC_{LA}} = \frac{\pi_{LGw}\pi_{GwGl}|S_t - LGwGl|}{(\pi_{LA} + \pi_{LGw}\pi_{GwA})} \frac{A_{Gl0}}{A_{A0}}. \quad (45)$$

Again, cancellation of common factors ($|S_t|^{-1}$ and $CL_{tL}^{-1}$) simplifies comparative risk expressions, lowers dimensionality and may improve parameter estimation.
Similarly, comparative kidney risk for stomach versus blood inputs is

$$CR_{KGl/KA}^{(8)} = \frac{AUC_{KGl}}{AUC_{KA}} = \frac{\pi_{KA}\pi_{AL}\pi_{LGw}\pi_{GwGl}}{\pi_{KA}|LGwGl|} \frac{A_{Gl0}}{A_{A0}} = \frac{\pi_{AL}\pi_{LGw}\pi_{GwGl}}{|LGwGl|} \frac{A_{Gl0}}{A_{A0}}, \quad (46)$$

where $|LGwGl| = (1 - \pi_{GlL}\pi_{LGw}\pi_{GwGl}) > 0$. These formulas are more complex than for P4 but yield similar comparative insights. For example, kidney risk is less from stomach than from blood inputs ($A_{Gl0} = A_{A0}$) if $\pi_{AL} \leq 1, \pi_{LGw}\pi_{GwGl} < 1/2$ and $\pi_{GlL} \leq 1$, e.g., when fractional bile clearance ($\pi_{GlL} = CL_{GlL}/CL_{tL}$) is small and GI absorption ($\pi_{GwGl} = CL_{GwGl}/CL_{tGl}$) is low. Thus, symbolic formulas help quantify qualitative insights and can sometimes correct faulty intuition.

**P8 Body-Burden, Experimental Design.** Optimal experimental designs for blood and urine measurements from stomach inputs can be derived from

$$CR_{UGl/AGl}^{(8)} \equiv \frac{AUC_{UGl}}{AUC_{AGl}} = \frac{CL_{tU}^{-1}(S_{ta})_{UGl}}{CL_{tA}^{-1}(S_{ta})_{AGl}} = \frac{CL_{tU}^{-1}}{CL_{tA}^{-1}} \frac{\pi_{UK}\pi_{KA}\pi_{AL}\pi_{LGw}\pi_{GwGl}}{\pi_{AL}\pi_{LGw}\pi_{GwGl}}$$

$$= \frac{CL_{tU}^{-1}}{CL_{tA}^{-1}} \pi_{UK}\pi_{KA} = \frac{CL_{tA}}{CL_{tU}} \frac{CL_{UK}}{CL_{tK}} \frac{CL_{KA}}{CL_{tA}} = \frac{CL_{UK}}{CL_{tU}} \frac{CL_{KA}}{CL_{tK}} = \frac{CL_{KA}}{CL_{tK}} \quad (47)$$

where $CL_{UK} = CL_{tU}$ for kidney, urine processes, yielding results analogous to the muscle, blood P4 liver input example discussed earlier (33), but extended to more widely separated monitored organs (a UKA vs. MA separation). Thus, poorly known kidney metabolic elimination, e.g., $CL_{eK}$ in $CL_{tK} = (CL_{UK} + CL_{AK} + CL_{eK})$, might be estimated from better known clinical parameters $CL_{AK} = Q_{AK}/R_{AK}$, $CL_{UK}$, $CL_{KA} = Q_{KA}$ and biomeasurements $AUC_{UGI}$ and $AUC_{AGI}$ after bolus input (or steady-state concentration measurements $C_{SS,U}$ and $C_{SS,A}$ after constant infusion). Other experimental designs yield less cost-effective $CL_{eK}$ estimates if they require biomonitoring of additional, less accessible organs.

Another comparative body burden application of (47) is written as

$$AUC_{AGI} = \frac{AUC_{UGI}}{\frac{CL_{KA}}{CL_{tK}}} = \frac{AUC_{UGI}CL_{tK}}{CL_{KA}}, \quad (48)$$

so that difficult to monitor internal doses ($AUC_{AGI}$) may be estimated from more easily monitored doses ($AUC_{UGI}$) if one knows intermediate clearance parameters ($CL_{tK}$, $CL_{KA}$) between target organ ($A$) and biomonitored compartment ($U$).

**P8 Subpathway Example.** P4 has single toxic pathways for all input, target organ pairs, but P8 allows multiple toxic subpathways for some IO pairs. Symbolic comparison of two subpathways helps identify dominant subpathways. This is useful for determining cost-effective physiological risk reduction or drug enhancement intervention strategies. For example, the liver target, blood input comparative risk for two toxic subpathways (40) is given by

$$CR_{LA^{(1)}/LA^{(2)}} = \frac{\pi_{LA}}{\pi_{LGw}\pi_{GwA}}. \quad (49)$$

Again comparative formulas are simpler than absolute formulas and provide more efficient estimators via cancellation of nonessential, biasing peripheral factors.

### Discussion and Future Directions

A linear 4-compartment P4 PBPK prototype has been reformulated using intuitive notation, physiologically consistent matrix factorizations and mathematically efficient matrix normalizations to clarify physiological structure, simplify symbolic solution, and facilitate extension to a larger P8 model. The classical 1-compartment formula, $AUC = CL^{-1}A_0$, has been generalized and SISO AUCs decomposed into 3 meaningful factors. Specific QRA and ED applications explored include: (i) extrapolation of target dose from bolus to infusion input; (ii) reconstruction of external exposure from internal dose; (iv) route-to-route exposure extrapolation of target dose; (v) comparative body burdens from single exposure routes; (vi) comparative subpathway risks for given input, output pairs; and (vii) efficient biomonitoring experimental designs for parameter estimation.

Future theoretical work on general linear PBPK models will include (1) graph-theoretic proof of the decomposability of SISO AUCs into sums of products of toxic pathway conductances, circulation multipliers and crude AUCs; (2) development of simple graph-theoretic determinant expansions to replace tedious row-column expansions; and (3) further model or computational validation efforts via AUC comparisons of symbolic and general purpose ODE solver results.

**Acknowledgments.** The author thanks J. Springer and R. Lorentzen (FDA) for support during early feasibility studies and J. Blancato, B. Hagstrom and J. Nocerino (EPA), and E. Pellazarri (RTI) for review comments.

# References

[1] Anderson,M.E., Clewell III,H.J., Gargas,M.L, Smith,F.A. and Reitz,R.H. (1987). 'Physiologically Based Pharmacokinetics and the Risk Assessment Process for Methylene Chloride.' *Toxicol. Appl. Pharmacol. 87*:185-205.

[2] Arnold.,V.I.(1992). *Ordinary Differential Equations.* Springer Verlag, New York, NY. pp 1-333.

[3] Brown,R.N. (1994). 'Analytic Solution of a Linear PBPK Model Prototype Useful in Risk Assessment.' *Biomarkers of Human Exposure to Pesticides* (M.A. Saleh, C.H. Nauman, J.N. Blancato, eds.). Vol. 542, ACS Symposium Series, Wash. D.C. pp. 301-317.

[4] Gear,C.W.(1971). *Numerical Initial Value Problems in Ordinary Differential Equations.* Prentice Hall, Englewood Cliffs, N.J. pp. 1-253.

[5] Gibaldi,M and Perrier,D.(1982). *Pharmacokinetics, 2nd Ed.* Marcel Dekker, New York, NY. pp. 1-494.

[6] Horn,R.A. and Johnson,C.R.(1988). *Matrix analysis.* Cambridge University Press, New York, NY. pp. 1-561.

[7] Nakashima,E. and Benet,L.Z.(1989). 'An Integrated Approach to Pharmacokinetic Analysis for Linear Mammillary Systems in Which Input and Exit May Occur in/from Any Compartment.' *J.Pharmaco.Biopharm.* 17:673-685.

# Chapter 19

# Comparison of Symbolic, Numerical Area Under the Concentration Curves of Small Linear Physiologically Based Pharmacokinetic Models

**Blaine L. Hagstrom, Robert N. Brown, and Jerry N. Blancato**

**Characterization Research Division, National Exposure Research Laboratory, U.S. Environmental Protection Agency, 944 East Harmon, Las Vegas, NV 89193–3478**

Using the area under the concentration curve (AUC) dose metric common in physiologically based pharmacokinetic (PBPK) models for human risk assessment, this paper compares recently developed exact symbolic solutions for linear PBPK models with conventional approximate numerical solutions. Comparisons are given for both 4 and 8-compartment PBPK models (P4 and P8). Relative error comparisons are presented for all exposure route, target organ, combinations for P4 and P8. AUC formulas are decomposed into the sum of independent toxic pathways, each pathway the product of 3 physiologically meaningful symbolic factors. The effect of extreme parameter perturbations upon relative error is explored and user options for recovering accuracy are discussed. The complementary role of symbolic and numerical solutions is emphasized.

To improve the understanding, validation, uncertainty analysis and general scientific basis of its risk assessment processes in exposure analysis, dose-response prediction, and risk characterization, the U.S. Environmental Protection Agency (EPA) is investigating new diagnostic tools for the analysis of physiologically based pharmacokinetic (PBPK) models describing the exposure and disposition of environmental toxicants in humans and animals. Such tools also help formulate cost-effective regulatory risk prevention and management strategies.

PBPK models examine toxicant dose at all physiological scales, including systemic exposure, target tissue dose and sub-cellular response. PBPK models help refine the qualitative formulation of biological processes, their quantitative mathematical computation, and their sensitivity to model, parameter and data uncertainties. PBPK models are described by mixed systems of linear and nonlinear ordinary differential equations (ODEs) which often reduce to purely linear systems at low concentrations typical at most human and some test animal exposures. Both toxicant concentrations and integrated concentrations are commonly computed using general purpose stiff numerical ODE algorithms such as Gear's [3]. In previous work with a 4-compartment PBPK prototype (P4) [1], symbolic analytic dose formulas were developed for the area under the concentration curve (AUC) dose metric common in quantitative risk assessment (QRA).

0097–6156/96/0643–0256$15.00/0

Concurrently in these Proceedings, a more intuitive terminology, consistent physiological formulation, and efficient symbolic AUC solution was presented for P4, and extended to selected inputs and outputs of an 8-compartment PBPK model (P8) [R.N. Brown; 'Symbolic AUC Solutions of Small Linear PBPK Models Useful in Risk Assessment and Experimental Design']. Those AUCs were observed to be decomposable into the product of three physiologically meaningful factors: a crude AUC, a fractional conductance, and a circulation multiplier.

The main purpose here is to compare, for all exposure route, target organ combinations for P4 and P8, the numerical AUC accuracy of approximate Gear-type ODE solvers with exact symbolic formulas. We further explore the physiological structure of symbolic AUCs. The complementary roles of symbolic and numerical solutions for risk assessment and management applications are emphasized. We also explore, via extreme fat partition coefficient perturbations, numerical solver breakdown and error-control options for accuracy recovery. All ODE solver and symbolic evaluations were performed using SimuSolv software [2], but P4 symbolic results were also verified by hand evaluation of symbolic formulas.

## Review of P4 Model and Symbolic AUCs

The P4 prototype consisted of (i) a main compartment ($A \equiv$ composite arterial-venous blood); (ii) a slowly perfused strong sink compartment ($F \equiv$ fat); (iii) a slowly perfused weak sink compartment ($M \equiv$ muscle); and (iv) a large flow systemic clearance compartment ($L \equiv$ liver). The mass balance equations were given previously in low concentration linear limit form (inputs suppressed) by

$$C'_A = \frac{A'_A}{V_A} = \frac{1}{V_A}\left(\frac{Q_{AF}C_F}{R_{AF}} + \frac{Q_{AM}C_M}{R_{AM}} + \frac{Q_{AL}C_L}{R_{AL}}\right) - \frac{1}{V_A}\left(\frac{Q_A C_A}{R_A} + V_A k_{eA} C_A\right)$$

$$C'_F = \frac{A'_F}{V_F} = \frac{1}{V_F}\frac{Q_{FA}C_A}{R_{FA}} - \frac{1}{V_F}\left(\frac{Q_{AF}C_F}{R_{AF}} + V_F k_{eF} C_F\right)$$

$$C'_M = \frac{A'_M}{V_M} = \frac{1}{V_M}\frac{Q_{MA}C_A}{R_{MA}} - \frac{1}{V_M}\left(\frac{Q_{AM}C_M}{R_{AM}} + V_M k_{eM} C_M\right)$$

$$C'_L = \frac{A'_L}{V_L} = \frac{1}{V_L}\frac{Q_{LA}C_A}{R_{LA}} - \frac{1}{V_L}\left(\frac{Q_{AL}C_L}{R_{AL}} + V_L k_{eL} C_L + \frac{V_{mL}^{(A)}}{K_{mL}}\frac{C_L}{1+\frac{C_L}{K_{mL}}}\right) \tag{1}$$

where clearance terms (in vol./time units) are given by $Q_{Ai}$; by $V_i k_{ei}$ where $k_{ei}$ is a kinetic transfer rate for volume $V_i$; or by $V_{mi}^{(A)}/K_{mi}$ when the nonlinear perturbation term $1 + \frac{C_L}{K_{mL}} \rightarrow 1$, where $V_{mi}^{(A)}$ is the amount form maximum metabolic elimination velocity (in mass/time units) and $K_{mi}$ is the Michaelis constant (in mass/vol. units). $R_{ij}$ is the unitless partition coefficient describing the ratio of toxicant concentration in compartment $j$ relative to $i$. In matrix form, the mass balance equations, for input vector $i$, are given by

$$C' = -V^{-1}(CL)C + i, \tag{2}$$

where $C' = V^{-1}A'$ is the concentration derivative vector (in mass/volume-time units); $A'$ the amount vector; $V^{-1}$ the diagonal matrix of inverse volumes; and $-CL$ the negative system clearance matrix given by

$$-CL = \begin{pmatrix} -\frac{Q_A}{R_A} - k_{eA}V_A & \frac{Q_{AF}}{R_{AF}} & \frac{Q_{AM}}{R_{AM}} & \frac{Q_{AL}}{R_{AL}} \\ \frac{Q_{FA}}{R_{FA}} & -\frac{Q_{AF}}{R_{AF}} - k_{eF}V_F & 0 & 0 \\ \frac{Q_{MA}}{R_{MA}} & 0 & -\frac{Q_{AM}}{R_{AM}} - k_{eM}V_M & 0 \\ \frac{Q_{LA}}{R_{LA}} & 0 & 0 & -\frac{Q_{AL}}{R_{AL}} - k_{eL}V_L - \frac{V_{mL}^{(A)}}{K_{mL}} \end{pmatrix}$$

$$= \begin{pmatrix} -CL_{tA} & CL_{AF} & CL_{AM} & CL_{AL} \\ CL_{FA} & -CL_{tF} & 0 & 0 \\ CL_{MA} & 0 & -CL_{tM} & 0 \\ CL_{LA} & 0 & 0 & -CL_{tL} \end{pmatrix}, \tag{3}$$

with off-diagonal elements ($CL_{ij}$) representing local circulatory clearance to compartment $i$ from $j$, and diagonal elements $CL_{ti}$ ($\equiv CL_{ii}$) denoting total clearances from $i$. Previously $CL$ was rewritten in t-normalized matrix form as

$$CL = S_t CL_t$$

$$= \begin{pmatrix} 1_A & -\pi_{AF} & -\pi_{AM} & -\pi_{AL} \\ -\pi_{FA} & 1_F & 0 & 0 \\ -\pi_{MA} & 0 & 1_M & 0 \\ -\pi_{LA} & 0 & 0 & 1_L \end{pmatrix} \begin{pmatrix} CL_{tA} & 0 & 0 & 0 \\ 0 & CL_{tF} & 0 & 0 \\ 0 & 0 & CL_{tM} & 0 \\ 0 & 0 & 0 & CL_{tL} \end{pmatrix} \tag{4}$$

where $S_t$ is $CL$ with $i^{th}$ column divided by $CL_{ti}$ so that diagonal entries are unity ($\pi_{ii} \equiv 1_i$) and so that off-diagonal entries represent the fractional clearance to compartment $i$ from $j$ ($\pi_{ij} = CL_{ij}/CL_{tj} = k_{ij}/k_{tj}$). $S_t$ is the system's total- or t-normalized unitless structural interconnection matrix, which emphasizes the importance of fractional flows between compartments in model formulation, understanding and solution. Figure 1 shows the P4 system graph useful in summarizing symbolic AUC solutions.

AUC, a useful internal dose refinement of external exposure for studying many intermediate or long term health effects, especially those independent of short-term exposure variations, was given previously as

$$AUC = \int_0^\infty C(\tau)d\tau = CL^{-1}A_0 = CL_t^{-1}S_t^{-1}A_0, \tag{5}$$

generalizing the classical 1-compartment formula $AUC = A_0/CL_e$ [4]. Using matrix row-column determinant expansions, the finite, nonnegative, t-normalized single-input, single-output (SISO) components for an $i^{th}$ compartment output from a $j^{th}$ compartment input were given [1] by

$$AUC_{ij} = CL_{ti}^{-1} \left( S_t^{-1} \right)_{ij} A_{j0} = \frac{(S_{ta})_{ij}}{|S_t|} \frac{A_{j0}}{CL_{ti}} \tag{6}$$

if $|S_t| = 1 - \pi_{AF}\pi_{FA} - \pi_{AM}\pi_{MA} - \pi_{AL}\pi_{LA} > 0$ (i.e., if $S_t$ invertible). For all P4 and selected P8 inputs and outputs, this reduced to the more intuitive

$$AUC_{ij} = \sum_{\pi_{i..j}} \pi_{i...j} \frac{|S_t - \pi_{i...j}|}{|S_t|} \frac{A_{j0}}{CL_{ti}} \tag{7}$$

where overall AUC reduced to the sum, over all possible toxic pathways from $j$ to $i$, of 3 physiologically meaningful factors: (i) where $\pi_{i...j}$, denoted the fractional conductance (i.e., product of individual fractional clearances) along a particular pathway; (ii) where the determinant ratio $|S_t - \pi_{i...j}| / |S_t|$ denoted the circulation multiplier, describing the effective average number of times the parent toxicant reentered the input compartment $j$ for repeat conductance down the toxic pathway to target organ $i$; and (iii) where $A_{j0}/CL_{ti}$ denoted the crude AUC, as if all input went directly to the target organ. Note that $|S_t - \pi_{i...j}| \equiv |S_t - (i...j)|$ represented the residual subdeterminant after subtracting from $S_t$ the $(i...j)$ rows and columns making up the toxic pathway.

## P4 Symbolic, Numerical AUC Comparisons

For comparing analytic and numerical AUCs, nondefault independent parameters for P4 are given in Table I (default partition coefficients = 1, default systemic clearances = 0). Dependent parameters for symbolic AUC construction are given in Table II. Numerical, symbolic AUC computations are given in Table III in the 'Approx.' and 'Exact' columns for all 4 SISO AUC outputs from an arterial bolus input ($A_{i0}$). Table truncation anomalies, involving apparent repeating decimals, are corrected using overbar notation in examples below. Interpretation is simplified by giving the relative error between approximate and exact AUCs, along with numerical evaluation of the 3 factors making up each AUC subpathway. Bolus muscle, fat and liver input results are given in Tables IV, V and VI.

Table I: P4 nondefault independent parameters.

| $Q_{FA} = 25$ | $Q_{MA} = 275$ | $V_A = 5$ | $V_F = 13$ | $R_{AL} = 10$ | $R_{AF} = 100$ |
|---|---|---|---|---|---|
| $Q_{LA} = 100$ | $Q_{Ai} = Q_{iA}$ | $V_L = 2$ | $V_M = 50$ | $R_{AM} = 10$ | $CL_{eL} = 45$ |

Table II: P4 dependent parameters.

| $\pi_{AL} = 0.18\overline{18}$ | $\pi_{LA} = 0.2500$ | $\pi_{AF} = 1.0000$ | $\pi_{FA} = 0.0625$ |
|---|---|---|---|
| $\pi_{AM} = 1.0000$ | $\pi_{MA} = 0.6875$ | $CL_{FA}=25.00$ | $CL_{MA}=275.00$ |
| $CL_{LA}=100.00$ | $CL_{AF}=0.25$ | $CL_{AM}=27.50$ | $CL_{AL}=10.00$ |
| $CL_{tM}=27.50$ | $CL_{tF}=0.25$ | $CL_{tA}=400.00$ | $CL_{tL}=55.00$ |

Table III: P4 $AUC_{iA}$ for bolus Arterial input. $RelErr. = \frac{APPROX. - EXACT}{EXACT}$, Cond. $= \pi_{i...A}$, Mult. $= \frac{|S_t - \pi_{i...A}|}{|S_t|}$, Crude $= \frac{A_{A0}}{CL_{ti}}$, $A_{A0} = 0.1$.

| i | Approx. | Exact | RelErr. | Cond. | Mult. | Crude |
|---|---|---|---|---|---|---|
| A | 0.00122222 | 0.00122222 | 5.7149E-7 | 1.000000 | 4.88889 | 0.00025 |
| F | 0.12222200 | 0.12222200 | 3.6576E-7 | 0.062500 | 4.88889 | 0.40000 |
| M | 0.01222220 | 0.01222220 | 2.2860E-7 | 0.687500 | 4.88889 | 0.00364 |
| L | 0.00222222 | 0.00222222 | 5.2387E-7 | 0.250000 | 4.88889 | 0.00182 |

For all P4 tables, the salient feature is the small relative error of the approximate AUCs. Looking closer at AUC structure, the baseline example of an arterial

AUC from an arterial input is given from (6, 7) and Table III as

$$AUC_{AA} = \pi_{AA} \frac{|S_t - A|}{|S_t|} \frac{A_{A0}}{CL_{tA}} = \frac{1}{|S_t|} \frac{A_{A0}}{CL_{tA}} = .00122\bar{2}, \qquad (8)$$

with fractional conductance $\pi_{AA} = 1$ since input is to target organ; circulation multiplier $1/|S_t| = 1/.204545 = 4.89$, where residual numerator subdeterminant $|S_t - A| = 1$ since $(S_t - A)$ is diagonal; and crude AUC small $(A_{A0}/CL_{tA} = .00025)$. For a weak sink (moderate partitioning), length-1 toxic pathway,

$$AUC_{MA} = \pi_{MA} \frac{|S_t - MA|}{|S_t|} \frac{A_{A0}}{CL_{tM}} = \pi_{MA} \frac{1}{|S_t|} \frac{A_{A0}}{CL_{tM}} = .0122\bar{2}, \qquad (9)$$

with crude AUC 15 times larger than baseline (at .00364 vs. .00025), conductance 1/3 smaller (at .6875 vs. 1), and AUC 10 times higher (at .01222 vs. .001222). For a strong sink example, the fat AUC from a bolus arterial input is given by

$$AUC_{FA} = \pi_{FA} \frac{|S_t - FA|}{|S_t|} \frac{A_{A0}}{CL_{tF}} = \pi_{FA} \frac{1}{|S_t|} \frac{A_{A0}}{CL_{tF}} = .122\bar{2}, \qquad (10)$$

with crude AUC about 1000 times higher (at .4) and conductance about 10 times smaller (at $\pi_{AF} = .0625$), to yield AUC 100 times higher than baseline. For a smaller multiplier, Table IV shows muscle AUC from a bolus muscle input as

$$AUC_{MM} = \frac{|S_t - M|}{|S_t|} \frac{A_{M0}}{CL_{tM}} = \frac{1 - \pi_{AF}\pi_{FA} - \pi_{AL}\pi_{LA}}{|S_t|} \frac{A_{M0}}{CL_{tM}} = .0159, \qquad (11)$$

with conductance unity $(\pi_{MM} = 1)$ and multiplier $4.36 = .89/.204$ (<max. of 4.89) since the numerator determinant is below unity. For a very low multiplier example, fat AUC from a bolus fat input is given in Table V by

$$AUC_{FF} = \frac{|S_t - F|}{|S_t|} \frac{A_{F0}}{CL_{tF}} = \frac{1 - \pi_{AM}\pi_{MA} - \pi_{AL}\pi_{LA}}{|S_t|} \frac{A_{F0}}{CL_{tF}} = .522\bar{2}, \qquad (12)$$

with conductance unity and multiplier $1.31 = .26/.204$ since the numerator subdeterminant is nearly as small as $|S_t|$. For a long pathway example, the muscle dose from a fat input is given by

$$AUC_{MF} = \pi_{MA}\pi_{AF} \frac{|S_t^{(4)} - MAF|}{|S_t|} \frac{A_{F0}}{CL_{tM}} = \pi_{MA}\pi_{AF} \frac{1}{|S_t^{(4)}|} \frac{A_{F0}}{CL_{tM}} = .0122\bar{2}, \qquad (13)$$

with large conductance (.69) dominated by $\pi_{MA} = .69$ since the $\pi_{AF}$ fractional clearance is unity, reflecting the default pure pass-through behavior of organs lacking competing clearances - by themselves, not affecting AUC.

On the other hand, Table VI indicates that AUC differences can be simply proportional to target organ partitioning: $AUC_{AL} = 00022\bar{2}$; 10-fold larger $AUC_{ML}$ and $AUC_{LL}$, due to muscle/blood and liver/blood partitioning of 10; and 100-fold larger $AUC_{FL}$, due to fat/blood partitioning of 100. Note here that muscle and liver AUC factors are completely different but $AUC_{ML}$ still equals $AUC_{LL}$ - an important fact apparently related to the pass-through behavior of

Table IV: P4 $AUC_{iM}$ for bolus Muscle input, $A_{M0} = 0.1$.

| i | Approx. | Exact | RelErr. | Cond. | Mult. | Crude |
|---|---------|-------|---------|-------|-------|-------|
| A | 0.00122222 | 0.00122222 | 7.6199E-7 | 1.000000 | 4.88889 | 0.00025 |
| F | 0.12222200 | 0.12222200 | 2.4384E-7 | 0.062500 | 4.88889 | 0.40000 |
| M | 0.01585860 | 0.01585860 | 1.1745E-7 | 1.000000 | 4.36111 | 0.00364 |
| L | 0.00222222 | 0.00222222 | 7.3342E-7 | 0.250000 | 4.88889 | 0.00182 |

Table V: P4 $AUC_{iF}$ for bolus Fat input, $A_{F0} = 0.1$.

| i | Approx. | Exact | RelErr. | Cond. | Mult. | Crude |
|---|---------|-------|---------|-------|-------|-------|
| A | 0.00122222 | 0.00122222 | 1.0477E-6 | 1.000000 | 4.88889 | 0.00025 |
| F | 0.52222200 | 0.52222200 | 2.2827E-7 | 1.000000 | 1.30555 | 0.40000 |
| M | 0.01222220 | 0.01222220 | 2.2860E-7 | 0.687500 | 4.88889 | 0.00364 |
| L | 0.00222222 | 0.00222222 | 3.1432E-7 | 0.250000 | 4.88889 | 0.00182 |

Table VI: P4 $AUC_{iL}$ for bolus Liver input, $A_{L0} = 0.1$.

| i | Approx. | Exact | RelErr. | Cond. | Mult. | Crude |
|---|---------|-------|---------|-------|-------|-------|
| A | 0.00022222 | 0.00022222 | 4.5839E-7 | 0.1818180 | 4.88889 | 0.00025 |
| F | 0.02222220 | 0.02222220 | 5.8673E-7 | 0.0113636 | 4.88889 | 0.40000 |
| M | 0.00222222 | 0.00222222 | 4.1910E-7 | 0.1250000 | 4.88889 | 0.00364 |
| L | 0.00222222 | 0.00222222 | -6.2864E-7 | 1.0000000 | 1.22222 | 0.00182 |

the nonliver organs that will be discussed elsewhere in the context of organ extraction efficiencies. Note also that arterial, fat and muscle compartment AUCs are greatly reduced when the drug or toxin is administered orally (via liver input for P4). This is a quantitative illustration of the liver's first pass effect.

## P8 Model and Symbolic-Numerical AUC Comparisons

The P8 extension of P4, incorporating GI Lumen (Gl), GI-wall (Gw), Kidney (K) and Urine (U), was given in these proceedings (inputs suppressed) as

$$C'_A = \frac{CL_{AF}C_F}{V_A} + \frac{CL_{AM}C_M}{V_A} + \frac{CL_{AK}C_K}{V_A} + \frac{CL_{AL}C_L}{V_A} - \left(\frac{CL_{\bullet A}C_A}{V_A} + \frac{CL_{eA}C_A}{V_A}\right)$$

$$C'_F = \frac{CL_{FA}C_A}{V_F} - \left(\frac{CL_{AF}C_F}{V_F} + \frac{CL_{eF}C_F}{V_F}\right)$$

$$C'_M = \frac{CL_{MA}C_A}{V_M} - \left(\frac{CL_{AM}C_M}{V_M} + \frac{CL_{eM}C_M}{V_M}\right)$$

$$C'_K = \frac{CL_{KA}C_A}{V_K} - \left(\frac{CL_{AK}C_K}{V_K} + \frac{CL_{UK}C_K}{V_K} + \frac{CL_{eK}C_K}{V_K}\right) \qquad (14)$$

$$C'_U = \frac{CL_{UK}C_K}{V_U} - \left(\frac{CL_{eU}C_U}{V_U}\right)$$

$$C'_{Gl} = \frac{CL_{GlL}C_L}{V_{Gl}} - \left(\frac{CL_{GwGl}C_{Gl}}{V_{Gl}} + \frac{CL_{eGl}C_{Gl}}{V_{Gl}}\right)$$

$$C'_{Gw} = \frac{CL_{GwA}C_A}{V_{Gw}} + \frac{CL_{GwGl}C_{Gl}}{V_{Gw}} - \left(\frac{CL_{LGw}C_{Gw}}{V_{Gw}} + \frac{CL_{eGw}C_{Gw}}{V_{Gw}}\right)$$

$$C'_L = \frac{CL_{LA}C_A}{V_L} + \frac{CL_{LGw}C_{Gw}}{V_L} - \left(\frac{CL_{AL}C_L}{V_L} + \frac{CL_{GlL}C_L}{V_L} + \frac{CL_{eL}C_L}{V_L}\right)$$

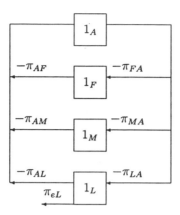

Figure 1: P4 model system diagram.

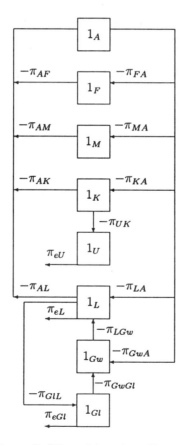

Figure 2: P8 model system diagram.

or, in matrix form, by $C' = -V^{-1}(CL)C = -V^{-1}(S_tCL_t)C$ where

$$S_t = \begin{pmatrix} 1_A & -\pi_{AF} & -\pi_{AM} & -\pi_{AK} & 0 & 0 & 0 & -\pi_{AL} \\ -\pi_{FA} & 1_F & 0 & 0 & 0 & 0 & 0 & 0 \\ -\pi_{MA} & 0 & 1_M & 0 & 0 & 0 & 0 & 0 \\ -\pi_{KA} & 0 & 0 & 1_K & 0 & 0 & 0 & 0 \\ 0 & 0 & 0 & -\pi_{UK} & 1_U & 0 & 0 & 0 \\ 0 & 0 & 0 & 0 & 0 & 1_{Gl} & 0 & -\pi_{GlL} \\ -\pi_{GwA} & 0 & 0 & 0 & 0 & -\pi_{GwGl} & 1_{Gw} & 0 \\ -\pi_{LA} & 0 & 0 & 0 & 0 & 0 & -\pi_{LGw} & 1_L \end{pmatrix}. \quad (15)$$

Figure 2 shows the system graph for P8 useful in summarizing symbolic AUC structure. Nondefault independent parameters are given in Table VII. Dependent parameters necessary to construct symbolic AUCs are given in Table VIII.

Table VII: P8 nondefault independent parameters.

| $Q_{FA} = 50$ | $Q_{MA} = 250$ | $Q_{LA} = 50$ | $Q_{KA} = 75$ |
|---|---|---|---|
| $Q_{GwA} = 75$ | $Q_{GwGl} = 0.2$ | $Q_{UK} = 0.075$ | $Q_{GlL} = 0.0625$ |
| $V_A = 5.5$ | $V_F = 15.0$ | $V_M = 44.5$ | $V_L = 2.2$ |
| $V_K = 0.3$ | $V_{Gw} = 1.0$ | $V_U = 0.5$ | $V = 1.0$ |
| $R_{AF} = 100$ | $R_{AM} = 10$ | $R_{AL} = 2$ | |
| $CL_{eL} = 12.5$ | $CL_{eGl} = 0.1$ | $CL_{eU} = Q_{UK}$ | $Q_{AL} = Q_{LA} + Q_{GwA}$ |

Table VIII: P8 dependent parameters.

| $\pi_{AF} = 1.000000$ | $\pi_{AK} = 0.999001$ | $\pi_{AL} = 0.832639$ | $\pi_{LA} = 0.1000000$ |
|---|---|---|---|
| $\pi_{GlL}=8.3264\text{E-}4$ | $\pi_{GwA} = 0.150000$ | $\pi_{GwGl} = 0.666667$ | $\pi_{AM} = 1.000000$ |
| $\pi_{LGw} = 1.000000$ | $\pi_{MA} = 0.500000$ | $\pi_{UK} = 0.000999$ | $\pi_{FA} = 0.100000$ |
| $\pi_{KA} = 0.150000$ | $CL_{tA} = 500$ | $CL_{tF} = 0.5$ | $CL_{tL} = 75.0625$ |
| $CL_{tM} = 25$ | $CL_{tK} = 75.075$ | $CL_{tU} = 0.075$ | $CL_{tGl} = 0.3$ |
| $CL_{tGw} = 75$ | | | |

Table IX: P8 $AUC_{iA}$ for bolus Arterial input, $A_{A0} = 0.1$.

| i | Approx. | Exact | RelErr. | Cond. | Mult. | Crude |
|---|---|---|---|---|---|---|
| A | 0.00477619 | 0.00477619 | -2.9249E-7 | $1.000000_1$ | 23.8810 | 0.00020000 |
| F | 0.47761900 | 0.47761900 | 1.8719E-7 | $0.100000_1$ | 23.8810 | 0.20000000 |
| M | 0.04776190 | 0.04776190 | 1.5599E-7 | $0.500000_1$ | 23.8810 | 0.00400000 |
| K | 0.00477142 | 0.00477142 | 1.9519E-7 | $0.150000_1$ | 23.8810 | 0.00133200 |
| U | 0.00477142 | 0.00477142 | 5.8556E-7 | $1.4985\text{E-}4_1$ | 23.8810 | 1.33333333 |
| L | 0.00795812 | 0.00795811 | 9.3623E-7 | $0.250000_2$ | 23.8942 | 0.00133222 |
| Gw | 0.00478062 | 0.00478061 | 5.8444E-7 | $0.150056_2$ | 23.8942 | 0.00133333 |
| Gl | 0.00165794 | 0.00165794 | -7.0217E-8 | $2.0816\text{E-}4_2$ | 23.8942 | 0.33333333 |

All P8 SISO AUC comparisons are given in the 7 Tables IX–XV. As with P4, the most salient feature is the low relative error in the numerical AUCs and will be discussed more fully later. P8 AUCs were derived by determinant expansions of the structural and adjoint matrices $(S_t, (S_{ta})_{ij})$ and reexpressed as the sum of toxic pathways, each the product of a fractional conductance, circulation multiplier and crude AUC. For example, liver dose from a GI lumen

Table X: P8 $AUC_{iF}$ for bolus Fat input, $A_{F0} = 0.1$.

| i | Approx. | Exact | Rel. Err. | Cond. | Mult. | Crude |
|---|---------|-------|-----------|-------|-------|-------|
| A | 0.00477619 | 0.00477619 | 4.8748E-7 | $1.000000_1$ | 23.88100 | 0.00020000 |
| F | 0.67761900 | 0.67761900 | 5.2777E-7 | $1.000000_1$ | 3.38809 | 0.20000000 |
| M | 0.04776190 | 0.04776190 | 4.6798E-7 | $0.500000_1$ | 23.88100 | 0.00400000 |
| K | 0.00477142 | 0.00477142 | -6.8316E-7 | $0.150000_1$ | 23.88100 | 0.00133200 |
| U | 0.00477142 | 0.00477142 | -3.9038E-7 | $1.4985\text{E-}4_1$ | 23.88100 | 1.33333333 |
| L | 0.00795810 | 0.00795811 | -9.3623E-7 | $0.250000_1$ | 23.89420 | 0.00133222 |
| Gw | 0.00478061 | 0.00478061 | -8.7666E-7 | $0.150056_1$ | 23.89420 | 0.00133333 |
| Gl | 0.00165794 | 0.00165794 | 6.3195E-7 | $2.0816\text{E-}4_1$ | 23.89420 | 0.33333333 |

Table XI: P8 $AUC_{iM}$ for bolus Muscle input, $A_{M0} = 0.1$.

| i | Approx. | Exact | RelErr. | Cond. | Mult. | Crude |
|---|---------|-------|---------|-------|-------|-------|
| A | 0.00477620 | 0.00477619 | 1.0725E-6 | $1.000000_1$ | 23.8810 | 0.00020000 |
| F | 0.47761900 | 0.47761900 | 1.2480E-7 | $0.100000_1$ | 23.8810 | 0.20000000 |
| M | 0.05176200 | 0.05176190 | 1.0795E-6 | $1.000000_1$ | 12.9405 | 0.00400000 |
| K | 0.00477142 | 0.00477142 | 8.7834E-7 | $0.150000_1$ | 23.8810 | 0.00133200 |
| U | 0.00477142 | 0.00477142 | 7.8075E-7 | $1.4985\text{E-}4_1$ | 23.8810 | 1.33333333 |
| L | 0.00795811 | 0.00795811 | 7.0217E-7 | $0.250000_2$ | 23.8942 | 0.00133222 |
| Gw | 0.00478062 | 0.00478061 | 5.8444E-7 | $0.150056_2$ | 23.8942 | 0.00133333 |
| Gl | 0.00165794 | 0.00165794 | 1.1937E-6 | $2.0816\text{E-}4_2$ | 23.8942 | 0.33333333 |

Table XII: P8 $AUC_{iK}$ for bolus Kidney input, $A_{K0} = 0.1$.

| i | Approx. | Exact | RelErr. | Cond. | Mult. | Crude |
|---|---------|-------|---------|-------|-------|-------|
| A | 0.00477142 | 0.00477142 | 4.8797E-7 | $0.9990010_1$ | 23.88100 | 0.00020000 |
| F | 0.47714200 | 0.47714200 | 4.9968E-7 | $0.0999001_1$ | 23.88100 | 0.20000000 |
| M | 0.04771420 | 0.04771420 | 8.5883E-7 | $0.4995010_1$ | 23.88100 | 0.00400000 |
| K | 0.00609866 | 0.00609865 | 1.4507E-6 | $1.0000000_1$ | 4.57856 | 0.00133200 |
| U | 0.00609866 | 0.00609865 | 6.8719E-7 | $9.9900\text{E-}4_1$ | 4.57856 | 1.33333333 |
| L | 0.00795016 | 0.00795016 | 7.0287E-7 | $0.2497500_2$ | 23.89420 | 0.00133222 |
| Gw | 0.00477584 | 0.00477584 | 0.0 | $0.1499060_2$ | 23.89420 | 0.00133333 |
| Gl | 0.00165628 | 0.00165628 | 8.4345E-7 | $2.0795\text{E-}4_2$ | 23.89420 | 0.33333333 |

input is given in Table XV by

$$AUC_{LGl}^{(8)} = \frac{\left(S_{ta}^{(8)}\right)_{LGl}}{|S_t|} \frac{A_{Gl0}}{CL_{tL}} = \pi_{LGw}\pi_{GwGl}\frac{|S_t - LGwGl|}{|S_t|}\frac{A_{Gl0}}{CL_{tL}} = .00531, \quad (16)$$

$$|S_t| = \left(\begin{array}{c} |LGwGl| - \pi_{AF}\pi_{FA}|LGwGl| - \pi_{AM}\pi_{MA}|LGwGl| \\ -\pi_{AK}\pi_{KA}|LGwGl| - \pi_{AL}\pi_{LA} - \pi_{AL}\pi_{LGw}\pi_{GwA} \end{array}\right) = .04185 \quad (17)$$

where $\pi_{LGw}\pi_{GwGl} = .66\bar{6}$ denotes the fractional conductance; $|S_t - LGwGl| / |S_t|$
$= 1 - \pi_{AF}\pi_{FA} - \pi_{AM}\pi_{MA} - \pi_{AK}\pi_{KA}/|S_t| = .25/.04185 = 5.98$ the circula-
tion multiplier; and $A_{Gl0}/CL_{tL} = .00133$ the crude AUC. Also, $|LGwGl| =$
$1 - \pi_{LGw}\pi_{GwGl}\pi_{GlL} = .000555$ is the $3 \times 3$ subdeterminant of $S_t$ generated by
the (L,Gw,Gl) row-column intersections, and $1/|S_t| = 23.9$ is the multiplier for
organs outside the GI tract, liver system.

A liver dose, blood input, dual pathway example is given in Table IX by

$$AUC_{LA}^{(8)} = \frac{\pi_{LA}|S_t - LA| + \pi_{LGw}\pi_{GwA}|S_t - LGwA|}{|S_t|}\frac{A_{A0}}{CL_{tL}}$$

$$= \frac{\pi_{LA} + \pi_{LGw}\pi_{GwA}}{|S_t|} \frac{A_{A0}}{CL_{tL}} = .00796 \tag{18}$$

since $|S_t - LA| = |S_t - LGwA| = 1$. When multipliers are equal for dual pathways, conductances are added together and indicated by a subindex 2 in the conductance column (a subindex 1 indicates a single path). In Table XIII, the dual path GI-wall AUC, from a liver input, is

$$AUC_{GwL}^{(8)} = \frac{\pi_{GwA}\pi_{AL}|S_t - LGwA| + \pi_{GwGl}\pi_{GlL}|S_t - LGwGl|}{|S_t|} \frac{A_{L0}}{CL_{tGw}},$$

but $|S_t - LGwA| \neq |S_t - LGwGl|$, so multipliers and conductances are separated.

Table XIII: P8 $AUC_{iL}$ for bolus Liver input, $A_{L0} = 0.1$.

| i | Approx. | Exact | RelErr. | Cond. | Mult. | Crude |
|---|---------|-------|---------|-------|-------|-------|
| A | 0.00397906 | 0.00397905 | 1.0533E-6 | $0.8326390_1$ | 23.894200 | 0.00020000 |
| F | 0.39790600 | 0.39790500 | 3.7449E-7 | $0.0832639_1$ | 23.894200 | 0.20000000 |
| M | 0.03979060 | 0.03979050 | 1.1235E-6 | $0.4163200_1$ | 23.894200 | 0.00400000 |
| K | 0.00397508 | 0.00397508 | 7.0287E-7 | $0.1248960_1$ | 23.894200 | 0.00133200 |
| U | 0.00397508 | 0.00397508 | 4.6858E-7 | $1.2477\text{E-}4_1$ | 23.894200 | 1.33333333 |
| L | 0.00796288 | 0.00796288 | 2.3392E-7 | $1.0000000_1$ | 5.977130 | 0.00133222 |
| Gw | 0.00398348 | 0.00398348 | 9.3519E-7 | $0.1248960_1$ | 23.894200 | 0.00133333 |
|  |  |  |  | $5.5509\text{E-}4_1$ | 5.977130 |  |
| Gl | 0.00165893 | 0.00165893 | 1.4035E-6 | $8.3264\text{E-}4_1$ | 5.977130 | 0.33333333 |

Table XIV: P8 $AUC_{iGw}$ for bolus Gut wall input, $A_{Gw0} = 0.1$.

| i | Approx. | Exact | RelErr. | Cond. | Mult. | Crude |
|---|---------|-------|---------|-------|-------|-------|
| A | 0.00397906 | 0.00397905 | 8.1920E-7 | $0.8326390_1$ | 23.89420 | 0.00020000 |
| F | 0.39790500 | 0.39790500 | -2.2469E-7 | $0.0832639_1$ | 23.89420 | 0.20000000 |
| M | 0.03979060 | 0.03979050 | 4.6811E-7 | $0.4163200_1$ | 23.89420 | 0.00400000 |
| K | 0.00397508 | 0.00397508 | 3.5144E-7 | $0.1248960_1$ | 23.89420 | 0.00133200 |
| U | 0.00397508 | 0.00397508 | -3.5144E-7 | $1.2477\text{E-}4_1$ | 23.89420 | 1.33333333 |
| L | 0.00796288 | 0.00796288 | 9.3566E-7 | $1.0000000_1$ | 5.97713 | 0.00133222 |
| Gw | 0.00531681 | 0.00531681 | 7.8825E-7 | $1.0000000_1$ | 3.98761 | 0.00133333 |
| Gl | 0.00165893 | 0.00165893 | 7.0175E-8 | $8.3264\text{E-}4_1$ | 5.97713 | 0.33333333 |

Looking for comparative toxicological insights for common arterial and gut lumen input routes (Tables IX, XV), it is clear that the liver dose from a given arterial bolus input is greater than from the same-sized gut lumen input. However, if relative body burdens to some non-liver system organs must be minimized to avoid undesirable toxic side effects, a gut lumen entry route may still be optimal. For example, if the dose to the gut lumen were increased by 50%, then $AUC_{LGl}$ would be the same as $AUC_{LA}$ for the smaller baseline arterial input, but AUCs for the 5 non-liver system organs would be only 83% of that for the baseline arterial input route; while $AUC_{GwGl}$ would be 107%, and $AUC_{GlGl}$ 30300% of baseline. For many drug therapies this would provide a reasonable trade off and justify drug delivery by ingestion rather than injection. Note also that urinary AUCs vary somewhat depending on the route of input; this may assist in the planning and interpretation of urinary monitoring studies.

Table XV: P8 $AUC_{iGl}$ for bolus Gut lumen input, $A_{Gl0} = 0.1$.

| i | Approx. | Exact | RelErr. | Cond. | Mult. | Crude |
|---|---|---|---|---|---|---|
| A | 0.00265270 | 0.00265270 | 3.5108E-7 | $0.5550930_1$ | 23.89420 | 0.00020000 |
| F | 0.26527100 | 0.26527000 | 1.0111E-6 | $0.0555093_1$ | 23.89420 | 0.20000000 |
| M | 0.02652700 | 0.02652700 | 4.9152E-7 | $0.2775460_1$ | 23.89420 | 0.00400000 |
| K | 0.00265005 | 0.00265005 | 7.0287E-7 | $0.0832639_1$ | 23.89420 | 0.00133200 |
| U | 0.00265005 | 0.00265005 | 4.3929E-7 | $8.3181E\text{-}5_1$ | 23.89420 | 1.33333333 |
| L | 0.00530859 | 0.00530858 | 6.1403E-7 | $0.6666670_1$ | 5.97713 | 0.00133222 |
| Gw | 0.00354454 | 0.00354454 | 9.1962E-7 | $0.6666670_1$ | 3.98761 | 0.00133333 |
| Gl | 0.33443900 | 0.33443900 | 2.6733E-7 | $1.0000000_1$ | 1.00332 | 0.33333333 |

Table XVI: P8, blood bolus model behavior for listed parameters.

| i | RelErr. | RelErr. | RelErr. | RelErr. |
|---|---|---|---|---|
| $R_{AF}$ | 100 | 1000 | 10000 | 100000 |
| TSTOP | 2000 | 20000 | 200000 | 2000000 |
| A | -2.9249E-7 | 7.7997E-7 | 8.7747E-7 | -1.7198E-4 |
| F | 1.8719E-7 | -1.9967E-7 | -3.1948E-7 | -1.7169E-4 |
| M | 1.5599E-7 | -1.5599E-7 | -5.4598E-7 | -1.8010E-4 |
| K | 1.9519E-7 | 8.7834E-7 | 5.8556E-7 | -1.7372E-4 |
| U | 5.8556E-7 | 1.9519E-7 | 5.8556E-7 | -1.4951E-4 |
| L | 9.3623E-7 | -4.6811E-7 | 1.1703E-7 | -1.7999E-4 |
| Gw | 5.8444E-7 | 8.7666E-7 | 1.9481E-7 | -1.7777E-4 |
| Gl | -7.0217E-8 | 0.0 | -2.1065E-7 | -1.7610E-4 |

## Accuracy Effects of Model Perturbations - User Recovery Options

The behavior of SimuSolv's stiff ODE solvers is explored through parameter perturbations of P8's fat partition coefficient. This coefficient is preferred over other parameters because it's baseline value reflects the system's strongest sink compartment. Sinks give rise to stiff ODE systems which are difficult to solve numerically because toxicant mean residence time or half-life in the sink tends to be orders of magnitude longer than in other organs. Increasing fat partitioning exaggerates stiffness. Moreover, the fat partition coefficient tends to be a sensitive and highly variable AUC parameter.

The coefficient was sequentially perturbed from its baseline value in 10-fold increments for a given blood input, and AUC behavior described in Table XVI. Within the range of plausible large values ($10^3$) and somewhat beyond ($10^4$), the table shows that the maximum relative AUC error over all organs remains small and consistent with the unperturbed baseline case (RelErr=9.4E-7). To cover the longer residence times in fat, 10-fold increases in integration time (TSTOP) were used for each 10-fold increase in partitioning.

Integration performance, however, was compromised at extremely high partitioning values. For example, at $R_{AF} = 10^5$, Table XVI shows a parallel increase in integration time (TSTOP=2.0E+6) to be insufficient since maximum relative error balloons to 1.8E-4. This error increase may result from initial and minimum algorithm step sizes being proportional to integration times, eventually causing significant degradation in global numerical error as $R_{AF}$ increases. To recover these 3 lost digits, other SimuSolv error-control options were adjusted from the perturbed baseline case ($R_{AF} = 10^5$): (i) simulation run times (from

TSTOP=2.0E-6); (ii) number of communication intervals (from Com. Int. = 480 per run); (iii) absolute error tolerance (from 1.0E-6); (iv) relative error tolerance (from 1.0E-4); (v) computation precision (from single to double); (vi) stiff integration algorithm (from Gear to LSODE, a modified-Gear algorithm). Table XVII lists the best accuracy obtained by modifying single error-control options. 'Best' here means the lowest obtained maximum relative AUC error over all 8 target sites (the minimax criterion). The results were obtained by increasing or decreasing the option value until results no longer improved - starting in most cases with 10-fold changes in the default option. To achieve near-optimal values, ever smaller option increments or decrements were tried using a variation of the method of bisection. In practice, this user-controlled option adjustment process was tedious since many intermediate adjustments resulted in accuracy stagnation, degradation, or in long integration times. We summarize these recovery results in Tables XVII, XVIII and discuss error-control strategies for the perturbed P8 model.

In phase one, single error-control options were adjusted from baseline perturbed case values until no further improvement could be achieved. After resetting each option back to its perturbed baseline value, the next option was similarly optimized. First, lengthening integration time beyond TSTOP = 2.0E+6 decreased the maximum relative error by 1 digit to 1.5E-5. Second, increasing the number of reporting intervals by 10-fold to 4800 also decreased the maximum relative error 1 digit to about $10^{-5}$. Third, decreasing the relative error criterion to $10^{-7}$ decreased error only 1/2 digit to 4.4E-5. Fourth, decreasing the absolute error tolerance to 1.5E-8 reduced error over 2 digits to 1.0E-6. Fifth, changing from single to double precision reduced error 1/2 digit to 6.9E-5. Sixth, changing from Gear to LSODE, which must be run in double precision in SimuSolv, reduced error over 2 1/2 digits to 1.17E-6. Tightening up the absolute error tolerance or switching from single precision Gear to double precision LSODE appears to produce maximum error reduction. For other exposure scenarios or models, different adjustments may be optimal (e.g., steady state computations may require tightening relative error).

Integration accuracy was further restored using several error-control user options in conjunction. The best performance results are shown in Table XVIII for case (a) where the maximum relative error is <6.0E-7 using 3 error-control option changes (TSTOP = 2.5E+6 , Com. Int. = 48000 and double precision Gear); and for case (b) with smaller error of <2.0E-7 using 3 other option changes (double precision LSODE and relative error tolerance = 1.0E-7). In summary, it appears that baseline model accuracy of highly perturbed models can be recovered or even exceeded (as above) via extensive, judicious, but rather tedious trial-and-error option selection and refinement strategies. However, the task is not simple and results not always predictable.

The reason why the numerical solvers break down at a fat partitioning near $10^5$ is not clear, Moreover, several unexpected, or even erratic, results were also noted while experimenting with the available error-control options. These consisted of sudden inexplicable accuracy degradations after a string of successive accuracy improvements. For example, successively longer simulation times, (increases in TSTOP), sometimes initially improved accuracy, but later degraded accuracy. In another case, successively lower absolute error settings initially increased accuracy, but later degraded accuracy. Perhaps more interesting, the perturbed baseline case simulation time frequently exceeded that for many option adjustments yielding lower minimax error. Indeed, for option changes in

Table XVII: Best achieved results for listed parameters.

| i | RelErr. | RelErr. | RelErr. | RelErr. | RelErr. | RelErr. |
|---|---|---|---|---|---|---|
| | TSTOP 3000000 | Com. int. 4800 | Rel. Er. 1.0E-7 | Abs. Er. 1.5E-8 | Dbl. Prec. | LSODE |
| A | -5.8498E-7 | -9.0672E-6 | 4.1728E-5 | -9.7496E-8 | -6.5323E-5 | 1.0725E-6 |
| F | 2.6197E-6 | -4.2810E-6 | 4.3832E-5 | -1.9169E-7 | -6.5556E-5 | 1.1501E-6 |
| M | 1.4897E-5 | -5.6158E-6 | 3.8843E-5 | 3.1199E-7 | -6.8871E-5 | 1.0140E-6 |
| K | 0.0 | -1.0735E-5 | 4.0111E-5 | 9.7594E-8 | -6.5388E-5 | 1.0735E-6 |
| U | 2.9278E-7 | -8.1979E-6 | 4.0013E-5 | 2.9278E-7 | -6.3241E-5 | 1.1711E-6 |
| L | 2.1065E-6 | -7.6068E-6 | 4.0141E-5 | -1.0533E-6 | -6.7291E-5 | 1.1703E-6 |
| Gw | -8.7666E-7 | -9.7406E-6 | 3.9742E-5 | -9.7406E-8 | -6.5554E-5 | 1.1689E-6 |
| Gl | 2.8087E-6 | -5.3365E-6 | 4.1849E-5 | -2.1065E-7 | -6.7830E-5 | 1.1235E-6 |

Table XVII, run time was significantly less than for the perturbed baseline case. This indicates that near-optimal error-controls should be expected to achieve both lower error and shorter simulation times. Other anomalies, such as fewer communication intervals sometimes providing lower errors, are difficult to explain without a deeper understanding of Gear-type algorithms.

One lesson to be learned is that without exact linear symbolic solutions, numerical accuracy degradations tend to go unnoticed, much less corrected, by the modeler. While minor accuracy degradations can be tolerated for many regulatory applications (e.g., 1-2 significant digits may be sufficient given other sources of risk assessment error), it is still discomforting to have accuracy degradations in the 3rd-5th digits. Even modelers with significant numerical or specific model experience, can easily fail to detect 3rd digit accuracy errors. Since finding optimal error-control options is tedious for one model, and more so for model changes, the availability of simple exact symbolic solutions is important.

Table XVIII: Resultant values for combination pertubation corrections.

| i | (a) RelErr. | (b) RelErr. | i | (a) RelErr. | (b) RelErr |
|---|---|---|---|---|---|
| A | -8.7747E-7 | -1.9499E-7 | U | -6.8316E-7 | -9.7594E-8 |
| F | -7.6674E-7 | -6.3895E-8 | L | -8.1920E-7 | -1.1703E-7 |
| M | -8.5797E-7 | -1.5599E-7 | Gw | -7.7925E-7 | -9.7406E-8 |
| K | -7.8075E-7 | -1.9519E-7 | Gl | -8.4260E-7 | -1.4043E-7 |

## Summary Discussion and Future Directions

It has been demonstrated that within realistic physiological parameter ranges for P4 and P8, SimuSolv's Gear and LSODE numerical algorithm performance is good, supporting their continued use as general purpose PBPK simulation tools. On the other hand, symbolic analytic AUC solutions for linear PBPK models (and some nonlinear models in their low concentration linear-limit range) were found to yield simple quantitative checks on the accuracy of conventional numerical solvers; thus they provide critical quality assurance and validation tools previously unavailable for physiological models and only rarely used used with smaller nonphysiological models. Symbolic AUC formulas also provide unique qualitative summarizations for exposure analysis, dose-response prediction, and, potentially, risk characterization - risk assessment tasks not readily accomplished using conventional numerical simulators alone. Furthermore, symbolic solutions appear to be physiologically insightful, mathematically exact, and computationally efficient and robust. It seems clear that symbolic analytic solutions provide diagnostic capabilities complementing and enhancing results from general purpose solvers of linear or linear-limit PBPK models.

Future work on linear PBPK models may include: (i) graph theoretical approaches to the description and proof of symbolic AUC decomposition; (ii) elucidation of the physiological and mathematical significance of system subdeterminants; (iii) and further validation of general ODE solver results against symbolic results.

**Notice.** The U.S. Environmental Protection Agency's Office of Research and Development performed this research. It has been subjected to Agency peer review and approved as an EPA publication.

# References

[1] Brown,R.N. In *Biomarkers of Human Exposure to Pesticides*; Editors, M.A. Saleh, C.H. Nauman, J.N. Blancato; ACS Symposium Series, Vol. 542; American Chemical Society; Washington, D.C., 1994; pp. 301-317.

[2] Dow Chemical Co. *SimuSolv Modeling and Simulation Software*; Midland, MI, 1990.

[3] Gear, C.W. *Numerical Initial Value Problems in Ordinary Differential Equations;* Prentice-Hall: Englewood Cliffs, NJ, 1971; pp.1-253.

[4] Gibaldi, M; Perrier, D. *Pharmacokinetics, 2nd Ed.*; Marcel Dekker: New York, NY, 1982; pp 1-494.

INDEXES

# Author Index

Abdel Rahman, Fawzia, 49,106
Abou Zeid, Mohamed, 106
Afify, Abdel Moneim, 114
Al-Bayati, Mohammed A., 206
Anderson, Jack W., 150
Bailey, Sandra L., 39
Baker, Sam, 39
Bieber, L. L., 140
Blancato, Jerry N., 206,256
Bothner, Kristen, 150
Brown, Robert N., 242,256
Chester, N. A., 191
Crofton, K., 70
Dary, Curtis C., 2
Davis, J. A., 169
Dong, Michael H., 229
Edelman, David, 150
Ehrich, Marion, 79
El-Baroty, Gamal, 114
El-Sebae, Abdel Khalek H., 49,114
Esteban, Emilio, 39
Famini, G. R., 191
Fry, D. M., 169
Gage, D. A., 140
Garst, John E., 126
Hagstrom, Blaine L., 256
Haley, M. V., 191
Head, Susan L., 39
Hern, Stephen C., 2
Hill, Robert H., Jr., 39

Hodge, Vernon, 180
Huang, Z. H., 140
Johannesson, Kevin, 180
Kamel, Alaa, 49,114
Knaak, James B., 206
Krieger, Robert I., 229
Kurnas, C. W., 191
Lassiter, L., 70
Mohamed, Zaher A., 49,106
Moser, V. C., 70
Nauman, Charles H., 2
Needham, Larry L., 24,39
Padilla, S., 70
Quackenboss, James J., 2
Raabe, Otto G., 206
Ragab, Awad, 114
Ross, John H., 229
Rubin, Carol, 39
Saleh, Mahmoud Abbas, 106,114
Seifert, Josef, 94
Shealy, Dana B., 39
Sterling, P. A., 191
Stetzenbach, Klaus, 180
Thongsinthusak, Thomas, 229
Tukey, Robert H., 150
Vincent, Steven, 150
Vu, Tien P., 150
Wilson, B. W., 169
Wilson, L. Y., 191

# Affiliation Index

California Department of Pesticide
   Regulation, 229
Columbia Analytical Services, 150
La Sierra University, 191
MEC Analytical Systems, Inc., 150
Michigan State University, 140
Occidental Chemical Corporation, 206

Texas Southern University, 49,106,114
University of California—Davis,
   169,206,229
University of California—San Diego, 150
University of Hawaii, 94
University of Nevada, 180
University of North Carolina, 70

U.S. Army Edgewood Research, Development and Engineering Center, 191
U.S. Department of Health and Human Services, 24,39
U.S. Environmental Protection Agency, 2,70,206,242,256
Virginia Polytechnic Institute and State University, 79

# Subject Index

**A**

Absorbed malathion doses in humans, spot urine sample results in physiologically based pharmacokinetic modeling, 229–239
Acetylate phenotype, biomarkers for health risk assessment, 53–54
Acetylcholinesterase, function, 70
Acetylcholinesterase inhibition, type of organophosphorus-induced neurotoxicity, 81
Acute bioassays, aquatic toxicity of chemical agent simulants, 191–203
Adverse health effect biomarkers, 60
Age, biomarkers for health risk assessment, 54
Agrochemicals, biomarkers for health risk assessment in exposed susceptible individuals, 49–62
Alkoxycoumarin substrates, use in fluorescence-based catalytic assays for cytochrome P450 induction measurement in birds, 172,174
Alkoxyresorufin substrates, use in fluorescence-based catalytic assays for cytochrome P450 induction measurement in birds, 172–175t
Aquatic toxicity of chemical agent simulants
experimental description, 192,195–197t
results, 198,199t
theoretical linear solvation energy relationship model, 193–195
TLSER vs. microtox assay, 198,201t,203
TOPKAT model, 192–193
TOPKAT vs. QSAR, 198,200t,202–203

Area under curves, See Symbolic, numerical area under curves of small linear physiologically based pharmacokinetic models
Arylhydrocarbon hydroxylase, biomarkers for health risk assessment, 54
Assessment of risk, See Risk assessment

**B**

Bioassays, aquatic toxicity of chemical agent simulants, 191–203
Biological marker, See Biomarkers
Biological monitoring of pesticide residues and metabolites, 107
Biologically effective dose, 25–27
Biomarkers
applications, 6–7
blood cholinesterase activity, 70–76
breast milk, 114–123
carnitine and its esters, 126–136,140–146
categories, 229
cytochrome P450 in birds, 169–176
definition, 6,86,229
environmental health, 11–15
exposure, See Exposure biomarkers
exposure, use in environmental legislation, 24–37
health risk assessment in susceptible individuals exposure to agrochemicals
education constraints, 61–62
epidemiological biomarkers, 57–60
extrinsic factors for susceptibility variation to toxic chemicals, 55–57
importance, 50
intrinsic factors for susceptibility variation to toxic chemicals, 53–55

Biomarkers—*Continued*
  health risk assessment in susceptible
    individuals exposure to
    agrochemicals—*Continued*
    regulatory and management
      constraints, 62
    socioeconomic status constraint, 61
  human 101L cells for risk assessment of
    environmental samples, 150–167
  human exposure, 7–8,10–11
  neurotoxic esterase inhibition, 79–89
  relationship to risk assessment and risk
    management, 2–18
  serum protein profile for insecticides,
    106–112
  trace elements in striped bass, 180–190
  urinary excretion of xanthurenic acid
    for organophosphorus insecticide
    exposure, 94–104
  use of carnitine as environmental-
    toxicological exposure to nongenotoxic
    tumorigens, 126–135
Biomonitoring, function, 39
Biomonitoring data applications
  biomarkers
    environmental health, 11–15
    human exposure, 7–8,10–11
    mathematical modeling on risk
      assessment, 13,16
    regulation, 7,9
Birds, fluorescence-based catalytic assays
  for cytochrome P450 induction
  measurement, 169–176
Blood, source of biomarkers, 106–107
Blood cholinesterase, use as biomarker, 8
Blood cholinesterase activity, 71–76
Blood type, biomarkers for health risk
  assessment, 50–53
Breast milk as marker for human exposure
  to environmental pollutants
  chlorinated insecticide levels
    1,1'-(dichloroethenylidene)bis(4-
      chlorobenzene), 118,121–122$t$
    hexachlorocyclohexanes, 121$t$,123
    1,1,1-trichloro-2,2-bis($p$-chlorophenyl)-
      ethane, 118,121$t$,123
  experimental procedure, 115–117
  lead concentration, 117–120$t$

C

Carbamate insecticides, inhibition of
  cholinesterase activity, 70–76
Carnitine
  biosynthesis, 126
  detoxicant of nonmetabolizable
    acyl-coenzyme As, 141–146
  roles in intermediary metabolism, 140
  structure, 126
Carnitine acyltransferases, 140–146
Carnitine and esters
  antiapoptotic properties, 132
  antioxidant properties, 131
  biomarkers of environmental-
    toxicological exposure, 134–135
  detoxification metabolism, 128,130
  DNA transcription effect, 133–134
  mitochondrial and peroxisomal
    metabolism, 128,129$f$
  nucleic acid transcription, 133
  protection against toxic substances, 127
  protein kinase C potentiation, 132
  protein translation effect, 132–133
  signal and regulatory roles, 130–131
Catalytic assays for cytochrome P450
  induction measurement in birds,
  fluorescence based, 171–175
Chemical agent simulants
  aquatic toxicity, 191–203
  toxicity assay methods, 191–192
  toxicity prediction, 13
Chlorinated insecticides
  breast milk as marker for exposure,
    118,121–123
  serum protein profile as biomarker, 106–112
Chlorpyrifos, environmental exposure, 46
Cholinesterase activity, blood, 71–76
Cholinesterase-inhibiting insecticides,
  70–71
Cytochrome P450 enzymes, biomarkers of
  pollutant exposure in wild animals, 169
Cytochrome P450 induction measurement
  in birds, fluorescence-based catalytic
  assays, 169–176
Cytochrome P450 reporter gene system,
  risk assessment of environmental
  samples, 150–167

D

*Daphnia magna* bioassay, aquatic toxicity of chemical agent simulants, 191–203
Detoxicant, carnitine for nonmetabolizable acyl-coenzyme As, 140–146
Development, biomarkers for health risk assessment, 54
Diazinon, role in urinary excretion of xanthurenic acid as biomarker of organophosphorus insecticide exposure, 98,101–104
*p*-Dichlorobenzene, environmental exposure, 46
1,1'-Dichloroethenylidene)bis(4-chlorobenzene), breast milk as marker for exposure, 118,121–122*t*
Digestive organs of oceanic fish, highest concentrations of natural radioactivity, 180
Dioxin
  human 101L cells as biomarker for risk assessment, 150–167
  human exposure biomarker use in environmental legislation, 28–29
DNA transcription, role of carnitine and esters, 133–134
Dose–response determination, 2


E

Ecological health outcome, 16
Endogenous response biomarkers, 12
Endosulfan, serum protein profile as biomarker, 106–112
Endrin, serum protein profile as biomarker, 106–112
Environmental approach for exposure assessment, 25–26
Environmental health paradigm
  activity patterns, 5
  applications, 16–18
  bioavailability, 5–6
  biomonitoring, 7–16
  body burden, 6–7
  challenge, 17–18
  environmental concentrations, 3,5

Environmental health paradigm—*Continued*
  exposure conditions, 5
  exposure routes, 5
  schematic representation, 3–4
  sources, 3
Environmental health studies, reference range concentration use, 39–47
Environmental legislation, human exposure biomarkers, 24–37
Environmental pollutants, breast milk as marker for exposure, 114–123
Environmental samples, human 101L cells as biomarker for risk assessment, 150–167
Environmental sampling, exposure index determination, 7–8
Environmental–toxicological exposure to nongenotoxic tumorigens, carnitine and esters as biomarkers, 126–135
Epidemiological biomarkers, types, 5
Ethoxyresorufin-*o*-deethylase assay, field application for cytochrome P450 induction measurement in birds, 169–176
Experimental design, symbolic solutions of small linear physiologically based pharmacokinetic models, 242–254
Exposure biomarkers
  description, 57–60
  time effectiveness, 106
  use in environmental legislation, *See* Human exposure biomarker use in environmental legislation
Exposure index
  basis, 25
  determination, 7–8,10–11
  disadvantages, 26
External biomarker, 11–12


F

Fast protein liquid chromatography
  blood contaminant detection, 10–11
  monitoring of serum protein profile as biomarker of insecticide exposure, 106–112
Federal Insecticide, Fungicide, and Rodenticide Act, 7

Fenvalerate, serum protein profile as biomarker, 106–112

Field applications, fluorescence-based catalytic assays for cytochrome P450 induction measurement in birds, fluorescence-based catalytic assays for induction measurement, 169–176

Fluorescence-based catalytic assay biomarkers of environmental health, 12 case study of wild avian embryos, 175–176
cytochrome P450 enzyme, 170–171
cytochrome P450 induction in birds, 169–176

Furan, human 101L cells as biomarker for risk assessment, 150–167

**G**

Genetics, biomarkers for health risk assessment, 50–53

Glucose-6-phosphate dehydrogenase deficiency, biomarkers for health risk assessment, 53

Good laboratory practice standards final rule, function, 17

**H**

Health outcome, influencing factors, 16–17

Health risk assessment in susceptible individuals, 49–62

Health studies, reference range concentration use, 39–47

Heptachlor, serum protein profile as biomarker, 106–112

Hexachlorocyclohexanes, breast milk as marker for exposure, 121$t$,123

Hormone disruption, biomarkers for health risk assessment, 54–55

Human(s)
exposure to chemicals in commercial procedure, 25
spot urine sample results in physiologically based pharmacokinetic modeling of absorbed malathion doses, 229–239

Human 101L cells as biomarker for risk assessment of environmental samples
advantages, 167
environmental samples, 166–167
experimental objectives, 152
mode of action, 151
responses
extracts of environmental samples, 157,160–164
standards, 153,157–159$f$
test substances, 153–156$t$
test system, 152–153
toxic equivalency, 162,165$t$

Human exposure
biomarkers, 7–8,10–11
to environmental pollutants, breast milk as marker, 114–123
to xenobiotics, assessment methods, 25

Human exposure biomarker use in environmental legislation
analytical methods, 33–34
assessment hierarchy, 35
collecting and banking of human specimens, 31–32
dioxin, 28–29
experimental description, 25
exposure index, 25–26
lead, 30
measurement currently performed, 35–37
prioritizing chemically, 32–33
types
biologically effective dose, 26–27
effect biomarkers, 27
internal dose, 27
response biomarkers, 27
susceptibility biomarkers, 27
volatile organic compounds, 30–31

Human health outcome, 16

Human lymphocyte antigens, biomarkers for health risk assessment, 53

Human milk, importance to newborns, 115

**I**

Immune dysfunctions, biomarkers for health risk assessment, 58–60

Immunoassays, analysis of human
exposure, 34
Immunodepression, role of pesticides, 50
Inductively coupled plasma–mass
spectrometric determination of trace
elements in organs of striped bass from
Lake Mead
element concentrations, 182–187t
experimental description, 180–181
rare earth element concentrations,
182–183,188–189
Infectious disease distribution, variation
in background, 55,56t
Insecticides
inhibition of cholinesterase activity, 70–76
serum protein profile as biomarker,
106–112
urinary excretion of xanthurenic acid as
biomarker of exposure, 94–104
Internal dose
advantages, 40
definition, 25
examples, 27
information requirements, 27
measures, 25
Intravenous administration, mass
balance, 211
Isofenphos
mathematical modeling in risk
assessment, 13,16
multipathway, multiroute physiologically
based pharmacokinetic and
pharmacodynamic model for toxicity
prediction, 207–227
*I*-Isopropylisofenphos oxon, mass
balance, 212–214

K

α-Keto acid metabolism, role in
L-carnitine as detoxicant of
nonmetabolizable acyl-coenzyme As,
141,144–145
L-Kynurenine pathway, L-tryptophan
metabolism, 95,96f

L

Lake Mead, inductively coupled plasma–
MS determination of trace elements in
organs of striped bass, 180–189
Lead
breast milk as marker for exposure,
117–120t
human exposure biomarker use in
environmental legislation, 30
Lindane, serum protein profile as
biomarker, 106–112
Liver kynurenine formamidase,
organophosphorus insecticide effect,
95,98–99

M

Malathion doses in humans, spot urine
sample results in physiologically
based pharmacokinetic modeling,
229–239
Management constraints, biomarkers for
health risk assessment, 62
Management of risk, *See* Risk
management
Mass spectrometric immunoassay, analysis
of human exposure, 34
Mass spectrometric methods, analysis of
human exposure, 34
Mass spectrometry–inductively coupled
plasma determination of trace elements
in organs of striped bass from Lake
Mead, *See* Inductively coupled
plasma–mass spectrometric
determination of trace elements in
organs of striped bass from Lake Mead
Mathematical modeling, risk assessment,
13,16
Metabolism, mass balance, 211
Methyl parathion, 40–46
Methyl salicylate, aquatic toxicity,
191–203
Milk, *See* Breast milk as marker for human
exposure to environmental pollutants

Multipathway, multiroute physiologically based pharmacokinetic and pharmacodynamic model for toxicity prediction
absorption, 222,224*t*
applications, 206–207
elimination, 223,224*t*
experimental conditions, 207,209*t*
inhibition
acetylcholinesterase and butyrylcholinesterase, 220,222–227
carboxylesterase, 220,222*t*
mass balance equations, 207–208,211–214
metabolic pathways, 214–219
metabolic rate constants
carboxylases, 219–220
cytochrome P450, 219,221–222*t*,226
metabolism, 223,224*t*
model, 207
organophosphate hydrolases, 219
output, 223,224*t*
partition coefficients, 214
physiological parameters, 214

**N**

National Health and Nutrition Examination Survey, 17–18,23
National Human Adipose Tissue Survey, 32
National Human Exposure Assessment Survey, 17–18,32
National Human Monitoring Program, 32
Neuropathy, organophosphorus-induced neuropathy, prediction of potential by neurotoxic esterase inhibition, 79–88
Neuropathy target esterase
function, 10
inhibition by organophosphorus compounds, 80
Neurotoxic esterase inhibition, 81–85
*p*-Nitrophenol, measurement using reference range concentrations, 41–46
Nongenotoxic tumorigens, carnitine and esters as biomarkers of environmental-toxicological exposure, 126–135
Nonmetabolizable acyl-coenzyme As, carnitine as detoxicant, 140–146

Nucleic acid transcription, role of carnitine and esters, 133
Numerical area under curves of small linear physiologically based pharmacokinetic models, symbolic, *See* Symbolic, numerical area under curves of small linear physiologically based pharmacokinetic models
Nutrition, biomarkers for health risk assessment, 54–55

**O**

Organophosphorus compounds, 79
Organophosphorus-induced delayed neuropathy
detection, 10
evaluation, 80–81
prediction of potential by neurotoxic esterase inhibition, 79–88
type of organophosphorus-induced neurotoxicity, 81
Organophosphorus-induced neurotoxicity, evaluation, 80–81
Organophosphorus insecticides
inhibition of cholinesterase activity, 70–76
mechanism of action, 94
urinary excretion of xanthurenic acid as biomarker of exposure, 94–104

**P**

Palmitoyl-coenzyme A role, 145–146
Percutaneous absorption, mass balance, 210
Peroxisomal proliferating agent induced tumorigenicity, role of carnitine and esters, 134–135
Pesticide(s), role in immunodepression, 50
*Pesticide Assessment Guidelines*, 7
Pharmacokinetic modeling of absorbed malathion doses in humans, spot urines, 229–239
*Photobacterium phosphoreum* bioassay, aquatic toxicity of chemical agent simulants, 191–203

Physiologically based pharmacokinetic and
   pharmacodynamic (PBPK/PD) models,
   toxicity prediction, 206–227
Physiologically based pharmacokinetic
   (PBPK) models
 absorbed malathion doses in humans,
   229–239
 applications, 230,256
 description, 243
 mathematical modeling in risk
   assessment, 13,16
 symbolic numerical area under curves,
   256–269
 symbolic solutions in risk assessment
   and experimental design, 242–254
Poisoning by agrochemicals, 50
Pollutants, breast milk as marker for
   exposure, 114–123
Polychlorinated biphenyls
 human 101L cells as biomarker for risk
   assessment, 150–167
 serum protein profile as biomarker,
   106–112
Polycyclic aromatic hydrocarbons, human
   101L cells as biomarker for risk
   assessment, 150–167
Potency of stressor, determination, 2
Protein kinase C potentiation, role of
   carnitine and esters, 132
Protein translation, role of carnitine and
   esters, 132–133


Q

Quantitative structure–activity
   relationship (QSAR) bioassays,
   toxicity of chemical agent simulants,
   13,191–203


R

Reference range concentration(s), 40–47
Regulatory constraints, biomarkers for
   health risk assessment, 62

Reporter gene system, measurement of
   biomarkers of environmental health, 12
Reporter gene system assay, risk
   assessment of environmental samples,
   150–167
Risk, description, 2
Risk assessment
 description, 2
 human 101L cells as biomarker of
   environmental samples, 150–167
 mathematical modeling, 13,16
 process, 2
 symbolic solutions of small linear
   physiologically based pharmacokinetic
   models, 242–254
Risk management, relationship to exposure
   biomarkers, 2–18


S

Serum $\alpha_1$-antitrypsin, biomarkers for
   health risk assessment, 54
Serum protein profile
 electrophoretic separation using sodium
   dodecyl sulfate–polyacrylamide gel
   electrophoresis after chlorinated
   insecticide exposure, 108–111
 experimental procedure, 107–108
 fast protein LC chromatograms after
   chlorinate insecticide exposure,
   111,112*f*
Short-chain metabolism, role in
   carnitine as detoxicant of
   nonmetabolizable acyl-coenzyme As,
   141,144–145
Skin, mass balance, 210
Small linear physiologically based
   pharmacokinetic models
 symbolic numerical area under curves,
   256–269
 use in risk assessment and experimental
   design, symbolic solutions, 242–254
Socioeconomic status constraint,
   biomarkers for health risk
   assessment, 61

Sodium dodecyl sulfate–polyacrylamide gel electrophoresis, monitoring of serum protein profile as biomarker of insecticide exposure, 106–112

Spot urine sample results in physiologically based pharmacokinetic modeling of absorbed malathion doses in humans
dermal absorption, 236–237
excretion rates, 236
experimental description, 230–233
experimental vs. simulated absorbed dose, 237–239
malathion excretion, 233–236
reproducibility, 236
skin permeability constants, 237

Striped bass, inductively coupled plasma–MS determination of trace elements in organs, 180–189

Susceptibility biomarkers, 60–61

Symbolic, numerical area under curves of small linear physiologically based pharmacokinetic models
accuracy effect of model perturbations, 266–268
area under curve description, 258–259
experimental description, 257
P4 model, 257–262f
P8 model, 261–266

Symbolic solutions of small linear physiologically based pharmacokinetic models useful in risk assessment and experimental design
experimental description, 243
P4 model
matrix factorizations, 243–247
quantitative risk assessment applications, 247–251
reasons for development, 242

T

Target dose, exposure index determination, 8,10–11

Theoretical linear solvation energy relationship (TLSER) assay, aquatic toxicity of chemical agent simulants, 191–203

Tissue–blood exchange, mass balance, 211

TOPKAT model, aquatic toxicity of chemical agent simulants, 191–203

Total human exposure relational data base and advanced simulation environment, mathematical modeling in risk assessment, 13

Toxaphene, serum protein profile as biomarker, 106–112

Toxic equivalent factors
function, 151
human 101L cells as biomarker for risk assessment of environmental samples, 150–167

Toxic Substances Control Act
function, 7
list of chemicals, 25
use of human exposure biomarkers in environmental legislation, 24–37

Toxicity prediction, multipathway, multiroute physiologically based pharmacokinetic and pharmacodynamic model, 207–227

Toxicological–environmental exposure to nongenotoxic tumorigens, carnitine and esters as biomarkers, 126–135

Trace elements, inductively coupled plasma–MS determination in organs of striped bass from Lake Mead, 180–189

1,1,1-Trichloro-2,2-bis(p-chlorophenyl)-ethane, breast milk as marker for exposure, 118,121t,123

Trichlorophenol, serum protein profile as biomarker, 106–112

3,5,6-Trichloro-2-pyridinol, measurement using reference range concentration, 46

L-Tryptophan metabolism, L-kynurenine pathway, 95,96f

Tumorigens, carnitine and esters as biomarkers of environmental-toxicological exposure, 126–135

U

Urinary excretion of xanthurenic acid as biomarker of organophosphorus insecticide exposure
animals, 95

Urinary excretion of xanthurenic acid as
   biomarker of organophosphorus
   insecticide exposure—*Continued*
endogenous and exogenous factors on
   diazinon-induced increase in
   xanthurenic acid urinary excretion,
   101,103
experimental procedure, 95,97*f*
L-kynurenine pathway of L-tryptophan
   metabolism, 95,96*f*
multiple treatment of diazinon effect
   on urinary xanthurenic acid,
   102*f*–104
organophosphorus insecticide effects
liver kynurenine formamidase,
   95,98–99
xanthurenic acid urinary excretion,
   100–101
Urine analysis, limitations, 229–230

Urine sample results in physiologically
   based pharmacokinetic modeling of
   absorbed malathion doses in humans,
   *See* Spot urine sample results in
   physiologically based pharmacokinetic
   modeling of absorbed malathion doses
   in humans

V

Volatile organic compounds, human
   exposure biomarker use in
   environmental legislation, 30–31

X

Xanthurenic acid, urinary excretion as
   biomarker of organophosphorus
   insecticide exposure, 94–104

# Highlights from ACS Books

# Bestsellers from ACS Books

*The ACS Style Guide: A Manual for Authors and Editors*
Edited by Janet S. Dodd
264 pp; clothbound ISBN 0–8412–0917–0; paperback ISBN 0–8412–0943–X

*Understanding Chemical Patents: A Guide for the Inventor*
By John T. Maynard and Howard M. Peters
184 pp; clothbound ISBN 0–8412–1997–4; paperback ISBN 0–8412–1998–2

*Chemical Activities* (student and teacher editions)
By Christie L. Borgford and Lee R. Summerlin
330 pp; spiralbound ISBN 0–8412–1417–4; teacher ed. ISBN 0–8412–1416–6

*Chemical Demonstrations: A Sourcebook for Teachers,*
*Volumes 1 and 2,* Second Edition
Volume 1 by Lee R. Summerlin and James L. Ealy, Jr.;
Vol. 1, 198 pp; spiralbound ISBN 0–8412–1481–6;
Volume 2 by Lee R. Summerlin, Christie L. Borgford, and Julie B. Ealy
Vol. 2, 234 pp; spiralbound ISBN 0–8412–1535–9

*Chemistry and Crime: From Sherlock Holmes to Today's Courtroom*
Edited by Samuel M. Gerber
135 pp; clothbound ISBN 0–8412–0784–4; paperback ISBN 0–8412–0785–2

*Writing the Laboratory Notebook*
By Howard M. Kanare
145 pp; clothbound ISBN 0–8412–0906–5; paperback ISBN 0–8412–0933–2

*Developing a Chemical Hygiene Plan*
By Jay A. Young, Warren K. Kingsley, and George H. Wahl, Jr.
paperback ISBN 0–8412–1876–5

*Introduction to Microwave Sample Preparation: Theory and Practice*
Edited by H. M. Kingston and Lois B. Jassie
263 pp; clothbound ISBN 0–8412–1450–6

*Principles of Environmental Sampling*
Edited by Lawrence H. Keith
ACS Professional Reference Book; 458 pp;
clothbound ISBN 0–8412–1173–6; paperback ISBN 0–8412–1437–9

*Biotechnology and Materials Science: Chemistry for the Future*
Edited by Mary L. Good (Jacqueline K. Barton, Associate Editor)
135 pp; clothbound ISBN 0–8412–1472–7; paperback ISBN 0–8412–1473–5

---

For further information and a free catalog of ACS books, contact:
American Chemical Society
Customer Service & Sales
1155 16th Street, NW, Washington, DC 20036
Telephone 800–227–5558